optimal control
of differential equations

PURE AND APPLIED MATHEMATICS

A Program of Monographs, Textbooks, and Lecture Notes

LECTURE NOTES IN PURE AND APPLIED MATHEMATICS

1. *N. Jacobson,* Exceptional Lie Algebras
2. *L.-Å. Lindahl and F. Poulsen,* Thin Sets in Harmonic Analysis
3. *I. Satake,* Classification Theory of Semi-Simple Algebraic Groups
4. *F. Hirzebruch, W. D. Newmann, and S. S. Koh, Differentiable Manifolds and Quadratic Forms*
5. *I. Chavel,* Riemannian Symmetric Spaces of Rank One
6. *R. B. Burckel,* Characterization of C(X) Among Its Subalgebras
7. *B. R. McDonald, A. R. Magid, and K. C. Smith,* Ring Theory: Proceedings of the Oklahoma Conference
8. *Y.-T. Siu,* Techniques of Extension on Analytic Objects
9. *S. R. Caradus, W. E. Pfaffenberger, and B. Yood,* Calkin Algebras and Algebras of Operators on Banach Spaces
10. *E. O. Roxin, P.-T. Liu, and R. L. Sternberg,* Differential Games and Control Theory
11. *M. Orzech and C. Small,* The Brauer Group of Commutative Rings
12. *S. Thomier,* Topology and Its Applications
13. *J. M. Lopez and K. A. Ross,* Sidon Sets
14. *W. W. Comfort and S. Negrepontis,* Continuous Pseudometrics
15. *K. McKennon and J. M. Robertson,* Locally Convex Spaces
16. *M. Carmeli and S. Malin,* Representations of the Rotation and Lorentz Groups: An Introduction
17. *G. B. Seligman,* Rational Methods in Lie Algebras
18. *D. G. de Figueiredo,* Functional Analysis: Proceedings of the Brazilian Mathematical Society Symposium
19. *L. Cesari, R. Kannan, and J. D. Schuur,* Nonlinear Functional Analysis and Differential Equations: Proceedings of the Michigan State University Conference
20. *J. J. Schäffer,* Geometry of Spheres in Normed Spaces
21. *K. Yano and M. Kon,* Anti-Invariant Submanifolds
22. *W. V. Vasconcelos,* The Rings of Dimension Two
23. *R. E. Chandler,* Hausdorff Compactifications
24. *S. P. Franklin and B. V. S. Thomas,* Topology: Proceedings of the Memphis State University Conference
25. *S. K. Jain,* Ring Theory: Proceedings of the Ohio University Conference
26. *B. R. McDonald and R. A. Morris,* Ring Theory II: Proceedings of the Second Oklahoma Conference
27. *R. B. Mura and A. Rhemtulla,* Orderable Groups
28. *J. R. Graef,* Stability of Dynamical Systems: Theory and Applications
29. *H.-C. Wang,* Homogeneous Branch Algebras
30. *E. O. Roxin, P.-T. Liu, and R. L. Sternberg,* Differential Games and Control Theory II
31. *R. D. Porter,* Introduction to Fibre Bundles
32. *M. Altman,* Contractors and Contractor Directions Theory and Applications
33. *J. S. Golan,* Decomposition and Dimension in Module Categories
34. *G. Fairweather,* Finite Element Galerkin Methods for Differential Equations
35. *J. D. Sally,* Numbers of Generators of Ideals in Local Rings
36. *S. S. Miller,* Complex Analysis: Proceedings of the S.U.N.Y. Brockport Conference
37. *R. Gordon,* Representation Theory of Algebras: Proceedings of the Philadelphia Conference
38. *M. Goto and F. D. Grosshans,* Semisimple Lie Algebras
39. *A. I. Arruda, N. C. A. da Costa, and R. Chuaqui,* Mathematical Logic: Proceedings of the First Brazilian Conference
40. *F. Van Oystaeyen,* Ring Theory: Proceedings of the 1977 Antwerp Conference
41. *F. Van Oystaeyen and A. Verschoren,* Reflectors and Localization: Application to Sheaf Theory
42. *M. Satyanarayana,* Positively Ordered Semigroups
43. *D. L Russell,* Mathematics of Finite-Dimensional Control Systems
44. *P.-T. Liu and E. Roxin,* Differential Games and Control Theory III: Proceedings of the Third Kingston Conference, Part A
45. *A. Geramita and J. Seberry,* Orthogonal Designs: Quadratic Forms and Hadamard Matrices
46. *J. Cigler, V. Losert, and P. Michor,* Banach Modules and Functors on Categories of Banach Spaces

47. *P.-T. Liu and J. G. Sutinen,* Control Theory in Mathematical Economics: Proceedings of the Third Kingston Conference, Part B
48. *C. Byrnes,* Partial Differential Equations and Geometry
49. *G. Klambauer,* Problems and Propositions in Analysis
50. *J. Knopfmacher,* Analytic Arithmetic of Algebraic Function Fields
51. *F. Van Oystaeyen,* Ring Theory: Proceedings of the 1978 Antwerp Conference
52. *B. Kadem,* Binary Time Series
53. *J. Barros-Neto and R. A. Artino,* Hypoelliptic Boundary-Value Problems
54. *R. L. Sternberg, A. J. Kalinowski, and J. S. Papadakis,* Nonlinear Partial Differential Equations in Engineering and Applied Science
55. *B. R. McDonald,* Ring Theory and Algebra III: Proceedngs of the Third Oklahoma Conference
56. *J. S. Golan,* Structure Sheaves Over a Noncommutative Ring
57. *T. V. Narayana, J. G. Williams, and R. M. Mathsen,* Combinatorics, Representation Theory and Statistical Methods in Groups: YOUNG DAY Proceedings
58. *T. A. Burton,* Modeling and Differential Equations in Biology
59. *K. H. Kim and F. W. Roush,* Introduction to Mathematical Consensus Theory
60. *J. Banas and K. Goebel,* Measures of Noncompactness in Banach Spaces
61. *O. A. Nielson,* Direct Integral Theory
62. *J. E. Smith, G. O. Kenny, and R. N. Ball,* Ordered Groups: Proceedings of the Boise State Conference
63. *J. Cronin,* Mathematics of Cell Electrophysiology
64. *J. W. Brewer,* Power Series Over Commutative Rings
65. *P. K. Kamthan and M. Gupta,* Sequence Spaces and Series
66. *T. G. McLaughlin,* Regressive Sets and the Theory of Isols
67. *T. L. Herdman, S. M. Rankin III, and H. W. Stech,* Integral and Functional Differential Equations
68. *R. Draper,* Commutative Algebra: Analytic Methods
69. *W. G. McKay and J. Patera,* Tables of Dimensions, Indices, and Branching Rules for Representations of Simple Lie Algebras
70. *R. L. Devaney and Z. H. Nitecki,* Classical Mechanics and Dynamical Systems
71. *J. Van Geel,* Places and Valuations in Noncommutative Ring Theory
72. *C. Faith,* Injective Modules and Injective Quotient Rings
73. *A. Fiacco,* Mathematical Programming with Data Perturbations I
74. *P. Schultz, C. Praeger, and R. Sullivan,* Algebraic Structures and Applications: Proceedings of the First Western Australian Conference on Algebra
75. *L Bican, T. Kepka, and P. Nemec,* Rings, Modules, and Preradicals
76. *D. C. Kay and M. Breen,* Convexity and Related Combinatorial Geometry: Proceedings of the Second University of Oklahoma Conference
77. *P. Fletcher and W. F. Lindgren,* Quasi-Uniform Spaces
78. *C.-C. Yang,* Factorization Theory of Meromorphic Functions
79. *O. Taussky,* Ternary Quadratic Forms and Norms
80. *S. P. Singh and J. H. Burry,* Nonlinear Analysis and Applications
81. *K. B. Hannsgen, T. L. Herdman, H. W. Stech, and R. L. Wheeler,* Volterra and Functional Differential Equations
82. *N. L. Johnson, M. J. Kallaher, and C. T. Long,* Finite Geometries: Proceedings of a Conference in Honor of T. G. Ostrom
83. *G. I. Zapata,* Functional Analysis, Holomorphy, and Approximation Theory
84. *S. Greco and G. Valla,* Commutative Algebra: Proceedings of the Trento Conference
85. *A. V. Fiacco,* Mathematical Programming with Data Perturbations II
86. *J.-B. Hiriart-Urruty, W. Oettli, and J. Stoer,* Optimization: Theory and Algorithms
87. *A. Figa Talamanca and M. A. Picardello,* Harmonic Analysis on Free Groups
88. *M. Harada,* Factor Categories with Applications to Direct Decomposition of Modules
89. *V. I. Istrăţescu,* Strict Convexity and Complex Strict Convexity
90. *V. Lakshmikantham,* Trends in Theory and Practice of Nonlinear Differential Equations
91. *H. L. Manocha and J. B. Srivastava,* Algebra and Its Applications
92. *D. V. Chudnovsky and G. V. Chudnovsky,* Classical and Quantum Models and Arithmetic Problems
93. *J. W. Longley,* Least Squares Computations Using Orthogonalization Methods
94. *L. P. de Alcantara,* Mathematical Logic and Formal Systems
95. *C. E. Aull,* Rings of Continuous Functions
96. *R. Chuaqui,* Analysis, Geometry, and Probability
97. *L. Fuchs and L. Salce,* Modules Over Valuation Domains

98. *P. Fischer and W. R. Smith,* Chaos, Fractals, and Dynamics
99. *W. B. Powell and C. Tsinakis,* Ordered Algebraic Structures
100. *G. M. Rassias and T. M. Rassias,* Differential Geometry, Calculus of Variations, and Their Applications
101. *R.-E. Hoffmann and K. H. Hofmann,* Continuous Lattices and Their Applications
102. *J. H. Lightbourne III and S. M. Rankin III,* Physical Mathematics and Nonlinear Partial Differential Equations
103. *C. A. Baker and L, M. Batten,* Finite Geometrics
104. *J. W. Brewer, J. W. Bunce, and F. S. Van Vleck,* Linear Systems Over Commutative Rings
105. *C. McCrory and T. Shifrin,* Geometry and Topology: Manifolds, Varieties, and Knots
106. *D. W. Kueker, E. G. K. Lopez-Escobar, and C. H. Smith,* Mathematical Logic and Theoretical Computer Science
107. *B.-L. Lin and S. Simons,* Nonlinear and Convex Analysis: Proceedings in Honor of Ky Fan
108. *S. J. Lee,* Operator Methods for Optimal Control Problems
109. *V. Lakshmikantham,* Nonlinear Analysis and Applications
110. *S. F. McCormick,* Multigrid Methods: Theory, Applications, and Supercomputing
111. *M. C. Tangora,* Computers in Algebra
112. *D. V. Chudnovsky and G. V. Chudnovsky,* Search Theory: Some Recent Developments
113. *D. V. Chudnovsky and R. D. Jenks,* Computer Algebra
114. *M. C. Tangora,* Computers in Geometry and Topology
115. *P. Nelson, V. Faber, T. A. Manteuffel, D. L. Seth, and A. B. White, Jr.,* Transport Theory, Invariant Imbedding, and Integral Equations: Proceedings in Honor of G. M. Wing's 65th Birthday
116. *P. Clément, S. Invernizzi, E. Mitidieri, and I. I. Vrabie,* Semigroup Theory and Applications
117. *J. Vinuesa,* Orthogonal Polynomials and Their Applications: Proceedings of the International Congress
118. *C. M. Dafermos, G. Ladas, and G. Papanicolaou,* Differential Equations: Proceedings of the EQUADIFF Conference
119. *E. O. Roxin,* Modern Optimal Control: A Conference in Honor of Solomon Lefschetz and Joseph P. Lasalle
120. *J. C. Díaz,* Mathematics for Large Scale Computing
121. *P. S. Milojević,* Nonlinear Functional Analysis
122. *C. Sadosky,* Analysis and Partial Differential Equations: A Collection of Papers Dedicated to Mischa Cotlar
123. *R. M. Shortt,* General Topology and Applications: Proceedings of the 1988 Northeast Conference
124. *R. Wong,* Asymptotic and Computational Analysis: Conference in Honor of Frank W. J. Olver's 65th Birthday
125. *D. V. Chudnovsky and R. D. Jenks,* Computers in Mathematics
126. *W. D. Wallis, H. Shen, W. Wei, and L. Zhu,* Combinatorial Designs and Applications
127. *S. Elaydi,* Differential Equations: Stability and Control
128. *G. Chen, E. B. Lee, W. Littman, and L. Markus,* Distributed Parameter Control Systems: New Trends and Applications
129. *W. N. Everitt,* Inequalities: Fifty Years On from Hardy, Littlewood and Pólya
130. *H. G. Kaper and M. Garbey,* Asymptotic Analysis and the Numerical Solution of Partial Differential Equations
131. *O. Arino, D. E. Axelrod, and M. Kimmel,* Mathematical Population Dynamics: Proceedings of the Second International Conference
132. *S. Coen,* Geometry and Complex Variables
133. *J. A. Goldstein, F. Kappel, and W. Schappacher,* Differential Equations with Applications in Biology, Physics, and Engineering
134. *S. J. Andima, R. Kopperman, P. R. Misra, J. Z. Reichman, and A. R. Todd,* General Topology and Applications
135. *P Clément, E. Mitidieri, B. de Pagter,* Semigroup Theory and Evolution Equations: The Second International Conference
136. *K. Jarosz*, Function Spaces
137. *J. M. Bayod, N. De Grande-De Kimpe, and J. Martínez-Maurica*, *p*-adic Functional Analysis
138. *G. A. Anastassiou*, Approximation Theory: Proceedings of the Sixth Southeastern Approximation Theorists Annual Conference
139. *R. S. Rees*, Graphs, Matrices, and Designs: Festschrift in Honor of Norman J. Pullman
140. *G. Abrams, J. Haefner, and K. M. Rangaswamy*, Methods in Module Theory

141. *G. L. Mullen and P. J.-S. Shiue*, Finite Fields, Coding Theory, and Advances in Communications and Computing
142. *M. C. Joshi and A. V. Balakrishnan*, Mathematical Theory of Control: Proceedings of the International Conference
143. *G. Komatsu and Y. Sakane*, Complex Geometry: Proceedings of the Osaka International Conference
144. *I. J. Bakelman*, Geometric Analysis and Nonlinear Partial Differential Equations
145. *T. Mabuchi and S. Mukai*, Einstein Metrics and Yang–Mills Connections: Proceedings of the 27th Taniguchi International Symposium
146. *L. Fuchs and R. Göbel*, Abelian Groups: Proceedings of the 1991 Curaçao Conference
147. *A. D. Pollington and W. Moran*, Number Theory with an Emphasis on the Markoff Spectrum
148. *G. Dore, A. Favini, E. Obrecht, and A. Venni*, Differential Equations in Banach Spaces
149. *T. West*, Continuum Theory and Dynamical Systems
150. *K. D. Bierstedt, A. Pietsch, W. Ruess, and D. Vogt*, Functional Analysis
151. *K. G. Fischer, P. Loustaunau, J. Shapiro, E. L. Green, and D. Farkas*, Computational Algebra
152. *K. D. Elworthy, W. N. Everitt, and E. B. Lee*, Differential Equations, Dynamical Systems, and Control Science
153. *P.-J. Cahen, D. L. Costa, M. Fontana, and S.-E. Kabbaj*, Commutative Ring Theory
154. *S. C. Cooper and W. J. Thron*, Continued Fractions and Orthogonal Functions: Theory and Applications
155. *P. Clément and G. Lumer*, Evolution Equations, Control Theory, and Biomathematics
156. *M. Gyllenberg and L. Persson*, Analysis, Algebra, and Computers in Mathematical Research: Proceedings of the Twenty-First Nordic Congress of Mathematicians
157. *W. O. Bray, P. S. Milojević, and Č. V. Stanojević*, Fourier Analysis: Analytic and Geometric Aspects
158. *J. Bergen and S. Montgomery*, Advances in Hopf Algebras
159. *A. R. Magid*, Rings, Extensions, and Cohomology
160. *N. H. Pavel*, Optimal Control of Differential Equations

Additional Volumes in Preparation

optimal control of differential equations

A Festschrift in Honor of Constantin Corduneanu

edited by

Nicolae H. Pavel
Ohio University
Athens, Ohio

Marcel Dekker, Inc. **New York • Basel • Hong Kong**

Library of Congress Cataloging-in-Publication Data

Optimal control of differential equations / edited by Nicolae H. Pavel.
p. cm. — (Lecture notes in pure and applied mathematics; v. 160)
"Based on an International Conference on Optimal Control of Differential Equations held in Athens, Ohio, a Festschrift to honor the 65th birthday of Constantin Corduneanu"—Galley.
Includes bibliographical references.
ISBN 0-8247-9234-3
1. Mathematical optimization—Congresses. 2. Control theory—Congresses. 3. Differential equations, Partial—Congresses. I. Pavel, N. H. (Nicolae H.) II. Corduneanu, C. III. International Conference on Optimal Control of Differential Equations (1993: Athens, Ohio) IV. Series.
QA402.5.058 1994
515'.64—dc20 94-11101
CIP

The publisher offers discounts on this book when ordered in bulk quantities. For more information, write to Special Sales/Professional Marketing at the address below.

This book is printed on acid-free paper.

MARCEL DEKKER, INC.
270 Madison Avenue, New York, New York 10016

Current printing (last digit):
10 9 8 7 6 5 4 3 2 1

PRINTED IN THE UNITED STATES OF AMERICA

To Constantin Corduneanu on the Occasion of His Sixty-Fifth Birthday

This volume is dedicated to Professor Constantin Corduneanu for his outstanding contribution to research in differential and integral equations and for his teaching and services to the mathematical community.

The main concentration of his research work has been on qualitative theory of differential and integral equations, including delay equations, and integrodifferential equations. The following problems have been studied:

a) Global existence of solutions.
b) Boundedness of solutions on a half-axis or on the entire real line.
c) Existence of almost periodic solutions (periodic solutions included).
d) Boundary value problems for ordinary differential equations.
e) Stability theory for various classes of equations (ordinary differential, delay, integrodifferential or integral). Professor Corduneanu's early contribution to the comparison method in stability theory is now widely known.
f) Admissibility theory for differential and integral equations. Some of his first results in admissibility theory for integral equations go back to 1965.
g) Equations with abstract Volterra operators. His first paper was published in 1966 (*Funkcialaj Ekvacioj* [Japan]); his further contributions since 1986 include a brief introduction to this class of functional/functional differential equations in his monograph *Integral Equations and Applications* (Cambridge University Press, 1991).
h) Control theory for various classes of equations: stability of feedback systems, existence of optimal controls in various systems, including systems described by abstract Volterra equations.
i) Varia: qualitative inequalities and their applications, singular perturbation for some classes of partial differential equations or in control problems.

His research work has influenced many mathematicians around the world. More than 120 textbooks and monographs refer to his papers, which now exceed 130.

Professor Corduneanu has published the very successful textbook *Principles of Differential and Integral Equations* (three English and two Romanian editions), used by many universities throughout the world. He has also published a monograph, *Almost Periodic Functions* (two English and one Romanian editions), and two monographs on integral equations (one English edition of each) as well as the one published in 1991.

Professor Corduneanu is an excellent teacher and doctoral student advisor. He has played a great role in the scientific formation of many young mathematicians. His contribution to the mathematical community is also significant. He has organized successful conferences of differential and integral equations, served on the editorial boards of many journals of mathematics and he has been a reviewer for *Mathematical Reviews* and *Zentralblatt für Mathematik* since the early 1960s. Professor Corduneanu is still very active, publishing papers and books and delivering invited talks. Currently he is a professor of mathematics at the University of Texas at Arlington and a member of the Romanian Academy of Sciences.

As one of his many former Ph.D. students, I would like to take this opportunity to wish Academician Professor Constantin Corduneanu and his distinguished and devoted wife Mrs. Alice Corduneanu long life, success and happiness in the years to come.

Nicolae H. Pavel

Constantin Corduneanu

Preface

The objectives of the international conference *Optimal Control of Differential Equations*, held at the Department of Mathematics, Ohio University, Athens, Ohio, March 25-27, 1993, were the following:

1. To bring together internationally recognized authorities from various branches of optimal control of differential equations for the purpose of exchanging information sharing recent results, ascertaining new directions, and identifying open problems relevant to optimal control and its applications.
2. To provide opportunities for young investigators, graduate students, and postdoctoral scientists early in their careers to actively participate in this very dynamic research and to promote more cooperation with leading researchers.

The topics in this volume, presented as papers at the conference, include periodic and antiperiodic control with applications to physical systems (parabolic heat equation, wave equation) involving the reversal of sign of state vectors, optimal control of some stochastic differential equations, Pareto optimality conditions, identification problems, boundary control, bang-bang principles, shape optimization via nonsmooth analysis with applications to fluid mechanics and construction of hydrodynamic surfaces, Riccati equations, and the geometric point of view in optimal control.

I would like to acknowledge the financial aid received from various offices of Ohio University and other sources in Athens which made this conference possible. My idea to organize this conference was ably supported by Professors V. Barbu and G. Da Prato beginning in the summer of 1991. I also would like to thank my colleagues A. R. Aftabizadeh, S. Aizicovici, and U. Ledzewicz as well as my Ph.D. students I. Hrinca and G. S. Wang for helping me in the organization of this conference.

Nicolae H. Pavel

Contents

Contributors

N. U. Ahmed Departments of Electrical Engineering and Mathematics, University of Ottawa, Ottawa, Ontario, Canada

Sergiu Aizicovici Department of Mathematics, Ohio University, Athens, Ohio

Viorel Barbu School of Mathematics, University of Iaşi, Iaşi, Romania

E. N. Barron Department of Mathematical Sciences, University of Chicago, Chicago, Illinois

Zhixiong Cai Department of Mathematics, Barton College, Wilson, North Carolina

Constantin Corduneanu Department of Mathematics, The University of Texas at Arlington, Arlington, Texas

Giuseppe Da Prato Department of Mathematics, Scuola Normale Superiore di Pisa, Pisa, Italy

H. O. Fattorini Department of Mathematics, University of California at Los Angeles, Los Angeles, California

Erik Hendrickson Department of Applied Mathematics, University of Virginia, Charlottesville, Virginia

Mary Ann Horn School of Mathematics, University of Minnesota, Minneapolis, Minnesota

Yong Kang Huang Department of Mathematics, The Ohio State University at Marion, Marion, Ohio

Dariusz Idczak Department of Differential Equations and Computer Sciences, Institute of Mathematics, Lódź University, Lódź, Poland

Karl Kunisch Institut für Mathematik, Technische Universität Berlin, Berlin, Germany

Irena Lasiecka Department of Applied Mathematics, University of Virginia, Charlottesville, Virginia

Urszula Ledzewicz Department of Mathematics and Statistics, Southern Illinois University at Edwardsville, Edwardsville, Illinois

Boris S. Mordukhovich Department of Mathematics, Wayne State University, Detroit, Michigan

T. Murphy Department of Mathematics, University of California at Los Angeles, Los Angeles, California

Andrzej Nowakowski Department of Differential Equations and Computer Science, Institute of Mathematics, Lódź University, Lódź, Poland

Nikolaos S. Papageorgiou Department of Mathematics, Florida Institute of Technology, Melbourne, Florida

Nicolae H. Pavel Department of Mathematics, Ohio University, Athens, Ohio

Heinz Schaettler Department of Systems Science and Mathematics, Washington University, St. Louis, Missouri

Srdjan Stojanovic Department of Mathematical Sciences, University of Cincinnati, Cincinnati, Ohio

H. J. Sussmann Department of Mathematics, Rutgers University, New Brunswick, New Jersey

Thomas P. Svobodny Department of Mathematics and Statistics, Wright State University, Dayton, Ohio

Daniel Tataru Department of Mathematics, Northwestern University, Evanston, Illinois

Roberto Triggiani Department of Applied Mathematics, University of Virginia, Charlottesville, Virginia

Stanislaw Walczak Department of Differential Equations and Computer Science, Institute of Mathematics, Lódź University, Lódź, Poland

G. S. Wang Department of Mathematics, Ohio University, Athens, Ohio

Shih-liang Wen Department of Mathematics, Ohio University, Athens, Ohio

Kaixia Zhang Department of Mathematics, Wayne State University, Detroit, Michigan

S. Q. Zhu Automation and Robotics Research Institute, The University of Texas at Arlington, Arlington, Texas

1 Optimal Relaxed Controls for Nonlinear Infinite Dimensional Stochastic Differential Inclusions

N. U. Ahmed University of Ottawa, Ottawa, Ontario, Canada

Abstract: In this paper we present some new results on the existence of optimal relaxed controls for systems governed by stochastic differential inclusions in Hilbert space..

1. INTRODUCTION

Questions of optimal relaxed controls for systems governed by deterministic Differential Inclusions in Banach space have been considered in recent years [Ahmed, 1-4]. Questions of admissibility of relaxations have been studied in [Ahmed and Xiang, 9]. Relaxed boundary controls for infinite dimensional systems have been studied in [Ahmed, 4]; and for linear stochastic systems in [Ahmed ,5]. Partially observed optimal relaxed feedback controls have been considered in [Ahmed, 6]. To the knowledge of the author, there seems to be a gap in the literature on question of optimal control of uncertain infinite dimensional stochastic systems modeled here as Stochastic Differential Inclusions. Here we study precisely this problem.

We consider the following stochastic differential inclusion as the model for the system dynamics.

$$\begin{aligned} &dx + A(t,x)dt \in G(t,x)dt + f(t,x,\mu_t)dt + \sigma(t)dW, t \in I \equiv [0,\tau], \\ &x(0) = x_0; \tau < \infty \end{aligned} \tag{1.1}$$

where A and f may represent nonlinear partial differential operators and σ a linear operator possibly stochastic, and G is a suitable multivalued map representing model uncertainty and μ is the control generally a measure valued process to be defined precisely later. The control problem is to find a control policy that minimizes the maximum risk defined as

$$J^o(\mu) \equiv Sup\{\mathcal{E}(\psi(x,\mu)), x \text{ any solution of } (1.1)\} \Longrightarrow Inf. \tag{1.2}$$

Here ψ is a suitable measure of performance of the system.

2. BASIC NOTATIONS

Let X be any Banach space and X^* the corresponding topological dual with the pairing $< x^*, x >$ denoting the value of x^* at x. For any pair of Banach spaces X, Y, $\mathcal{L}(X, Y)$ will denote the space of linear operators from X to Y. For any topological space $\mathcal{Z}$, $2^{\mathcal{Z}} \setminus \emptyset$ will denote the space of nonempty subsets of $\mathcal{Z}$, $cc(\mathcal{Z})$ the class of nonempty closed convex subsets of $\mathcal{Z}$, $cbc(\mathcal{Z})$ the class of nonempty closed bounded convex subsets of $\mathcal{Z}$, and kc(Z) the class of compact convex subsets of $\mathcal{Z}$.
Let $(\Omega, \mathcal{F}, \mathcal{F}_t \uparrow\subset \mathcal{F}, t \geq 0, P)$ denote a complete probability space furnished with an increasing family of right continuous complete sub σ-algebras $\mathcal{F}_t \subset \mathcal{F}$ having left limits. Let $\mathcal{E}(.)$ denote the integration with respect to the P measure. All random processes considered in the paper will be assumed to be strongly $\mathcal{F}_t$-progressively measurable processes unless stated otherwise. Let F be a separable Hilbert space, $\{W(t), t \geq 0\}$ an $\mathcal{F}_t$ Brownian motion (Wiener process) with values in F with the covariance operator $Q \in \mathcal{L}_n^+(F)$ where $\mathcal{L}_n^+(F)$is the space of positive nuclear operators in F. The operator Q is given by $\mathcal{E}((W(t), h)^2) \equiv t(Qh, h)$. To emphasize measurability, we shall use $L_p(\mathcal{F}, X), 1 \leq p \leq \infty$, to denote the Banach space of strongly $\mathcal{F}-$measurable, X-valued random variables satisfying $\mathcal{E} \parallel x \parallel_X^p < \infty$. Since for each $t \geq 0$ the subsigma algebras $\mathcal{F}_t$ are complete, $L_p(\mathcal{F}_t, X)$ are closed subspaces of $L_p(\mathcal{F}, X)$ and hence they are also Banach spaces. Similarly, $L_p^e(X) \equiv L_p^e(I, X)$ will denote the Banach space of $\mathcal{F}_t-$progressively measurable random processes defined on I, taking values from X satisfying $\mathcal{E} \int_I \parallel x(t) \parallel_X^p dt < \infty$. Further notations will be introduced as required.

3. SYSTEM MODEL AND SOME BASIC ASSUMPTIONS

For all practical purposes we can combine the single valued map f and the multivalued map G of the evolution inclusion (1.1) into one multivalued map $G(t, x, \mu) \equiv f(t, x, \mu) + G(t, x)$ and rewrite the equation as

$$\begin{aligned} &dx + A(t, x)dt \in G(t, x, \mu)dt + \sigma(t)dW \\ &x(0) = x_0. \end{aligned} \tag{3.1}$$

Let H be a separable Hilbert space and V a subset of H having the structure of a reflexive Banach space with the embedding $V \hookrightarrow H$ being continuous and dense. Identifying H with it's dual we have $V \hookrightarrow H \hookrightarrow V^*$. We shall denote the pairing between V^* and V by $< v^*, v >$ for $v^* \in V^*$ and $v \in V$. Let $1 < q \leq 2 \leq p < \infty$ with $(1/p) + (1/q) = 1$. Let $L_p^e(V)$ denote the Banach space of all $\mathcal{F}_t-$ progressively measurable V-valued stochastic processes $\{x\}$ satisfying $\mathcal{E} \int_I \parallel x(t) \parallel_V^p dt < \infty$, and let $L_q^e(V^*)$ denote it's topological dual. Note that these are reflexive Banach spaces. Similarly $L_\infty^e(H)$ will denote the Banach space of $\mathcal{F}_t-$adapted $H-$valued processes satisfying $ess-sup\{\mathcal{E} \parallel x(t) \parallel_H^2, t \in I\} < \infty$.

Let M_τ denote the space of $\mathcal{F}_t-$ adapted H valued random processes $\{x\}$ satisfying $Sup\{\mathcal{E} \parallel x(t) \parallel_H^2, t \in I\} \equiv \parallel x \parallel_{M_\tau}^2 < \infty$. It is easy to verify that M_τ, furnished with the norm topology as defined above, is a Banach space. Note that $M_\tau \subset L_\infty^e(H)$.
For admissible controls we need the following notions. Let U be a compact Polish space and $\mathcal{M} \equiv \mathcal{M}(U)$ be the Banach space of all bounded Radon measures on the Borel fields of U with the total variation norm. Since the controls are required to be non anticipative , we let $\mathcal{P}$ denote

the σ-algebra of progressively measurable subsets of the set $I \times \Omega$ and $L_\infty(\mathcal{P}, \mathcal{M})$ the Banach space of $w^* - \mathcal{P}-$ measurable processes with values in $\mathcal{M}$. Note that by Dunford-Pettis theorem, $L_\infty(\mathcal{P}, \mathcal{M})$ is the dual of $L_1(\mathcal{P}, C(U))$. To emphasize the time variable we shall write these as $L^e_\infty(I, \mathcal{M}(U))$ and $L^e_1(I, C(U))$ respectively.

Let $\mathcal{U}_r$ denote the class of admissible controls consisting of $w^* - \mathcal{P}$ measurable, also called $w^* - \mathcal{F}_t$ measurable, processes on I with values in $\mathcal{M}$ satisfying the following properties:

$$
\begin{aligned}
&(1) : t \longrightarrow \int_U g(t,u)\mu_{t,\omega}(du) \ \text{ is } \ \mathcal{F}_t \ \text{ adapted for all } \ g \in L^e_1(I, C(U)),\\
&(2) : \mu_{t,\omega}(U) = 1 \quad \text{for all} \quad t \in I, \omega \in \Omega,\\
&(3) : \int_{I\times U} \chi_{J\times U}(t,u)\mu_{t,\omega}(du)dt = \ell(J) \ \ P - a.s., \ \ \ell \equiv \ \text{Lebesgue measure},\\
&(4) : \int_{I\times U} g(t,u)\mu_{t,\omega}(du)dt \geq 0 \ \ P - a.s. \ \text{ for all } \ g \geq 0.
\end{aligned}
\tag{3.2}
$$

A sequence $\mu^n \in \mathcal{U}_r$ is said to converge in the Young topology τ_Y to $\mu \in \mathcal{U}_r$ if

$$
\mathcal{E}\int_{J\times\Gamma} g(t,u)\mu^n_{t,\omega}(du)dt \longrightarrow \mathcal{E}\int_{J\times\Gamma} g(t,u)\mu_{t,\omega}(du)dt
$$

for every $g \in L^e_1(I, C(U))$ and for every $J \times \Gamma \in B(I) \times B(U)$.

Furnished with the Young topology, $\mathcal{U}_r$ is a locally convex sequentially complete topological vector space. In fact it is also sequentially compact. Since U is a compact Polish space and the σ-algebras are separable, $L_1(\mathcal{P}, C(U))$ is separable and hence by theorem V.5.1 [13, p426], $\mathcal{U}_r \subset L_\infty(\mathcal{P}, \mathcal{M})$ is metrizable, and with respect to this metric it is a complete separable metric space and hence a Polish space.

For the proof of existence of solutions we shall introduce the following assumptions:

(A1): The operator $A : I \times V \longrightarrow V^*$ is measurable on I and strictly monotone and hemicontinuous on V.

(A2): There exist constants $\gamma \geq 0, \alpha > 0$ such that

$$
< A(t,\xi), \xi > +\gamma \parallel \xi \parallel^2_H \geq \alpha \parallel \xi \parallel^p_V \ \text{for all}\ t \in I, \xi \in V.
$$

(A3): There exists a $\delta > 0$ such that

$$
\parallel A(t,\xi) \parallel_{V^*} \leq \delta(1 + \parallel \xi \parallel^{p-1}_V), \ \text{for all}\ t \in I, \xi \in V.
$$

(G1): $G : I \times H \times \mathcal{M} \longrightarrow cc(H)$ so that for each fixed $x \in H$ and $\mu \in \mathcal{U}_r$, $t \longrightarrow G(t, x, \mu_t)$ is a measurable set valued map on I and that there exists a $k \in L_2(I)$ such that for each $\beta \in G(t, x, \mu_t)$

$$
\parallel \beta \parallel^2_H \leq k(t)^2(1 + \parallel x \parallel^2_H), \ \text{ for } \ (t,x) \in I \times H
$$

uniformly with respect to $\mu \in \mathcal{U}_r$. Further, for almost all $t \in I$, $\omega \longrightarrow G(t, \xi(\omega), \mu_{t,\omega})$ is $\mathcal{F}_t$ measurable for every H-valued $\mathcal{F}_t$ measurable ξ.

(G2): For each $\mu \in \mathcal{U}_r$, $(t,x) \longrightarrow G(t,x,\mu_t)$ is weakly upper semicontinuous on $I \times H$ with respect to inclusion.

(G3): $\mu \longrightarrow G(.,x,\mu_{(.)})$ is lower semicontinuous with respect to inclusion on $\mathcal{U}_r$ in the Young topology uniformly with respect to $x \in H$.

4 EXISTENCE OF SOLUTIONS AND REGULARITY PROPERTIES

DEFINITION 4.1. *For any $\mu \in \mathcal{U}_r$ and $x_0 \in L_2(\mathcal{F}_0, H)$, an element $x \in M_r$ is said to be a solution of the evolution inclusion (3.1) if there exists a $g \in L_2^e(H)$ such that x is a weak solution of the stochastic evolution equation,*

$$\begin{aligned} dx + A(t,x)dt &= g(t)dt + \sigma(t)dW \\ x(0) &= x_0, \end{aligned} \tag{4.1}$$

and that

$$g(t) \in G(t, x(t), \mu_t) \quad a.e, P - a.s. \tag{4.2}$$

In order to prove the existence of solutions we shall need some a-priori bounds as given in the following lemma.

LEMMA 4.2. *Suppose the assumptions (A1)-(A3) and (G1) hold and σ is $\mathcal{F}_t$ adapted satisfying*

$$\mathcal{E} \int_I tr(\sigma Q \sigma^*)dt < \infty.$$

Let x be any solution of the differential inclusion (3.1) corresponding to $x_0 \in L_2(\mathcal{F}_0, H)$ and $\mu \in \mathcal{U}_r$. Then there exists a constant c such that

$$Sup\{\mathcal{E} \parallel x(t) \parallel_H^2, t \in I\} \leq c, \ \mathcal{E} \int_I \parallel x(t) \parallel_V^p dt \leq c;$$

and further, $x \in C(I, H_w)$ P-a.s. and the estimates are independent of the choice of $\mu \in \mathcal{U}_r$.

Proof. Let $\{v_i\}$ be a common basis for the triple $\{V, H, V^*\}$ such that $\{v_i\}$ is orthogonal in V and orthonormal in H. Given that x is a solution of the system (3.1) in the sense of the definition (4.1), there exists a $g \in L_2^e(H)$ such that $g(t) \in G(t, x(t), \mu_t)$ a.e- P-a.s. and x satisfies the stochastic evolution equation

$$dx = -A(t,x)dt + g(t)dt + \sigma(t)dW, \quad \text{with} \quad x(0) = x_0;$$

in the weak sense, that is, for each $v \in V$,

$$(x(t), v) = (x_0, v) - \int_0^t < A(s, x(s)), v > ds + \int_0^t (g(s), v)ds + \int_0^t (dW(s), \sigma(s)^* v) \tag{4.3}$$

for each $t \in I$ with probability one. Then by applying Ito's formula to $\phi(t) \equiv (x(t), v_i)$ and summing over all the indices one can easily arrive at the following expression:

$$d(\| x(t) \|_H^2) + 2 < A(t, x(t)), x(t) > dt = 2(g(t), x(t))dt + tr(\sigma Q \sigma^*)dt + 2(x(t), \sigma dW). \quad (4.4)$$

Now integrating and taking the expectation and using the assumptions (A1),(A2) and (G1), we obtain the following estimate

$$(1 + \mathcal{E}(\| x(t) \|_H^2) + 2\alpha \int_0^t \mathcal{E} \| x(s) \|_V^p \, ds \leq (1 + \mathcal{E} \| x_0 \|_H^2) + \mathcal{E} \int_I tr(\sigma Q \sigma^*) ds + \int_0^t |K(s)|^2 (1 + \mathcal{E} \| x(s) \|_H^2) ds, \quad (4.5)$$

where $|K(t)|^2 \equiv |k(t)|^2 + (1 + 2|\gamma|)$. Hence, for any $t \in I$, by Gronwall inequality we have,

$$\left(1 + \mathcal{E}(\| x(t) \|_H^2)\right) \leq \left((1 + \mathcal{E} \| x_0 \|_H^2) + \mathcal{E} \int_I tr(\sigma Q \sigma^*) ds\right) exp\left(\int_0^t |K(s)|^2 ds\right).$$

Since $\sigma \in L_2^e(\mathcal{L}_Q(F, H))$ and $x_0 \in L_2(\mathcal{F}_0, H)$, it follows from the above inequality and the inequality (4.5) that there exists a constant c such that

$$Sup\{\mathcal{E} \| x(t) \|_H^2\} \leq c, \quad \text{and} \quad \mathcal{E} \int_I \| x(t) \|_V^p \, dt \leq c. \quad (4.6)$$

Thus $x \in L_\infty^e(H) \cap L_p^e(V)$. Clearly, for each $s, t \in I$ satisfying $t \geq s$, and $v \in V$, it follows from (4.3) that

$$(x(t) - x(s), v) = -\int_s^t < A(r, x(r)), v > dr + \int_s^t (g(r), v) dr + \int_s^t (dW(r), \sigma(r)^* v) \quad P-a.s.$$

Define $\eta(t) \equiv A(t, x(t)), t \in I$. By virtue of (A3) and (4.6), $\eta \in L_q^e(V^*)$. Since the martingale term is continuous P-a.s and $g \in L_2^e(H)$ it follows from the above expression that $Lim_{t \to s}(x(t) - x(s), v) = 0$ with probability one. Since V is dense in H, this implies that $x \in C(I, H_w)$ P -a.s. □

For the proof of existence of solutions, we shall use Galerkin approach. As usual we project the infinite dimensional system into a sequence of finite dimensional ones as follows. Since $x_0 \in L_2(\mathcal{F}_0, H)$, there exists a sequence of $\mathcal{F}_0$-measurable square integrable random variables $\{\gamma_i\}$which are the Fourier coefficients of x_0 such that the partial sums

$$x_0^n \equiv \sum_{1 \leq i \leq n} \gamma_i v_i \xrightarrow{s} x_0 \text{ in } L_2(\mathcal{F}_0, H) \quad \text{as } n \to \infty. \quad (4.7)$$

Define, for $t \in I$ and $\zeta \in R^n$,

$$(1): E_n(t, \zeta) \equiv \{z \in R^n : z_i = (y, v_i), i = 1, 2, \ldots.n, y \in G(t, \sum_{1 \leq i \leq n} \zeta_i v_i, \mu_t)\}.$$

$$(2): a^n \equiv \{a_j^n, j = 1, 2, \ldots n\}, a_j^n(t, \zeta) \equiv < A(t, \sum_{1 \leq i \leq n} \zeta_i v_i), v_j > \quad (4.8)$$

$$(3): M^n \equiv \{M_j^n, j = 1, 2, \ldots n\}, dM_j^n \equiv (\sigma dW, v_j).$$

Note that $E_n : I \times R^n \longrightarrow cc(R^n)$ since , for a fixed μ, $G : I \times H \longrightarrow cc(H)$. Then we can approximate the infinite dimensional stochastic differential inclusion (3.1) by the following sequence of finite dimensional stochastic differential inclusions

$$\begin{aligned} &d\xi^n(t) + a^n(t, \xi^n(t))dt \in E_n(t, \xi^n(t))dt + dM^n, t \in I \\ &\xi^n(0) = \gamma^n, \gamma^n \equiv \{\gamma_i, i = 1, 2, ...n\}. \end{aligned} \tag{4.9$_n$}$$

We shall prove that the sequence $\{x^n\}$ given by $x^n(t) \equiv \sum_{1\leq i\leq n} \xi_i^n(t)v_i$, where ξ^n are the solutions of the finite dimensional inclusions $(4.9)_n$ if they exist , converge in an appropriate sense to a solution of (3.1). Let $\eta \in L_2^e(R^n)$ and consider the finite dimensional stochastic differential equation

$$\begin{aligned} &d\xi + a^n(t, \xi)dt = \eta dt + dM^n, t \in I, \\ &\xi^n(0) = \gamma^n \equiv \xi_0, \end{aligned} \tag{4.10}$$

By virtue of assumption (A1), the function $a^n(t, \xi)$ is measurable in $t \in I$ and continuous in $\xi \in R^n$. Using this fact and the monotonicity and the growth property of A, one can prove the existence of a unique solution for (4.10) which is continuous with probability one and has finite second moment. Further one can easily verify that the solution map $\eta \longrightarrow \xi(t) \equiv S_t(\eta)$ is Lipschitz.

Define the multivalued map $\hat{E}_n$ with values,

$$\hat{E}_n(\eta) \equiv \{\beta \in L_2^e(R^n) : \beta(t) \in E_n(t, S_t(\eta)), a.e, P - a.s.\},$$

for each $\eta \in L_2^e(R^n)$. By virtue of the apriori estimates (see Lemma 4.1), the set valued map $t \longrightarrow E_n(t, S_t(\eta))$ is (square) integrably bounded with closed values and hence has measurable selections [Wagner 15, theorem. 4.1, theorem 8.1; pp 859-903, Ahmed and Teo 1, theorem 1.4.4, theorem 1.4.5 , pp37-40]. Thus $\hat{E}_n$ is nonempty. Then the question of existence of a solution of the system $(4.9)_n$ is equivalent to the question of existence of a fixed point of the multivalued map $\hat{E}_n$.

In the following lemma we prove the existence of solutions for the system $(4.9)_n$.

LEMMA 4.3. *Consider the system $(4.9)_n$ and define*

$$\Sigma \equiv \{\eta \in L_2^e(R^n) : \mathcal{E} \int_I \| \eta(t) \|_{R^n}^2 dt \leq \tilde{c} \equiv \| k \|^2 (1 + c)\},$$

where the constant c is as given in lemma 4.2. Suppose $\hat{E}_n(\eta) \cap \Sigma \neq \emptyset$ for all $\eta \in \Sigma$. Then the system $(4.9)_n$ has solutions which belong to $C(I, R^n)$ P-a.s and have finite second moments.

Proof. Furnished with the weak topology, $(L_2^e(R^n), \tau_w)$ is a locally convex sequentially complete topological vector space and Σ is a compact convex subset of this LCTVS. By virtue of assumption (G2), the multifunction $\eta \longrightarrow \hat{E}_n(\eta)$ is upper semicontinuous with values from the space of nonempty closed convex (bounded) subsets of $(L_2^e(R^n), \tau_w)$. Hence it follows from corollary 2 of theorem 6.3 of Browder [12, pp 75] that $Fix\hat{E}_n \neq \emptyset$. This means that $(4.9)_n$ has solutions.

The regularity properties as stated in the lemma, follow from those of the solutions of the finite dimensional system (4.10) with standard proof. □

Now we are prepared to prove the existence of solutions of the infinite dimensional stochastic differential inclusion (3.1). Let $\mathcal{X}_\mu$ (possibly an empty set) denote the family of solutions of the evolution inclusion (3.1) corresponding to the control $\mu \in \mathcal{U}_r$. We prove the existence by showing that this set is nonempty. Since, for the next theorem, $\mu \in \mathcal{U}_r$ is fixed, for convenience of notation we write $E(t,x) \equiv G(t,x,\mu_t)$.
For the existence, we shall introduce a more technical, though reasonable, hypothesis on the multifunction G. Let B_c denote the subset of $M_\tau \cap L_p^e(V)$ that satisfies the apriori estimates of Lemma 4.2 including the almost sure weak continuity in H. Define

$$S(E) \equiv \{g \in L_2^e(H) : g(t) \in E(t,x(t)) a.e. P - a.s. \quad \text{for some} \quad x \in B_c\}$$

THEOREM 4.4. *Suppose the assumptions (A1)-(A3), (G1)-(G3), and those of Lemma 4.3 hold for each $n \in N_+$. Further assume that*

(G4): the set $S(E)$, considered as a subset of $L_q^e(V^*)$, is conditionally compact. Then the evolution inclusion (3.1) has a nonempty set of solutions.

Proof. By Lemma 4.3, $Fix(\hat{E}_n) \neq \emptyset$ for each $n \in N_+$. Hence there exists an $\eta^n \in L_2^e(R^n)$ and $\xi^n \in C(I,R^n) - P - a.s.$ satisfying $Sup\{\mathcal{E} \parallel \xi^n(t) \parallel_{R^n}^2, t \in I\} < \infty$ so that $\eta^n(t) \in E_n(t,\xi^n(t))P - a.s$ for almost all $t \in I$ and that ξ^n solves the problem (4.10) with $\eta = \eta^n$ and $\xi^n(0) = \gamma^n$. Define

$$g^n(t) \equiv \sum_{1 \leq i \leq n} \eta_i^n(t) v_i$$

$$x^n(t) \equiv \sum_{1 \leq i \leq n} \xi_i^n(t) v_i.$$

Clearly $g^n(t) \in E(t,x^n(t)) a.e.P - a.s$ and x^n satisfies the equation:

$$d(x^n(t),v_i) + < A(t,x^n(t)), v_i > dt = (g^n(t),v_i)dt + (\sigma^* v_i, dW)_F, 1 \leq i \leq n$$

in its integral form

$$\begin{aligned}(x^n(t),v_i) = (x^n(0),v_i) - \int_0^t < A(s,x^n(s)), v_i > ds + \int_0^t (g^n(s),v_i)ds \\ + \int_0^t (\sigma^*(s)v_i, dW(s).\end{aligned} \tag{4.11}$$

By the apriori estimates (see Lemma 4.2), we have

$$Sup\{\mathcal{E} \parallel x^n(t) \parallel_H^2, t \in I\} \leq c, \quad \mathcal{E}\int_I \parallel x^n(t) \parallel_V^p dt \leq c;$$

and, since $g^n(t) \in E(t, x^n(t))$, we have $\mathcal{E} \int_I \| g^n(t) \|_H^2 \leq \| k \|^2 (1+c)$ for all $n \in N_+$. Since $L_2^e(H)$ is a Hilbert space and $L_p^e(V)$ is a reflexive Banach space , there exists a subsequence of the sequence $\{g^n, x^n\}$ relabeled as such and a pair $g^* \in L_2^e(H), x^* \in L_\infty^e(H) \cap L_p^e(V)$ such that

$$
\begin{aligned}
&g^n \xrightarrow{w} g^* \quad \text{in } L_2^e(H), \\
&x^n \xrightarrow{w^*} x^* \quad \text{in} \quad L_\infty^e(H) \subset L_\infty(I, L_2(\Omega, H)) \\
&x^n \xrightarrow{w} x^* \quad \text{in} \quad L_p^e(V).
\end{aligned}
\tag{4.12}
$$

The difficulty arises in showing that $A(x^n) \to A(x^*)$ where $A(x)$ denotes the Nemytski operator with values $A(t, x(t)), t \in I$. By virtue of the growth assumption (A3), the sequence $\{A(x^n)\}$ is bounded in $L_q^e(V^*)$. Hence there exists an $\eta \in L_q^e(V^*)$ so that, through a subsequence if necessary, $A(x^n) \to \eta$ in the weak star topology of $L_q^e(V^*)$. We show below that under the assumption (G4) we have $\eta = A(x^*)$. For convenience of notation we shall use $<< .,. >>$ to denote the duality pairing between $L_q^e(V^*)$ and $L_p^e(V)$ without distinguishing the order.
For an arbitrary $x \in L_p^e(V) \cap M_\tau$, that belongs to $C(I, H_w)$ P-a.s, define

$$
\begin{aligned}
Z^n &\equiv << A(x^n) - A(x), x^n - x >> + (1/2)\mathcal{E}|x^n(\tau) - x(\tau)|_H^2 \\
&\equiv Z_1^n + Z_2^n + Z_3^n,
\end{aligned}
$$

where

$$
\begin{aligned}
Z_1^n &= << A(x^n), x^n >> + (1/2)\mathcal{E}|x^n(\tau)|_H^2 \\
Z_2^n &= << A(x), x - x^n >> - << A(x^n), x >> \\
Z_3^n &= -\mathcal{E}(x^n(\tau), x(\tau)) + (1/2)\mathcal{E}|x(\tau)|_H^2.
\end{aligned}
$$

Using equation (4.4) it is easy to verify that

$$
Z_1^n = (1/2)\mathcal{E}|x_0|_H^2 + (1/2)\, \mathcal{E} \int_0^T tr(\sigma Q \sigma^*)dt + \mathcal{E} \int_0^T < g^n, x^n > dt.
$$

Since $\{g^n\} \in S(E)$, by hypothesis (G4), g^n converges strongly, through a subsequence if necessary, to g^* in $L_q^e(V^*)$; further, x^n converges to x^* weakly in $L_p^e(V)$. Hence

$$
\begin{aligned}
Lim Z_1^n &= Z_1 \equiv (1/2)\mathcal{E}|x_0|_H^2 + (1/2)\mathcal{E} \int_0^T tr(\sigma Q \sigma^*)dt + \mathcal{E} \int_0^T < g^*, x^* > dt, \\
Lim Z_2^n &= Z_2 \equiv << A(x), x - x^* >> - << \eta, x >>, \\
Lim Z_3^n &= Z_3 \equiv -\mathcal{E}(x^*(\tau), x(\tau)) + (1/2)\mathcal{E}|x(\tau)|_H^2.
\end{aligned}
$$

Since $Z^n \geq 0$ we have $Lim Z^n = Z \equiv Z_1 + Z_2 + Z_3 \geq 0$. Again by equation (4.4), applied to

$$
dx^* + \eta dt = g^* dt + \sigma dW,
$$

for the same initial condition we obtain

$$
\begin{aligned}
(1/2)\mathcal{E}|x^*(\tau)|_H^2 + << \eta, x^* >> = (1/2)\mathcal{E}|x_0|_H^2 + (1/2)\, \mathcal{E} \int_0^T tr(\sigma Q \sigma^*)dt \\
+ \mathcal{E} \int_0^T < g^*, x^* > dt.
\end{aligned}
$$

Substituting this for Z_1 in the expression for Z we obtain

$$Z = (1/2)\mathcal{E}|x^*(\tau) - x(\tau)|_H^2 + << A(x) - \eta, x - x^* >> \geq 0$$

for any $x \in L_p^e(V) \cap M_\tau$ which is weakly continuous in H with probability one. Choosing $x \equiv x^* + \theta y$ for $\theta \geq 0$ and y satisfying the same regularity as that of x, and substituting in the above expression we obtain

$$(1/2)\theta^2 \mathcal{E}|y(\tau)|_H^2 + \theta << A(x^* + \theta y) - \eta, y >> \geq 0$$

for all $\theta \geq 0$. Dividing through by θ and letting $\theta \downarrow 0$, and using the fact that A is hemicontinuous we obtain

$$<< A(x^*) - \eta, y >> \geq 0,$$

for all $y \in L_p^e(V) \cap M_\tau$ satisfying the weak continuity as mentioned above. Hence $\eta = A(x^*)$. Thus multiplying either side of (4.11) by a random variable $z \in L_\infty(\mathcal{F})$ and taking the limit in (4.11) and recalling that $\{v_i\}$ is a basis of V and z is arbitrary, it follows from (4.7) and (4.12) that

$$\begin{aligned}(x^*(t), v) = (x_0, v) - \int_0^t < A(s, x^*(s)), v > ds + \int_0^t (g^*(s), v) ds \\ + \int_0^t (\sigma^*(s)v, dW(s),\end{aligned} \tag{4.13}$$

for each $v \in V$ and for all $t \in I$, P-a.s. By virtue of strict monotonicity of A one can conclude that x^* is the unique solution of the stochastic evolution equation,

$$dx + A(t, x)dt = g^*(t)dt + \sigma(t)dW, x(0) = x_0,$$

in the weak sense. Now, in order to complete the proof, we must show that $g^*(t) \in E(t, x^*(t))$ a.e- P-a.s. Note that for each $t \in I$, $x^n(t)$ converges in the weak topology of H to $x^*(t)$ P-a.s. where, by Lemma 4.2, $x^* \in C(I, H_w)$- P-a.s. By assumption (G2), E is weakly upper semi continuous and hence , for each $t_0 \in I$ and $\epsilon > 0$, there exists a $\delta > 0$ such that

$$E(t, x^n(t)) \subset E_\epsilon(t_0, x^n(t_0)) \quad \text{for all} \quad n \in N_+, t \in I_\delta$$

where $I_\delta \equiv [t_0 - \delta, t_0] \cap I$; and E_ϵ denotes the epsilon-neighborhood of E. Again, since $x^n(t_0) \xrightarrow{w} x^*(t_0)$ in H P-a.s., for the given $\epsilon > 0$, there exists an integer $n_0 = n(t_0, \epsilon)$ such that

$$E_\epsilon(t_0, x^n(t_0)) \subset E_{2\epsilon}(t_0, x^*(t_0)) \quad \text{for all } n \geq n_0.$$

Hence for $n \geq n_0$, we have $g^n(t) \in E(t, x^n(t)) \subset E_{2\epsilon}(t_0, x^*(t_0))P - a.s.$ for all $t \in I_\delta$. Choose any $\Gamma \in \mathcal{F}_{t_0}$ and a sequence $\{\Gamma_\delta\} \in \mathcal{F}_{t_0 - \delta}$ so that $\Gamma_\delta \longrightarrow \Gamma$ as $\delta \rightarrow 0$. Since $E(t, x) \in cc(H)$, it is clear that for all $n \geq n_0$,

$$(1/\ell(I_\delta)) \int_{I_\delta \times \Gamma_\delta} g^n(s) \, ds \, dP \in \int_{\Gamma_\delta} E_{2\epsilon}(t_0, x^*(t_0)) \, dP.$$

Recall that, by the integral of a measurable set valued map one means the integral of it's measurable selections.

Therefore, by weak convergence of g^n to g^* in $L_2^e(H)$, we have

$$(1/\ell(I_\delta)) \int_{I_\delta \times \Gamma_\delta} g^*(s)ds \in C\ell \int_{\Gamma_\delta} E_{2\epsilon}(t_0, x^*(t_0))\, dP.$$

Letting $\delta \downarrow 0$, we obtain $\int_\Gamma g^*(t_0)\, dP \in C\ell \int_\Gamma E_{2\epsilon}(t_0, x^*(t_0))\, dP$. Now by a theorem on the integrals of measurable multifunctions with nonempty closed values [11, Theorem 8.6.4], we have

$$C\ell \int_\Gamma E_{2\epsilon}(t_0, x^*(t_0))\, dP = \int_\Gamma C\ell Co\; E_{2\epsilon}(t_0, x^*(t_0))\, dP.$$

But since $E_{2\epsilon}$ is convex valued, we have

$$C\ell \int_\Gamma E_{2\epsilon}(t_0, x^*(t_0))\, dP = \int_\Gamma C\ell\; E_{2\epsilon}(t_0, x^*(t_0))\, dP.$$

Hence

$$\int_\Gamma g^*(t_0)\, dP \in \int_\Gamma C\ell\; E_{2\epsilon}(t_0, x^*(t_0))\, dP.$$

This is true for all $\Gamma \in \mathcal{F}_{t_0}$ and consequently $g^*(t_0) \in C\ell E_{2\epsilon}(t_0, x^*(t_0))P - a.s.$ But since both $t_0 \in I$ and $\epsilon > 0$, are otherwise arbitrary, and E takes values from $cc(H)$, we have $g^*(t) \in E(t, x^*(t))\; a.e.P - a.s.$ This completes the proof. □

REMARK 4.5. Setting $\sigma \equiv 0$ in equation (3.1) and letting x_0 be a fixed element of H we obtain a deterministic system:

$$dx/dt + A(t,x) \in G(t,x,\mu), x(0) = x_0, \tag{4.14}$$

provided we reinterpret $\mathcal{U}_r$ as being the class of (deterministic) Radon probability measure valued functions on I. In this case the assumption (G2) of section 3 can be relaxed by replacing weak upper semi continuity by simple upper semi continuity. Here the solutions belong to the well known Banach space

$$W_{p,q}(I) \equiv \{\xi : \xi \in L_p(V), (d\xi/dt) \in L_q(V^*)\}$$

furnished with the standard norm topology

$$\| \xi \|_{W_{p,q}} \equiv \| \xi \|_{L_p(V)} + \| (d\xi/dt) \|_{L_q(V^*)} .$$

Since the injection $W_{p,q}(I) \hookrightarrow C(I,H)$ is continuous, the solution set, whenever it is nonempty, belongs to $C(I,H)$. Further, if the injection $V \hookrightarrow H$ is compact then the injection $W_{p,q}(I) \hookrightarrow L_p(H)$ is also compact. Hence the Glalerkin sequence $x^n(t) \xrightarrow{s} x^*(t)$ in H a.e. Thus in the later part of the proof of Theorem 4.4 showing $g^*(t) \in E(t, x^*(t))$ one needs only upper semi continuity of G with respect to $x \in H$. Hence we have the following result.

THEOREM 4.6. *Under the relaxed hypothesis, for each $\mu \in \mathcal{U}_r$, the solution set X_μ of the differential inclusion (4.14) is a nonempty subset of $W_{p,q}$.*

REMARK 4.7. (ASSUMPTION (G4)): Note that the set $S(E)$ is a conditionally weakly compact subset of $L_2^e(H)$ without the assumption (G4). Unfortunately the injection $L_2^e(H) \hookrightarrow L_q^e(V^*)$ is not compact even though $H \hookrightarrow V^*$ is compact. Thus the assumption (G4) can not be dropped unless the operator$A$ is continuous from $L_p^e(V)$ in the weak topology to $L_q^e(V^*)$ in the weak star topology.

5. EXISTENCE OF OPTIMAL CONTROLS

For the proof of existence of optimal controls we shall need few basic lemmas. For each $\mu \in \mathcal{U}_r$, let $\mathcal{X}_\mu$ denote the family of solutions of the controlled stochastic differential inclusion (3.1).

LEMMA 5.1. *For each control $\mu \in \mathcal{U}_r$, the solution set $\mathcal{X}_\mu \subset L_p^e(V) \cap L_\infty^e(H) \cap M_r$. Considered as a subset of $L_p^e(V)$, it is bounded, weakly closed, weakly sequentially compact and hence weakly compact. Considered as a subset of $L_\infty^e(H) \cap M_r$, it is bounded, w^*-closed and w^*-sequentially compact and hence w^*-compact.*

Proof. Note that by virtue of assumption (G1), the apriori bounds given in Lemma 4.1 are independent of the choice of $\mu \in \mathcal{U}_r$. Hence $\mathcal{X}_\mu$ is a bounded subset of $L_p^e(V) \cap L_\infty^e(H) \cap M_r$. We shall prove the closedness. Let $\{x^n\} \in \mathcal{X}_\mu$ and suppose

$$\begin{aligned} &x^n \xrightarrow{w^*} x^0 \quad \text{in} \quad L_\infty^e(H), \\ &x^n \xrightarrow{w} x^0 \quad \text{in} \quad L_p^e(V), \\ &x^n(t) \xrightarrow{w} x^0(t) \quad \text{in} \quad H_w - a.e, P - a.s. \end{aligned} \tag{5.1}$$

Then we show that $x^0 \in \mathcal{X}_\mu$. Since $x^n \in \mathcal{X}_\mu$, there exists a sequence $f^n \in L_2^e(H)$ such that $f^n(t) \in E(t, x^n(t)) a.e. P - a.s$ and x^n is a weak solution of the stochastic evolution equation

$$\begin{aligned} &dx + A(t, x(t))dt = f^n(t)dt + \sigma(t)dW, t \in I, \\ &x(0) = x^n(0) = x_0. \end{aligned}$$

That is, for each $t \in I$ and each $v \in V$,

$$(x^n(t), v) = (x_0, v) - \int_0^t < A(s, x^n(s)), v > ds + \int_0^t (f^n(s), v) ds + \int_0^t (\sigma^*(s)v, dW(s))$$

where the pairing under the stochastic integral is the scalar product in the Hilbert space F. By virtue of the apriori estimates (see Lemma 4.1) and the fact that $f^n(t) \in E(t, x^n(t)) a.e.\ P - a.s$, it follows from assumption (G1) that

$$\mathcal{E} \int_I \| f^n(t) \|_H^2 \, dt \le (1 + c) \| k \|^2$$

independent of $n \in N_+$. Hence there exists a subsequence $\{f^{n_k}, x^{n_k}\}$ of the sequence $\{f^n, x^n\}$, relabeled as $\{f^n, x^n\}$ and a pair $\{f^0, x^0\} \subset L_2^e(H) \times (L_\infty^e(H) \cap L_p^e(V) \cap M_\tau)$ such that

$$\begin{aligned} f^n &\xrightarrow{w} f^0 \quad \text{in} \quad L_2^e(H), \\ x^n &\xrightarrow{w^*} x^0 \quad \text{in} \quad L_\infty^e(H), \\ x^n &\xrightarrow{w} x^0 \quad \text{in} \quad L_p^e(V). \end{aligned} \tag{5.2}$$

We must prove that the pair $\{x^0, f^0\}$ satisfies the equation (4.1) and the inclusion (4.2). But this follows from similar arguments as given in the proof of Theorem 4.4. Hence $x^0 \in \mathcal{X}_\mu$ proving the closure. A bounded weakly sequentially closed subset of a reflexive Banach space is weakly sequentially compact. Since $L_p^e(V)$ is reflexive, $\mathcal{X}_\mu$, considered as a subset of $L_p^e(V)$, is weakly sequentially compact.

By Alaoglu's theorem, a bounded w^*- sequentially closed subset of $L_\infty^e(H)$ is w^*- sequentially compact. Hence $\mathcal{X}_\mu$, considered as a subset of $L_\infty^e(H)$, is w^* sequentially compact. Hence by Eberlien-Smulian theorem, $\mathcal{X}_\mu$ is a weakly compact subset of $L_p^e(V)$ and a w^*-compact subset of $L_\infty^e(H)$. This completes the proof. □

Before we can prove the existence of optimal controls we need one more lemma.

LEMMA 5.2. *Consider the stochastic evolution equation (4.1) and let N denote the map $x \equiv N(g)$ where $x(t) \equiv N_t(g), t \in I \equiv [0, \tau]$ is the solution of the evolution equation (4.1) corresponding to the process $g \in L_2^e(H)$. The map N is lipschitz from $L_2^e(H) \longrightarrow M_\tau$.*

Proof. Let $x_0, y_0 \in L_2(\mathcal{F}_0, H)$ and $f, g \in L_2^e(H)$ and let $x, y \in M_\tau$ denote the corresponding solutions of equation (4.1). Then by using the monotonicity of the operator A, one can easily verify that

$$\mathcal{E}(\| x(t) - y(t) \|_H^2) \leq \mathcal{E}(\| x_0 - y_0 \|_H^2) + \mathcal{E} \int_I \| f(s) - g(s) \|_H^2 \, ds + \int_0^t \mathcal{E} \| x(s) - y(s) \|_H^2 \, ds$$

for $t \in I$. The conclusion of the Lemma then follows from Gronwall inequality by setting $x_0 = y_0$. □

First we shall prove the existence of generalized random controls.

THEOREM 5.3. *Suppose the assumptions of Theorem 4.4 hold and the functional $x, \mu \longrightarrow \mathcal{E}\psi(x, \mu)$ is lower semi continuous in x in the w^*- topology of $L_\infty^e(H)$ and lower semi continuous in μ in τ_Y topology of $\mathcal{U}_r$. Then the control problem : (3.1) and (1.2) has a solution , that is, there exists a control $\mu^0 \in \mathcal{U}_r$ such that $J^0(\mu^0) \leq J^0(\mu)$ for all $\mu \in \mathcal{U}_r$.*

Proof. First we shall prove that $\mu \longrightarrow J^0(\mu)$ is lower semi continuous in the Young topology τ_Y of $\mathcal{U}_r$. With this end in view, we prove that the map $\mu \longrightarrow X_\mu$ is lower semi continuous with respect to inclusion from $\mathcal{U}_r$ to $2^{L_\infty^e(H)} \setminus \emptyset$ in their respective w^* topologies, where X_μ denotes the solution set corresponding to the control strategy μ. Note that by virtue of Lemma 4.2 and assumption

(G1), all solutions of the evolution inclusion (3.1) corresponding to any admissible control must satisfy the a priori bound given there. Hence, for any $\eta \in L_2^e(H)$ satisfying $\eta(t) \in G(t, x(t), \mu_t)$ for any $\mu \in \mathcal{U}_r$ and any $x \in \mathcal{X}_\mu$, we have

$$\mathcal{E} \int_I \| \eta(t) \|_H^2 \, dt \leq \| k \|^2 (1+c).$$

Define

$$\Sigma \equiv \{\beta \in L_2^e(H) : \mathcal{E} \int_I \| \beta(t) \|_H^2 \, dt \leq (1+c) \| k \|^2\}.$$

For each $\xi \in \Sigma$, define $\hat{G}_\mu(\xi)$ to be the set of measurable selections of the multi function $G_\mu(t, N_t(\xi)) \equiv G(t, N_t(\xi), \mu_t)$. Since these multifunctions are measurable and (square) integrably bounded with closed values , by a generalized Aumann selection theorem as mentioned earlier, they are nonempty and have selections from $L_2^e(H)$. Let $\{\mu^n\}$ be a sequence from $\mathcal{U}_r$ converging in the w^*-sense to μ^0.

Then by virtue of assumption (G3), for every $\epsilon > 0$, there exists an integer n_0 such that for all $n \geq n_0$,

$$\hat{G}_{\mu^0}(\xi) \subset \hat{G}_{\mu^n}(\xi) + \epsilon B_1(L_2^e(H))$$

for all $\xi \in \Sigma$ where $B_1(Z)$ denotes the unit ball in any normed space Z. Let $Fix(\hat{G}_\mu) \equiv \{\xi \in \Sigma : \xi \in \hat{G}_\mu(\xi)\}$ denote the set of fixed points of the multi function $\hat{G}_\mu$. Clearly by Theorem 4.4 this is a nonempty set. Hence it follows from the above inclusion that, for the same $\epsilon > 0$, there exists an integer $n_1 \geq n_0$, such that

$$Fix(\hat{G}_{\mu^0}) \subset Fix(\hat{G}_{\mu^n}) + \epsilon B_1(L_2^e(H)),$$

for all $n \geq n_1$. Note that the solution set is given by $X_\mu = N(Fix(\hat{G}_\mu))$. Hence, due to the operator N being Lipschitz, there exists an integer $n_2 \geq n_1$ such that

$$X_{\mu^0} \equiv N(Fix(\hat{G}_{\mu^0})) \subset N(Fix(\hat{G}_{\mu^n})) + \epsilon B_1(M_\tau) \equiv X_{\mu^n} + \epsilon B_1(M_\tau),$$

for all $n \geq n_2 \geq n_1$. This shows that the map $\mu \longrightarrow X_\mu$ is lower semicontinuous with respect to inclusion. We show that J^0 is lower semi continuous. Let $\mu^* \in \mathcal{U}_r$ and suppose $\{\mu^m\} \subset \mathcal{U}_r$ is a sequence such that

$$\mu^m \xrightarrow{\tau_Y} \mu^*$$

as $m \longrightarrow \infty$. We show that $J^0(\mu^*) \leq \underline{Lim}_{m\to\infty} J^0(\mu^m)$. Let d be any real number satisfying $d < J^0(\mu^*)$. Then there exists an $x^* \in X_{\mu^*}$ such that $d \leq \mathcal{E}(\psi(x^*, \mu^*))$. Since $\mu \longrightarrow X_\mu$ is lower semi continuous with respect inclusion there exists a sequence $\{x^m\} \in X_{\mu^m}$ such that $x^m \xrightarrow{w^*} x^*$ in $L_\infty^e(H)$ and, owing to lower semi continuity of $\mathcal{E}\psi(x.\mu)$,

$$\mathcal{E}\psi(x^*, \mu^*) \leq \underline{Lim}_{m\to\infty} \mathcal{E}\psi(x^m, \mu^m).$$

Clearly, by definition of J^0, we have $\mathcal{E}\psi(x^m, \mu^m) \leq J^0(\mu^m)$. Hence

$$d \leq \mathcal{E}\psi(x^*, \mu^*) \leq \underline{Lim}_{m\to\infty} \mathcal{E}\psi(x^m, \mu^m) \leq \underline{Lim}_{m\to\infty} J^0(\mu^m).$$

Since $d < J^0(\mu^*)$ is arbitrary, we can let $d \uparrow J^0(\mu^*)$ thereby obtaining

$$J^0(\mu^*) \leq \underline{lim}_{m\to\infty} J^0(\mu^m).$$

Hence J^0 is lower semi continuous in the Young topology. Since $\mathcal{U}_r$ is τ_Y-compact, J^0 attains it's infimum in $\mathcal{U}_r$. This completes the proof. □

REMARK 5.4. Suppose the injection $V \hookrightarrow H$ is compact. Then by a well known result, [Zeidler,16, p 450], the injection $W_{p,q} \hookrightarrow L_p(H)$ is compact. Hence in the deterministic case, according to remark 4.5 and Theorem 4.6, the existence result of Theorem 5.3 can be improved as stated below.

THEOREM 5.5. *Consider the system (4.14) with the cost functional*

$$J^0(\mu) \equiv Sup\{\Psi(x,\mu), x \in X_\mu\}, \tag{5.3}$$

where Ψ is lower semi continuous in x on $L_p(H)$ and w^*-lower semi continuous in μ on $L_\infty(I, \mathcal{M}(U))$. Then the control problem (4.14), (5.3) has a solution.

6. FEED BACK CONTROLS FOR PARTIALLY OBSERVED SYSTEM

Again we let U be a compact Polish space and $\mathcal{M}(U)$ the Banach space of bounded signed measures on the σ-algebra of Borel subsets of the set U furnished with the total variation norm. Let $Y = C(I, R^m)$ be furnished with the metric topology induced by the usual sup norm. Consider the topological space of continuous maps from Y to $L_\infty(I, \mathcal{M}(U))$ denoted by

$$\mathcal{C}(Y, L_\infty(I, \mathcal{M}(U)))$$

furnished with the topology of w^*-convergence uniformly on compact subsets of Y. That is,

$$\mu_y^n \xrightarrow{w^*} \mu_y \text{ in } L_\infty(I, \mathcal{M}(U))$$

uniformly with respect to y on compact subsets of Y. Since, by Dunford-Pettis theorem, the dual of $L_1(I, C(U))$ is $L_\infty(I, \mathcal{M}(U))$ it follows from the above discussion that, for every $f \in L_1(I, C(U))$,

$$\text{Lim}_{n\to\infty} \int_{I\times U} f(t,\sigma) \ \mu_y^n(t, d\sigma)dt \longrightarrow \int_{I\times U} f(t,\sigma) \ \mu_y(t, d\sigma)dt$$

uniformly with respect to y on every compact subset of Y. We shall denote this topology by τ_{co}. Let $\mathcal{C}_p(Y, L_\infty(I, \mathcal{M}(U)))$ denote the closure, in this topology, of the class of nonanticipative policies in the sense that for any $\mu \in \mathcal{C}_p(Y, L_\infty(I, \mathcal{M}(U)))$ and any pair $y,\ z \in Y$ satisfying $y(s) = z(s), s \leq t$ for some $t \in I$, we have $\mu_y(s,\Gamma) = \mu_z(s,\Gamma)$ for any $\Gamma \in \mathcal{B}(U)$ and all $s \leq t$.

For control applications it is natural to replace $\mathcal{M}(U)$ by the space of probability measures $P\mathcal{M}(U)$ and it is necessary that the control policies be nonanticipative. Hence, let $\mathcal{U}_f$, a closed subset of $\mathcal{C}_p(Y, L_\infty(I, \mathcal{M}(U)))$, denote the class of admissible relaxed feedback controls satisfying the following natural conditions:

$(1): \mu_y(t, U) = 1$ for all $t \in I, y \in Y$,

$(2): \int_{D\times U} \chi_{D\times U}(t,u)\ \mu_y(t,du)dt = \ell(D)$ for all $y \in Y$, where $\chi_{D\times U}$ is the characteristic function of the set $D \times U$ with D being any Borel suset of I,

$(3): \int_{I\times U} f(t,u)\ \mu_y(t,du)dt \geq 0$, for $f \geq 0$ and $f \in L_1(I, C(U))$ for all $y \in Y$.

For convenience, we shall often write, with slight abuse of notation, $\mu(t, \Pi_t y)$, for the values of μ_y for each $\mu \in \mathcal{U}_f$ where $\Pi_t y \equiv \{y(s), s \leq t\}$. Note $\mu(t, \Pi_t y) \in \mathcal{M}(U)$ for each fixed $t \in I$ and $y \in Y$.

For each $\mu \in \mathcal{U}_f$, consider the partially observed system

$$\begin{aligned} &dx + A(t,x)dt \in G(t,x,\mu_y)dt + \sigma(t)dW; x(0) = x_0 \\ &dy = h(x,y)dt + \sigma_0(x,y)dW_0, y(0) = 0; \end{aligned} \tag{6.1}$$

where y is the observed (measured) output process taking values from R^m, $h : H \times R^m \longrightarrow R^m$, and W_0 is a d-dimensional standard Brownian motion and $\sigma_0 : H \times R^m \longrightarrow \mathcal{L}(R^d, R^m)$. In other words only a finite number of noisy outputs are available for controlling the system.

The control problem is to find a feedback control law from the class $\mathcal{U}_f$ that minimizes the cost functional

$$\begin{aligned} J^0(\mu) \equiv Sup\{\mathcal{E}\Psi(y_\tau) : \{x,y\} &\text{ any solution of } (6.1) \\ &\text{corresponding to a fixed contol law } \mu\}. \end{aligned} \tag{6.2}$$

We shall use the following standard assumptions for the coefficients of the finite dimensional system.

(H1): there exists a constant k_0 such that

$$\begin{aligned} &|z(x_1,y_1) - z(x_2,y_2)|^2 \leq k_0^2(\| x_1 - x_2 \|_H^2 + |y_1 - y_2|^2) \text{ for all } x_1, x_2 \in H \\ &\qquad\qquad \text{and } y_1, y_2 \in R^m \\ &|z(x,y)|^2 \leq k_0^2(1 + \| x \|^2 + |y|^2) \text{ for all } x \in H, y \in R^m; \end{aligned}$$

where z stands both for h and σ_0, and $|.|$ denotes the Euclidean norm in any finite dimensional space.

Note that, given $x \in L_2^{\mathfrak{e}}(H)$, the equation (6.1b) is independent of (6.1a) and can be solved by standard methods.

LEMMA 6.1. *Under the assumption (H1), for each given $x \in L_2^{\mathfrak{e}}(H)$, the finite dimensional system (6.1b) has a unique solution $y \in M_\tau(R^m)$ and that $y \in Y \equiv C(I, R^m)$ P-a.s. Further, the map $F : M_\tau(H) \longrightarrow M_\tau(R^m)$, denoting the state to output map $x \longrightarrow y$ with* *values $y(t) = F_t(x)$, is Lipschitz, and maps bounded subsets of $M_\tau(H)$ into bounded subsets of $M_\tau(R^m)$.*

Proof. The proof is standard. □

Again, the question of existence of a solution of (6.1) reduces to the question of existence of a fixed point of the multifunction $\hat{G}_\mu$ defined by

$$\hat{G}_\mu(g) \equiv \{f \in L_2^e(H) : f(t) \in G(t, N_t(g), \mu(t, \Pi_t((FoN)(g)))a.e.Pa.s; \} \quad (6.3)$$

for $g \in L_2^e(H)$ where $\mu \in \mathcal{U}_f$ is any fixed control law and FoN is the composition of the maps F and N. Note that the a-priori bounds given by Lemma 4.2 remain valid for this case also and hence by virtue of the growth assumption (G1), the fixed points of $\hat{G}_\mu$ must belong, as before, to the same set

$$\Sigma \equiv \{\beta \in L_2^e(H) : \mathcal{E}\int_I \| \beta(t) \|_H^2 \, dt \leq (1+c) \| k \|^2\}.$$

The questions of existence of solutions of this class of systems have been studied in [7,8,10] by other methods. Here we shall consider the question of existence of optimal feedback controls. For this we shall need the following lemmas.

LEMMA 6.2. *Consider $\mathcal{U}_f$ to be the class of admissible relaxed feedback controls. Suppose the set $\mathcal{U}_f$ is pointwise closed in the Young topology and equicontinuous on compact subsets of Y. Then $\mathcal{U}_f$ is compact in the compact-open topology. Further, restricted to any compact subset of Y, $\mathcal{U}_f$ is sequentially compact.*

Proof. For proof see [6, Lemma 2].

LEMMA 6.3. *Suppose for each $\mu \in \mathcal{U}_f$, $\hat{G}_\mu(g) \cap \Sigma \neq \emptyset$ for all $g \in \Sigma$, then the set of output processes $\mathcal{Y}_0 \equiv \{y^x, x \in X_\mu, \mu \in \mathcal{U}_f\}$ is a compact family of stochastic processes with sample paths belonging to Y P-a.s.*

Proof. Replacing $(L_2^e(R^n), \tau_w)$ by $(L_2^e(H), \tau_w)$ and following similar arguments as in Lemma 4.3, we can prove that for each $\mu \in \mathcal{U}_f$, $Fix(\hat{G}_\mu) \neq \emptyset$. Hence for each $\mu \in \mathcal{U}_f, X_\mu \neq \emptyset$ and it is given by

$$X_\mu = N(Fix(\hat{G}_\mu)). \quad (6.4)$$

Define $Y_\mu \equiv F(X_\mu)$. This is the output set corresponding to a fixed control law μ. Then the set of all realizable outputs is given by

$$\mathcal{Y}_0 = \bigcup\{Y_\mu, \mu \in \mathcal{U}_f\}.$$

By virtue of the apriori bound as given in Lemma 4.2, it follows from Lemma 6.1, that this is a bounded subset of $M_\tau(R^m) \subset C(I, L_2(\Omega, R^m))$, with the sample paths belonging to Y P-a.s. In fact for all $x \in \mathcal{X} \equiv \bigcup\{X_\mu, \mu \in \mathcal{U}_f\}$, one can easily verify that there are constants c_1, c_2, independent of x, (depending only on k_0, τ and c of Lemma 4.2), such that

$$\mathcal{E}|y^x(t)|^2 \leq c_1 + c_2 \int_0^t \mathcal{E}|y^x(s)|^2 ds.$$

Hence by Gronwall inequality, $Sup\{\mathcal{E}|y^x(t)|^2, t \in I, x \in \mathcal{X}\} < \infty$. Thus

$$Lim_{N\to\infty}Sup_{x\in\mathcal{X}}Sup_{t\in I}P\{|y^x(t)| > N\} = 0. \tag{6.5}$$

Again we can prove that there exists a constant c_3 such that

$$\mathcal{E}|y^x(t) - y^x(s)|^2 \leq c_3|t-s|, s, t \in I,$$

uniformly with respect to $x \in \mathcal{X}$. Thus

$$Lim_{h\to 0}Sup_{x\in\mathcal{X}}Sup_{|t-s|<h}P\{|y^x(t) - y^x(s)| > \epsilon\} = 0 \tag{6.6}$$

for all $\epsilon > 0$. Hence it follows from a probabilistic analog of Arzela-Ascoli theorem [Gihmann-Skorohod,14, theorem 3.11, pp147,] that the set $\mathcal{Y}_0$ is sequentially compact in the sense that every sequence $\{y^k\} \in \mathcal{Y}_0$ has a subsequence that converges in probability in the usual topology of $Y = C(I, R^m)$ to an element $y^0 \in \mathcal{Y}_0$. This completes the proof. □

Now we are prepared to prove the existence of optimal controls.

THEOREM 6.4. *Consider the control problem (6.1)-(6.2) and suppose the function $\Psi : R^m \longrightarrow R \cup \{+\infty\}$ is lower semi continuous satisfying $\Psi(\zeta) \geq -a|\zeta|^2 - b$ for some $a, b \in R$ and for all $\zeta \in R^m$. Then there exists an optimal relaxed feedback control.*

Proof. Consider the cost functional J^0 as defined by (6.2). Since the elements of $\mathcal{Y}_0$ are continuous with probability one, y_τ as well as $\Psi(y_\tau)$ are well defined and hence $\mu \longrightarrow J^0(\mu)$ is a well defined functional on $\mathcal{U}_f$ taking values from the extended real number system. If $J^0(\mu) \equiv +\infty$ for all $\mu \in \mathcal{U}_f$, there is nothing to prove. So we may assume the contrary. Let $\{\mu^n \in \mathcal{U}_f\}$ be a minimizing sequence so that

$$Lim_{n\to\infty}J^0(\mu^n) = Inf\{J^0(\mu), \mu \in \mathcal{U}_f\} = \varpi. \tag{6.7}$$

Since $\mathcal{Y}_0$ is a bounded subset of $M_\tau(R^m)$, and Ψ majorizes a quadratic function, $\varpi > -\infty$. Again $\mathcal{U}_f$ being sequentially compact, there exists a subsequence $\{\mu^{n_k}\} \subset \{\mu^n\}$ and $\mu^0 \in \mathcal{U}_f$ such that

$$\mu^{n_k} \xrightarrow{\tau_{co}} \mu^0. \tag{6.8}$$

Since $\mu \longrightarrow X_\mu$ is lower semi continuous, for every $\epsilon > 0$, there exists an integer $n_0 = n_0(\epsilon)$ such that

$$X_{\mu^0} \subset X_{\mu^n} + \epsilon B_1(M_\tau(H))$$

for all $n \geq n_0$. By Lemma 6.1, $F : M_\tau(H) \longrightarrow M_\tau(R^m)$ is Lipschitz and maps bounded sets into bounded sets, and hence, for the given $\epsilon > 0$, there exists an integer $n_1 \geq n_0$ such that

$$Y_{\mu^0} = F(X_{\mu^0}) \subset F(X_{\mu^n}) + \epsilon B_1(M_\tau(R^m)) = Y_{\mu^n} + \epsilon B_1(M_\tau(R^m))$$

for all $n \geq n_1$. Thus $\mu \longrightarrow Y_\mu$ is also lower semicontinuous with respect to inclusion. Choose any $d < J^0(\mu^0)$.

Then there exists a $y^0 \in Y_{\mu^0}$ such that

$$d \leq \mathcal{E}\Psi(y_T^0). \tag{6.9}$$

By lower semi continuity of the map $\mu \longrightarrow Y_\mu$, it follows from (6.8) that, for k sufficiently large, say $k \geq k_0$, we have

$$Y_{\mu^0} \subset Y_{\mu^{n_k}} + \epsilon B_1.$$

Hence, by virtue of Lemma 6.3, we can find a subsequence $\{y^{n_{k_r}}\} \subset Y_{\mu^{n_{k_r}}}$ of the sequence $\{y^{n_k}\} \subset Y_{\mu^{n_k}}$ such that

$$y_T^{n_{k_r}} \longrightarrow y_T^0 \; P - a.s,$$

as $r \longrightarrow \infty$.

Then by lower semicontinuity of Ψ, we obtain

$$\Psi(y_T^0) \leq \underline{Lim}_{r\to\infty}\Psi(y_T^{n_{k_r}}) \; P - a.s. \tag{6.10}$$

Using Fatou's Lemma and the definition of J^0, we conclude that

$$\mathcal{E}\Psi(y_T^0) \leq \underline{Lim}_{r\to\infty}\mathcal{E}\Psi(y_T^{n_{k_r}}) \leq \underline{Lim}_{r\to\infty}J^0(\mu^{n_{k_r}}). \tag{611}$$

Since $d < J^0(\mu^0)$ is arbitrary, we can let $d \uparrow J^0(\mu^0)$ and use (6.9) and (6.11) to obtain

$$J^0(\mu^0) \leq \underline{Lim}_{r\to\infty}J^0(\mu^{n_{k_r}}) = Lim_{r\to\infty}J^0(\mu^{n_{k_r}}) = \varpi. \tag{6.12}$$

Since $\mu^0 \in \mathcal{U}_f$, $\varpi \leq J^0(\mu^0)$. Hence μ^0 is the optimal feedback control. □

REFERENCES

[1] Ahmed , N.U, Teo, K.L; Optimal Control of Distributed Parameter Systems, North Holland, New York, Oxford, 1981.

[2] Ahmed, N.U., Existence of Optimal Controls for a class of Systems Governed by Differential Inclusions on a Banach Space., Journal of Optimization Theory and Applications (JOTA), 50,2, (1986), 213-237.

[3] Ahmed, N.U., Existence of Optimal Relaxed Controls for Differential Inclusions in Banach Space. Nonlinear Analysis and Application (Ed.V.Lakshmikantham), Lect. Notes in Pure and Applied Math.109,39-49, Marcel Dekker,New York and Basel,1987.

[4] Ahmed, N.U.,1988, Optimization and Identification of Systems Governed by Evolution Equations on Banach Space, Pitman Res.Notes.in Math.Series, vol 184, Longman Scientific & Technical, U.K; Co-Publisher John Wiley & Sons, New York.

[5] Ahmed, N.U, (1990), Relaxed Controls for Stochastic Boundary Value Problems in Infinite Dimension, Lect. Notes in Control and Information Sciences, (Eds. K.H.Hoffmann, W.Krabs), Optimal Control of Partial Differential Differential Equations, Proc. of the IFIF WG 7.2 International conference, Irsee, Springer-Verlag, 149, 1-10.

[6] Ahmed, N. U., (1991), Partially Observed Optimal Relaxed Feedback Controls for Infinite Dimensional Stochastic Systems, IMACS Trans. (to Appear); also presented at the 13th IMACS World Congress on Computation and Applied Mathematics, Dublin, Ireland.

[7] Ahmed,N.U.,(1992), An Existence Theorem for Stochastic Nonlinear Evolution Equations on Banach Space. J.Stochastic Analysis and Applications,10,4 . 379-385.

[8] Ahmed, N. U, (1992), An Existence Theorem for Differential Inclusions on Banach Space, J. of Applied Math. and Stochastic Analysis, Vol.5,No.2 pp123-130.

[9] Ahmed, N. U. , Xiang. X., (1992), Admissible Relaxation in Optimal Control Problems for Infinite Dimensional Uncertain Systems, J. of Applied Math. and Stochastic Analysis, Vol5,No3,pp227-236.

[10] Ahmed .N.U., Existence of Solutions of Nonlinear Stochastic Differential Inclusions in Banach Space, Proceedings of the World Congress on Nonlinear Analysis and Applications (Ed. V. Lakshmikantham), Walter de Gruyter and Co., Berlin, New York, (to appear).

[11] Aubin.J.P., Frankowska.H, (1990), Set-Valued Analysis, Birkhauser, Boston.Basel.Berlin.

[12] Browder.F.E,(1976), Nonlinear Operators and Nonlinear Equations of Evolution in Banach Spaces, Proc.of Symp.in Pure and Applied Math.,Vol XVIII,part 2., AMS,Providence,Rhode Island.

[13] Dunford.N.,Schwartz.J.T, (1964), Linear Operators, Part 1, John Wiley & Sons, New York. London.

[14] I.I. Gihman, A.V. Skorohod, Controlled Stochastic Systems, Springer-Verlag; New York, Heidelberg, Berlin; 1977.

[15] Wagner, D.H.,(1977), Survey of Measurable Selection Theorems. SIAM J. Contr. Optim., 15,5 ,859-903.

[16] Zeidler. E. (1990), Nonlinear Functional Analysis and its Applications II, Springer, New York.

2 Optimal Control Problems Governed by Volterra Integral Inclusions

Sergiu Aizicovici Ohio University, Athens, Ohio

Nikolaos S. Papageorgiou Florida Institute of Technology, Melbourne, Florida

1 INTRODUCTION AND PRELIMINARIES

The purpose of this note is to summarize some recent results on the variational stability of a class of nonlinear, infinite dimensional optimal control problems, monitored by Volterra integral equations. A more detailed discussion with complete proofs will appear elsewhere [2].

We consider the parametrized control problem

$$\eta(x,\lambda) \to \inf = m(\lambda), \text{ over all } x \in C([0,T];X) \text{ satisfying}$$

$$x(t) + \int_0^t k(t-s)[A(\lambda)x(s) + f(s,x(s),\lambda)u(s)]\,ds \ni g(t,\lambda),\ t \in [0,T], \quad (\mathrm{P}_\lambda)$$

$$u(t) \in U(t,\lambda), \text{ a.e. on } (0,T),\ u - \text{measurable.}$$

Here X is a real Banach space, $\lambda \in \Lambda$ (a complete metric space), $A(\lambda)$ ($\lambda \in \Lambda$) is an m-accretive operator in X, $U : [0,T] \times \Lambda \to 2^Y$, where Y is another Banach space (the control space), $k : [0,T] \to \mathbb{R}$, $f : [0,T] \times X \times \Lambda \to \mathcal{L}(Y,X)$ (the space of all bounded linear operators from Y to X), $g : [0,T] \times \Lambda \to X$, and

$\eta : C([0,T];X) \times \Lambda \to \mathbb{R}$. Of main concern are the continuity properties of the function $\lambda \to m(\lambda)$ (the value of the minimization problem (P_λ)). In particular, we will see that (P_λ) is well-posed. Our analysis is based on a preliminary study of a related abstract parametrized Volterra integral inclusion.

We now recall some basic definitions and notations from the theory of multifunctions and set convergence. For more details, see [3, 6, 7].

Throughout this paper, $(X, \|\cdot\|)$ denotes a real separable Banach (or Hilbert) space. We will use $P_{f(c)}(X)$ and $P_{(w)k(c)}(X)$ to designate the collection of all nonempty, closed (convex), and respectively (weakly) compact (convex), subsets of X. On $P_f(X)$ one can define the so called Hausdorff generalized metric h, by:

$$h(A,B) = \max\left\{ \sup_{a\in A} d(a,B), \sup_{b\in B} d(b,A) \right\} \quad (A, B \in P_f(X)),$$

where 'd' stands for distance. If Ω is a measure space and $F : \Omega \to P_f(X)$, we say that F is measurable if the function $\omega \in \Omega \to d(x, F(\omega))$ is measurable for all $x \in X$. If V and W are Hausdorff topological spaces, then a multifunction $F : V \to 2^W$ is called continuous when it is both upper semicontinuous (i.e., $F^{-1}(\mathcal{O}) = \{v \in V : F(v) \subset \mathcal{O}\}$ is open for all open $\mathcal{O} \subset W$) and lower semicontinuous (i.e., $F^{-1}(C)$ is closed for all closed $C \subset W$).

Finally, let $\{A_n\}_{n\geq 1}$ be a sequence of nonempty subsets of X. Set:

$$\underline{\lim}\, A_n := \{x \in X : x = \lim_{n\to\infty} x_n, x_n \in A_n\},$$

$$\overline{\lim}\, A_n := \{x \in X : x = \lim_{k\to\infty} x_{n_k}, \text{ for some } x_{n_k} \in A_{n_k}\}.$$

Clearly, $\underline{\lim}\, A_n \subseteq \overline{\lim}\, A_n$. If $\underline{\lim}\, A_n = \overline{\lim}\, A_n$, then A is said to converge to A in Kuratowski's sense, and we write $A_n \xrightarrow{\mathrm{K}} A$.

2 MULTIVALUED VOLTERRA EQUATIONS

The study of (P_λ) is directly related to that of the Volterra integral inclusion

$$x(t) + \int_0^t k(t-s)(A(\lambda)x(s) + F(s, x(s), \lambda))\, ds \ni g(t,\lambda), t \in [0,T], \qquad (\mathrm{V}_\lambda)$$

in the Banach space X, where λ varies in a complete metric space Λ. The following minimal assumptions are needed:

(H_A) For each $\lambda \in \Lambda$, $A(\lambda) : D(\lambda) \subset X \to 2^X$ is an m-accretive operator such that $-A(\lambda)$ generates a compact contraction semigroup on $\overline{D(\lambda)}$,

(H_k) $k \in AC[0,T]$, $k(0) = 1$, $\dot{k} \in BV[0,T]$,

(H_F) $F : [0,T] \times X \times \Lambda \to P_{fc}(X)$ satisfies

(i) $t \to F(t,x,\lambda)$ is measurable for all $x \in X$, $\lambda \in \Lambda$,

(ii) $h(F(t,x,\lambda), F(t,y,\lambda)) \le \xi_\lambda(t)\|x-y\|$, a.e. on $(0,T)$, $\forall x,y \in X$, $\lambda \in \Lambda$, for some $\xi_\lambda \in L^1_+(0,T)$,

(iii) $|F(t,x,\lambda)| := \sup\{\|y\| : y \in F(t,x,\lambda)\} \le \alpha_\lambda(t) + \beta_\lambda(t)\|x\|$, a.e. on $(0,T)$, $\forall x \in X, \lambda \in \Lambda$, with $\alpha_\lambda, \beta_\lambda \in L^1_+(0,T)$,

(H_g) $g(\cdot,\lambda) \in W^{1,1}([0,T];X)$, $g(0,\lambda) \in \overline{D(\lambda)}$ $(\lambda \in \Lambda)$.

Using an idea of Crandall and Nohel [5], one can rewrite (V_λ) as a functional differential equation of the form

$$\dot{x}(t) + A(\lambda)x(t) + F(t,x(t),\lambda) \ni G(x,\lambda)(t), \quad t \in (0,T),$$
$$x(0) = g(0,\lambda),$$

where $G : C([0,T];X) \times \Lambda \to X$ is given by

$$G(x,\lambda)(t) = \dot{g}(t,\lambda) + r * \dot{g}(t,\lambda) - r(0)x(t) + r(t)g(0,\lambda) - x * \dot{r}(t)$$
$$\left(\text{with } x * \dot{r}(t) = \int_0^t x(t-s)\,dr(s)\right) \tag{1}$$
$$r + \dot{k} * r = -\dot{k}.$$

Definition. A generalized solution of (V_λ) is a function $x \in C([0,T];X)$ for which there exists $f(\cdot,\lambda) \in L^1(0,T;X)$, with $f(t,\lambda) \in F(t,x(t),\lambda)$, a.e. on $(0,T)$, such that x is an integral solution (in the sense of [4]) of $\dot{x} + A(\lambda)x(t) = G(x,\lambda)(t) - f(t)$, $t \in [0,T]$; $x(0) = g(0,\lambda)$, where G is defined by (1).

It turns out that assumptions (H_A), (H_k), (H_F) and (H_g) guarantee the existence of a generalized solution to (V_λ). Specifically, if $S(\lambda)$ denotes the solution set of (V_λ), we have, as a consequence of the results in [1]:

Theorem 1. *Suppose that X is reflexive and that conditions (H_A), (H_k), (H_F) and (H_g) are satisfied. Then $S(\lambda)$ is a nonempty compact subset of $C([0,T];X)$.*

We next examine the continuous dependence on λ of the solution set $S(\lambda)$. To this end, we need the following stronger hypotheses:

(H'_A) $A(\lambda)$ $(\lambda \in \Lambda)$ satisfy (H_A), and $(I + \mu A(\lambda_n))^{-1}x \to (I + \mu A(\lambda))^{-1}x$ for all $\mu > 0$, $x \in H$, as soon as $\lambda_n \to \lambda$ in Λ $(n \to \infty)$,

(H'_F) $F : [0,T] \times X \times \Lambda \to P_{wkc}(X)$ satisfies (H_F) with ξ_B, α_B, β_B in place of $\xi_\lambda, \alpha_\lambda, \beta_\lambda$, respectively, whenever $\lambda \in B \subset \Lambda$ with B compact; in addition $\lambda \to F(t,x,\lambda)$ is d-continuous (i.e., $\forall z \in X$, $\lambda \to d(z, F(t,x,\lambda))$ is continuous),

(H'_g) g satisfies H_g; also, $\lambda \to g(\cdot,\lambda)$ is continuous from Λ into $W^{1,1}([0,T];X)$ and $\|\dot{g}(t,\lambda)\| \le \theta_B(t)$, a.e. on $(0,T)$.

The following result can be viewed as a generalization of [1, Theorem 4.3].

Theorem 2. *Assume that X^* is uniformly convex and that* (H'_A), (H_k), (H'_F) *and* (H'_g) *hold. If $\lambda_n \to \lambda$ in Λ as $n \to \infty$, then $S(\lambda_n) \xrightarrow{K} S(\lambda)$ in $C([0,T];X)$.*

The conclusion of Theorem 2 can be strengthened when X is a Hilbert space and $A(\lambda)$ are cyclically maximal monotone in X for all $\lambda \in \Lambda$. More precisely, we impose the following condition (in place of (H'_A)):

$(\mathrm{H}\varphi)$

- (i) $A(\lambda)(\cdot) = \partial\varphi(\,\cdot\,,\lambda)$ (∂ = subdifferential), where $\varphi : X \times \Lambda \to (-\infty, +\infty]$ is proper, convex and lower semicontinuous,
- (ii) if $\lambda_n \to \lambda$ in Λ, as $n \to \infty$, then $(I + \mu A(\lambda_n))^{-1}x \to (I + \mu A(\lambda))^{-1}x$, for all $x \in X$ and $\mu > 0$,
- (iii) for all compact $B \subseteq \Lambda$ and $\varepsilon > 0$, the set $\bigcup_{\lambda \in B}\{x \in X : \|x\|^2 + \varphi(x,\lambda) \le \varepsilon\}$ is precompact in X.

We now obtain

Theorem 3. *Let X be a real separable Hilbert space and let* (H_k), (H_φ) *be satisfied. In addition assume that* (H'_F) *and* (H'_g) *hold with $\alpha_B, \beta_B, \theta_B \in L^2_+(0,T)$, and that $g(0,\lambda) \in \operatorname{dom}\varphi(\,\cdot\,,\lambda) := \{x \in H : \varphi(x,\lambda) < \infty\}$, with $\sup_{\lambda \in B}\varphi(g(0,\lambda),\lambda) < \infty$ for each compact $B \subset \Lambda$. Then $S : \Lambda \to P_f(C[0,T];X)$ is continuous.*

Remark. It is worth noting that in the setup of Theorem 3, $S(\lambda)$ consists of strong solutions of (V_λ), for each $\lambda \in \Lambda$.

3 THE CONTROL PROBLEM

We consider the problem (P_λ), where X is a real separable Hilbert space, Y is a real separable Banach space, and $A(\lambda)$, k and g satisfy at least (H_A), (H_k) and (H_g), respectively. As regards f, U and η, we assume that

(H_f) $f : [0,T] \times X \times \Lambda \to \mathcal{L}(Y,X)$ satisfies

- (i) $t \to f(t,x,\lambda)$ is measurable,
- (ii) $\|f(t,x,\lambda) - f(t,z,\lambda)\|_{\mathcal{L}} \le \theta_B(t)\|x - z\|$, a.e. on $(0,T)$, for all $\lambda \in B \subseteq \Lambda$, with B compact, and some $\theta_B \in L^1_+(0,T)$,
- (iii) $\|f(t,x,\lambda)\|_{\mathcal{L}} \le \alpha_B(t) + \beta_B(t)\|x\|$, a.e. on $(0,T)$, $\forall\lambda \in B \subseteq \Lambda$ (B compact), with $\alpha_B, \beta_B \in L^2_+(0,T)$,
- (iv) $\lambda \to f(t,x,\lambda)$ and $\lambda \to f^*(t,x,\lambda)$ are continuous from Λ into $\mathcal{L}(Y,X)$ and $\mathcal{L}(X,Y^*)$, respectively.

(H_U) $U : [0,T] \times \Lambda \to P_{wkc}(Y)$ is such that

- (i) $t \to U(t,\lambda)$ is measurable,

(ii) for each compact $B \subset \Lambda$, there exists $W_B \in P_{wkc}(Y)$ with $U(t, \lambda) \subset W_B$, a.e. on $(0, T)$, $\forall \lambda \in B$,

(iii) $\lambda \to U(t, \lambda)$ is continuous as a multifunction from Λ to Y.

(H_η) $\eta : C([0, T]; X) \times \Lambda \to \mathbb{R}$ is continuous.

Define $F : [0, T] \times X \times \Lambda \to P_{wkc}(X)$ by

$$F(t, x, \lambda) = f(t, x, \lambda) U(t, \lambda). \tag{2}$$

On account of (H_f) and (H_U), one can show that F, as defined by (2), satisfies (H'_F), and that (P_λ) can equivalently be rewritten as

$$\begin{aligned} &\eta(x, \lambda) \to \inf = m(\lambda), \text{ such that} \\ &x(t) + \int_0^t k(t-s)[A(\lambda)x(s) + F(s, x(s), \lambda)]\, ds \ni g(t, \lambda), \ 0 \le t \le T. \end{aligned} \tag{P'_λ}$$

By Theorem 1, it then follows that (P'_λ), and hence (P_λ) are well-posed. Furthermore, by strengthening the assumptions on $A(\lambda)$ and g, and invoking Theorems 2 and 3, we can establish the following result.

Theorem 4. *Let conditions (H'_A), (H_k), (H'_g), (H_g), (H_U) and (H_η) be satisfied. Then $m : \Lambda \to \mathbb{R}$ is upper semicontinuous. If also $A(\lambda)$ is cyclically maximal monotone for each $\lambda \in \Lambda$ and (H_φ) holds, then m is continuous.*

As an application of theorem 4, we study a terminal (Meyer type) optimal control problem. Let $Z \subset \mathbb{R}^N$ be a bounded domain with smooth boundary Γ, and let $D_i = \frac{\partial}{\partial z_i}, i = 1, 2, \dots, N$ ($z \in Z$). We consider the problem ($\lambda \in \Lambda, p \ge 2$)

$$\begin{aligned} &\int_Z l(z, x(T, z), \lambda)\, dz \to \inf = m(\lambda), \text{ such that} \\ &x(t, z) - \int_0^t k(t-s)\Big[\sum_{i=1}^N D_i(v_1(z, \lambda)|D_i x|^{p-2} D_i x) + v_2(z, x) x |x|^{p-2} \\ &\qquad - f_1(s, z, x(s, z), \lambda) u(s, z)\Big]\, ds \ni g_1(t, z, \lambda), \ t \in [0, T], \ z \in Z, \\ &x|_{[0,T]\times\Gamma} = 0, \quad |u(t, z)| \le \delta(t, z, \lambda), \quad u\text{-measurable}. \end{aligned} \tag{3}$$

We impose the following restrictions on the data in (3):

(H_v) $v_i : Z \times \Lambda \to \mathbb{R}$ ($i = 1, 2$) is measurable in z and continuous in λ; in addition, for each compact $B \subset \Lambda$, there are constants $c_{iB} > 0$ ($i = 1, 2$) such that $c_{1B} \le v_i(z, \lambda) \le c_{2B}$ for all $(z, \lambda) \in Z \times B$,

(H_{f_1}) $f_1 : [0, T] \times Z \times \mathbb{R} \times \Lambda \to \mathbb{R}$ satisfies

(i) $(t, z) \to f_1(t, z, x, \lambda)$ is measurable,
(ii) $|f_1(t, z, x, \lambda) - f_1(t, z, y, \lambda)| \leq \xi_B(t, z)|x - y|$ a.e. on $(0, T) \times Z$, $\lambda \in B \subset \Lambda$, and with $\xi_B \in L^1_+((0, T) \times Z)$,
(iii) $\lambda \to f_1(t, z, x, \lambda)$ is continuous,
(iv) $|f_1(t, z, x, \lambda)| \leq \alpha_B(t, z) + \beta_B(t, z)|x|$, a.e. on $(0, T) \times Z$, $\lambda \in B \subset \Lambda$ (compact), and with $\alpha_B, \beta_B \in L^2((0, T) \times Z)$,

(H_{g_1}) $\lambda \to g_1(\cdot, \cdot, \lambda)$ is continuous from Λ into $W^{1,2}(0, T; L^2(Z))$, $\|g_{1t}\|_{L^2(Z)} \leq \theta_B(t)$, a.e. on $(0, T)$, $\lambda \in B \subset \Lambda$ (compact), with $\theta_B \in L^2_+(0, T)$; in addition $g_1(0, \cdot, \lambda) \in W_0^{1,p}(Z)$ with $\sup_{\lambda \in B} \|g_1(0, \cdot, \lambda)\|_{W_0^{1,p}(Z)} < \infty$,

(H_δ) $\delta(\cdot, \cdot, \lambda) \in L^\infty((0, T) \times Z)$ with $\sup_{\lambda \in B} \|\delta(\cdot, \cdot, \lambda)\|_{L^\infty} < \infty$, for each compact $B \subset \Lambda$; in addition, $\lambda \to \delta(t, z, \lambda)$ is continuous,

(H_l) $l : Z \times \mathbb{R} \times \Lambda \to \mathbb{R}$ satisfies
(i) $z \to l(z, x, \lambda)$ is measurable,
(ii) $(x, \lambda) \to l(z, x, \lambda)$ is continuous,
(iii) $|l(z, x, \lambda)| \leq \gamma_B(z)(1 + |x|)$, a.e. on Z, with $\gamma_B \in L^2_+(Z)$, where $B \subset \Lambda$ is compact.

Let $X = Y = L^2(Z)$ and define $\varphi : X \times \Lambda \to (-\infty, +\infty)$, $f : [0, T] \times X \times \Lambda \to X$, $U : [0, T] \times \Lambda \to P_{wkc}(Y)$, $g : [0, T] \times \Lambda \to X$, and $\eta : C([0, T]; X) \to \mathbb{R}$ by

$$\varphi(x, \lambda) = \begin{cases} p^{-1}\left[\sum_{i=1}^N \int_Z v_1(z, \lambda)|D_i x|^p dz + \int_Z v_2(z, \lambda)|x|^p dz\right], & \text{if } x \in W_0^{1,p}(Z) \\ +\infty & \text{otherwise,} \end{cases}$$

$$f(t, x, \lambda)(\cdot) = f_1(t, \cdot, x(\cdot), \lambda),$$

$$U(t, \lambda) = \{u \in L^2(Z) : |u(z)| \leq \delta(t, z, \lambda), \text{ a.e.}\},$$

$$g(t, \lambda)(z) = g_1(t, z, \lambda),$$

$$\eta(x, \lambda) = \int_Z l(z, x(T, z), \lambda)\, dz,$$

respectively. Then the problem (3) can be rewritten in the form (P_λ) with $A(\lambda)(\cdot) = \partial\varphi(\cdot, \lambda)$, and conditions (H'_A), (H'_g) (H_f), (H_U) and (H_η) are verified. Applying Theorem 4 yields

Theorem 5. *Let assumptions* (H_v), (H_{f_1}), (H_{g_1}), (H_δ), (H_l) *and* (H_k) *be satisfied, and let* $m(\cdot)$ *denote the value of the minimization problem* (3). *Then* $m : \Lambda \to \mathbb{R}$ *is continuous.*

ACKNOWLEDGMENT

The research of the first author was supported in part by the National Science Foundation under Grant No. DMS-91-11794.

REFERENCES

[1] S. Aizicovici and N. S. Papageorgiou, Multivalued Volterra integral equations in Banach spaces, *Funkcial. Ekvac.*, to appear.
[2] S. Aizicovici and N. S. Papageorgiou, A sensitivity analysis of Volterra integral inclusions with applications to optimal control problems, *J. Math. Anal. Appl.,* to appear.
[3] H. Attouch, Variational Convergence for Functions and Operators, Pitman, Boston, 1984.
[4] P. Bénilan, Equations d'Evolution dans un Espace de Banach Quelconque et Applications, Thesis, University of Paris XI, 1972.
[5] M. G. Crandall and J. A. Nohel, An abstract functional differential equation and a related Volterra equation, *Israel J. Math.* 29 (1978), 313–328.
[6] E. Klein and A. Thompson, Theory of Correspondences, J. Wiley, New York, 1984.
[7] D. Wagner, Survey of measurable selection theorems, *SIAM J. Control Optim.* 15 (1977), 859–903.

3 Identifying the Nonlinearity in a Parabolic Boundary Value Problem

Viorel Barbu* University of Iaşi, Iaşi, Romania

Karl Kunisch Technische Universität Berlin, Berlin, Germany

1 INTRODUCTION

We shall present here some preliminary results on the identification of nonlinear function $a\colon \mathbb{R}^n \to \mathbb{R}^n$, $b\colon \mathbb{R} \to \mathbb{R}$ into the parabolic boundary value problem

$$\begin{aligned} &y_t - \operatorname{div} a(\nabla y) = f && \text{in } Q = \Omega \times (0,T) \\ &y = 0 && \text{in } \tau = \partial\Omega \times (0,T) \\ &y(x,0) = y_0(x) && \text{in } \Omega, \end{aligned} \tag{1.1}$$

where Ω is a bounded open subset of $\mathbb{R}^n$ with a smooth boundary $\partial\Omega$ and $f \in L^2(Q)$, $y_0 \in L^2(\Omega)$ are given.

The function a is to be chosen from a certain class of potential function on $\mathbb{R}^n$. For the identification we are given a distributed observation $y^0 \in L^2(Q)$ and one must find via least square quadratic technique the function a which best fits to y^0. Problems of this type arise in identification of conductivity in nonlinear thermic fields and were studied in [4], [5]. Our appoach is different and it relies on an approximating procedure developed in [2] (see also [3]) for the identification of nonlinear elliptic boundary value problems.

* Research supported by NSF Grant DMS-91-11794.

2 THE LEAST SQUARE APPROACH AND THE APPROXIMATING PROBLEM

We consider here system (1.1) for $a\colon \mathbb{R}^n \to \mathbb{R}^n$ in the class

$$A = \{a = \partial j, j \in K\}, \tag{2.1}$$

where

$$K = \{j\colon \mathbb{R}^n \to \mathbb{R} \text{ convex, continuous, } \beta_1|r|^2 + \gamma_1 \leq j(r) \leq \beta_2|r|^2 + \gamma_2, \forall r \in \mathbb{R}\} \tag{2.2}$$

for some $0 < \beta_1 < \beta_2 < \infty$, $\gamma_1 \leq \gamma_2$ and ∂j is the subdifferential of j.

Consider the problem: *Mimimize*

$$\text{(P)} \qquad \|y - y^0\|^2_{L^2(Q)}$$

on all $(y, j) \in L^2(0, T; H_0^1(\Omega)) \cap W^{1,2}([0, T]; H^{-1}(\Omega)) \times K$ *subject to* (1.1).

We recall (see e.g. [1]) that for every $y_0 \in L^2(\Omega)$, $f \in L^2(Q)$ and $a = \partial j$, $j \in K$ the boundary value problems (1.1) has a unique solution $y \in C([0, T]; L^2(\Omega)) \cap L^2(0, T; H_0^1(\Omega))$ with $y \in L^2(\delta, T; L^2(\Omega))$ for every $0 < \delta < T$. If $y_0 \in H_0^1(\Omega)$ then $y_t \in L^2(Q)$. In the sequel we shall assume that $y_0 \in H_0^1(\Omega)$ and $f \in L^2(Q)$. We may equivalently write Eq. (1.1) as

$$\begin{aligned} &z \in a(\nabla y) \text{ in } Q = \Omega \times (0, T) \\ &\operatorname{div} z = y_t - f \text{ in } Q \\ &y \in H_0^1(\Omega),\ y_t \in L^1(Q),\ z \in (L^2(Q))^n \end{aligned} \tag{2.3}$$

i.e.,

$$\begin{aligned} &j(\nabla y) + j^*(z) = z \cdot \nabla y \text{ in } Q \\ &\operatorname{div} z = y_t - f \text{ in } Q, \end{aligned} \tag{2.4}$$

where j^* is the conjugate of j. This suggests to approximate (P) by the following family of minimization problems

$$\begin{aligned} (\text{P}_\varepsilon) \qquad &\inf\{\|y - y^0\|^2_{L^2(Q)} + \varepsilon^{-1} \int_Q (j(\nabla y) + j^*(z) - z \cdot \nabla y)\, dx\, dt \\ &y \in L^2(0, T; H_0^1(\Omega)) \cap W^{1,2}([0, T]; H^{-1}(\Omega)) \\ &y(x, 0) = y_0(x); z \in (L^2(Q))^n, \operatorname{div} z = y_t - f \text{ in } Q; j \in K\}. \end{aligned}$$

We have

THEOREM 2.1. *For every $\varepsilon > 0$ problem (P_ε) has at least one solution $(y_\varepsilon, z_\varepsilon, j_\varepsilon) \in L^2(0,T;H_0^1(\Omega)) \times (L^2(Q))^n \times K$.*

Proof. Let d be the infimum in problem (P_ε) and let $\{y_k, z_k, j_k)\} \subset L^2(0,T;H_0^1(\Omega)) \times (L^2(Q))^n \times K$ be a minimizing sequence for (P_ε), i.e.,

$$d \le \|y_k - y^0\|^2_{L^2(Q)} + \varepsilon^{-1}\int_Q (j_k(\nabla y_k) + j_k^*(z_k) + y_k(y_k)_t - f))\,dx\,dt \le k^{-1} + d. \tag{2.5}$$

By (2.2) it follows that

$$\|\nabla y_k\|_{L^2(Q)} + \|z_k\|_{(L^2(Q))^n} + \|y_k\|_{L^2(Q)} \le C$$

and therefore on a subsequence, again denoted k,

$$\begin{aligned} &y_k \to y \text{ weakly in } L^2(0,T;H_0^1(\Omega)) \\ &\nabla y_k \to \nabla y \text{ weakly in } (L^2(Q))^n \\ &z_k \to z \text{ weakly in } (L^2(Q))^n. \end{aligned} \tag{2.6}$$

Moreover, by the equation $\operatorname{div} z_k = (y_k)_t - f$ in $D'(Q)$ and by (2.6) it follows that $\{(y_k)_t\}$ is bounded in $L^2(0,T;H^{-1}(\Omega))$ and

$$(y_k)_t \to y_t \text{ weakly in } L^2(0,T;H^{-1}(\Omega)). \tag{2.7}$$

In particular this implies that $\{y_k\}$ is compact in $L^2(0,T;L^2(\Omega)) \cap C([0,T];H^{-1}(\Omega)) \cap C_w([0,T];L^2(\Omega))$. We have therefore

$$\begin{aligned} &y_k \to y \text{ strongly in } L^2(Q) \\ &y_k(t) \to y(t) \text{ weakly in } L^2(\Omega) \text{ for all } t \in [0,T]. \end{aligned} \tag{2.8}$$

In particular this implies that

$$\begin{aligned} &\liminf_{k\to\infty} \int_Q y_k((y_k)_t - f)\,dx\,dt \\ &= \liminf_{k\to\infty} \left\{ \frac{1}{2}\|y_k(T)\|^2_{L^2(\Omega)} - \frac{1}{2}\|y_0\|^2_{L^2(\Omega)} - \int_Q y_k f\,dx\,dt \right. \\ &\ge \int_Q y(y_t - f)\,dx\,dt. \end{aligned} \tag{2.9}$$

Next we note that since $j_k, j_k^* \in K$ we have

$$|\partial j_k(r)| + |\partial j_k^*(r)| \le C(|r|^2 + 1) \quad \forall r \in \mathbb{R}^n,$$

where C is independent of k. Then by the Arzela theorem there exists $j \in K$ such that for $k \to \infty$

$$j_k \to j, \qquad j_k^* \to j^* \tag{2.10}$$

uniformly on compacta.

To conclude the proof it suffices to prove that

$$\liminf_{k\to\infty} \int_Q j_k(\nabla y_k)\,dx\,dt \ge \int_Q j(\nabla y)\,dx\,dt \tag{2.11}$$

$$\liminf_{k\to\infty} \int_Q j_k^*(z_k)\,dx\,dt \ge \int_Q j^*(z)\,dx\,dt \tag{2.12}$$

and to tend to zero in (2.5). This follows by an argument similar to that used in [2]. □

We shall formulate now the convergence theorem.

THEOREM 2.2. *The family of solutions $(y_\varepsilon, z_\varepsilon, j_\varepsilon)_{\varepsilon>0}$ of (P_ε) has at least one cluster point $(\bar{y}, \bar{z}, \bar{j})$ in $(L^2(0,T;H_0^1(\Omega)) \cap W^{1,2}([0,T];H^{-1}(\Omega)))_w \times (L^2(Q))_w^n \times K$ for $\varepsilon \to 0$,*

$$\operatorname{div} \bar{z} = \bar{y}_t - f \text{ in } D'(Q)$$

$(\bar{y}, \bar{z}, \bar{j})$ is a solution to problem (P) and

$$\liminf_{\varepsilon\to 0} f(\mathrm{P}_\varepsilon) = \inf(\mathrm{P}). \tag{2.13}$$

Moreover, if $\tilde{y}_\varepsilon$ is the solution to problem (1.1) with $a = \partial j_\varepsilon$ we have

$$\lim_{\varepsilon\to 0} \|\tilde{y}_\varepsilon - y^0\|_{L^2(Q)}^2 = \inf(P). \tag{2.14}$$

Proof. For every $\varepsilon > 0$ and $j \in K$ we have

$$\|y_\varepsilon - y^0\|_{L^2(Q)}^2 + \varepsilon^{-1} \int_Q (j_\varepsilon(\nabla y_\varepsilon) + j_\varepsilon^*(z_\varepsilon) + y_\varepsilon((y_\varepsilon)_t - f)\,dx \le \|y^j - y_0\|_{L^2(Q)}^2,$$

where y^j is the solution to (1.1) corresponding to j.

We have therefore

$$\limsup_{\varepsilon\to 0} \|y_\varepsilon - y^0\|_{L^2(Q)}^2 \le \inf\{\|y^j - y^0\|_{L^2(Q)}^2; j \in K\} = \inf(\mathrm{P}) \tag{2.15}$$

and

$$\lim_{\varepsilon\to 0} \int_\Omega (j_\varepsilon(\nabla y_\varepsilon) + j_\varepsilon^*(z_\varepsilon) - z_\varepsilon \cdot \nabla y_\varepsilon)\,dx\,dt = 0.$$

Since $j_\varepsilon(\nabla y_\varepsilon) + j^*_\varepsilon(z_\varepsilon) - z_\varepsilon \cdot \nabla y_\varepsilon \geq 0$, a.e. on Q we have

$$j_\varepsilon(\nabla y_\varepsilon) + j^*_\varepsilon(z_\varepsilon) - z_\varepsilon \nabla y_\varepsilon \to 0 \text{ strongly in } L^1(Q). \tag{2.16}$$

Then arguing as in the proof of Theorem 1 we infer that there exists a sequence $\varepsilon \to 0$ and $(\bar{y}, \bar{z}, \bar{j}) \in (L^2(0,T; H^1_0(\Omega)) \cap W^{1,2}([0,T]; H^{-1}(\Omega))) \times (L^2(Q))^n \times K$ such that

$$\begin{aligned}
&y_\varepsilon \to \bar{y} \text{ weakly in } L^2(0,T;H^1_0(\Omega)) \text{ and strongly in } L^2(Q)\\
&y_\varepsilon(t) \to \bar{y}(t) \text{ weakly in } L^2(\Omega) \quad \forall t \in [0,T]\\
&(y_\varepsilon)_t \to \bar{y}_t \text{ weakly in } L^2(0,T;H^{-1}(\Omega))\\
&z_\varepsilon \to \bar{z} \text{ weakly in } (L^2(Q))^n\\
&j_\varepsilon \to \bar{j} \text{ uniformly on campacta}\\
&j^*_\varepsilon \to \bar{j}^* \text{ uniformly on campacta,}
\end{aligned}$$

where $\operatorname{div} \bar{z} = \bar{y}_t - f$ in $D'(Q)$. Moreover, we have as above

$$\liminf_{\varepsilon \to 0} \int_Q (j_\varepsilon(\nabla y_\varepsilon) + (j^*_\varepsilon(z_\varepsilon))\,dx\,dt \geq \int_Q (j(\nabla \bar{y}) + \bar{j}^*(\bar{z}))\,dx\,dt, \tag{2.17}$$

while

$$\begin{aligned}
\liminf_{\varepsilon \to 0} \left(-\int_Q z_\varepsilon \nabla y_\varepsilon\,dx\,dt\right) &= \liminf_{\varepsilon \to 0} \int_Q y_\varepsilon((y_\varepsilon)_t - f)\,dx\,dt\\
&= \liminf_{\varepsilon \to 0} \left(\frac{1}{2}\int_Q (|y_\varepsilon(x,T)|^2 - |y_0(x)|^2)\,dx - \int_Q y_\varepsilon f\right)\\
&\geq \frac{1}{2}\int_\Omega (y^2(x,T) - y_0^2(x))\,dx - \int_Q \bar{y} f = -\int_Q \bar{z} \cdot \nabla \bar{y}\,dx dt.
\end{aligned}$$

In virtue of (2.16), (2.17) this yields therefore

$$\int_Q (\bar{j}(\nabla \bar{y}) + \bar{j}^*(z) - \nabla \bar{y} \cdot z)\,dx\,dt = 0$$

i.e.,

$$j(\nabla \bar{y}) + \bar{j}^*(z) - \nabla \bar{y} \cdot z = 0 \text{ a.e. in } Q.$$

We may therefore conclude that $(\bar{y}, \bar{z}, \bar{j})$ is optimal in problem (P) and so (2.13) follows. The proof of (2.14) follows by similar argument.

3 THE APPROXIMATING IDENTIFICATION PROCESS IN ONE DIMENSION

We shall study here problem (P) in the case $n = 1$ $\Omega = (0,1)$ and with the constraint set K replaced by

$$\tilde{K} = \{j = \varphi_0 + \varphi, \varphi \in H_0^1(\tau) \cap H^2(\tau), \alpha \leq \varphi'' \leq \beta \text{ a.e. in } \tau\}, \tag{3.1}$$

where τ is an interval $[-2\delta, 2\delta]$ to be precised later, $0 < \alpha < \beta < \infty$ and $\varphi_0 \in H^2(\tau)$ is a given convex function. Here $H_0^1(\tau)$ and $H^2(\tau)$ are usual Sobolev spaces on τ.

The interval τ is a priori determined by the condition that $y_x \in \tau$ for all solution y to boundary value problem

$$\begin{aligned} &y_t - (a(y_x))_x = f \text{ in } Q = \Omega \times (0,T); y = 0 \text{ in } \partial\Omega \times (0,T) \\ &y(x,0) = y_0(x) \qquad \forall x \in \Omega, \end{aligned} \tag{3.2}$$

where $a' \geq \alpha > 0$ in $\mathbb{R}$.

We shall assume that

$$f, f_t \in L^2(Q), \quad y_0 \in H_0^1(0,1) \cap H^2(0,1).$$

Then the solution y to (3.1) belongs to $H^{2,1}(Q) \cap W^{1,\infty}([0,T]; L^2(\Omega))$ i.e., $y_t \in L^\infty(0,T; L^2(\Omega))$. Moreover, the following estimate holds

$$\begin{aligned} \|y_t(t)\|_{L^2(\Omega)} \leq \|(a((y_0)_x))_x\|_{L^2(\Omega)} + \left(\int_0^1 f^2(x,0)\,dx\right)^{1/2} \\ + \int_0^T \left(\int_0^1 f_t^2(x,t)\,dx\right)^{1/2} dt, \forall t \in [0,T]. \end{aligned}$$

Hence

$$\begin{aligned} &y_{xx} = g \qquad \text{in } Q \\ &y(0,t) = y(1,t) = 0 \qquad \forall t \in [0,T], \end{aligned}$$

where

$$\begin{aligned} &\|g(t)\|_{L^2(\Omega)} \leq \alpha^{-1}(\|(a((y_0)_x))_x\|_{L^2(\Omega)} \\ &\quad + \left(\int_0^1 f^2(x,0)\,dx\right)^{1/2} \int_0^T \left(\int_0^1 f_t^2 dx\right)^{1/2} dt), \forall t \in [0,T]. \end{aligned} \tag{3.3}$$

This yields

$$|y_x(x,t)| \leq 2^{-1}C(f,y_0) \quad \forall (x,t) \in Q, \tag{3.4}$$

where $C(f,y_0)$ is the constant defined by the right hand side of estimate (3.3). In the following we shall take $\delta = 2^{-1}C(f,y_0)$ and $\tau = (-2\delta, 2\delta)$. The approximating problem (P_ε) has in this case the following form

$$(\tilde{\mathrm{P}}_\varepsilon) \qquad \inf\{\|y-y^0\|^2_{L^2(Q)} + \varepsilon^{-1}\int_Q (j(y_x)+$$

$$+ j^*\Big(\int_0^x y_t(\xi,t)\,d\xi - F(x,t) + v(t)) + y(y_t - f)\Big)\,dx\,dt,$$

$$y \in L^2(0,T;H_0^1(\Omega)) \cap W^{1,2}([0,T];H^{-1}(\Omega)), y(x,0) = y_0(x)$$

$$j \in \tilde{K}, v \in L^2(0,T);\ |y_x| \leq 2\delta \text{ a.e. in } Q\}.$$

Here $F(x,t) = \int_0^x f(\xi,t)\,d\xi$.

If extend j by $+\infty$ outside interval $\tau = [-2\delta, 2\delta]$ and redefine $j^*(p) = \sup\{p\cdot v - j(v); v \in \tau\}$ we may delete the constraint $|y_x| \leq 2\delta$ into problem (3.3).

Since all solutions to (3.2) with $a = \partial j$, $j \in \tilde{K}$ satisfy the a priori estimate (3.4) it follows as in Theorem 2.1 the existence and the convergence of solutions $(y_\varepsilon, j_\varepsilon)$ to problem $(\tilde{\mathrm{P}}_\varepsilon)$. To solve problem $(\tilde{\mathrm{P}}_\varepsilon)$ we may use the following steps

1° For fixed $y \in L^2(0,T;H_0^1(\Omega)) \cap W^{1,2}([0,T];H^{-1}(\Omega))$ with $|y_x| \leq 2\delta$ a.e. in Q solve the minimization problem

$$\inf\Big\{\int_Q (j(y_x) + j^*\Big(\int_0^x y_t d\xi - F + v(t))\Big)\,dx\,dt \\ j = \varphi_0 + \varphi;\ 0 < \alpha \leq \varphi'' \leq \beta \text{ a.e. in } \tau\Big\}, \tag{3.5}$$

where $j^*(p) = \sup\{p\cdot v - j(v); |v| \leq 2\delta\}$, $F_x = f$.

2° For fixed $(j,v) \in K \times L^2(0,T)$ solve

$$\inf\Big\{\|y-y_0\|^2_{L^2(Q)} + \varepsilon^{-1}\int_Q (j(y_x) + j^*\Big(\int_0^x y_t d\xi - F + v) + y(y_t - f)\Big)\,dx\,dt, \\ y \in L^2(0,T;H_0^1(\Omega)) \cap W^{1,2}([0,T];H^{-1}(\Omega); y(x,0) = y_0, |y_x| \leq 2\delta\Big\}. \tag{3.6}$$

It is readily seen that problem (3.6) is equivalent to boundary value problem

$$\begin{aligned} &z_t - (a(y_x))_x + 2\varepsilon(y - y^0) + N_K(y(t)) \ni f \text{ in } Q \\ &y = 0 \quad \text{in } \Sigma; \quad y(x,0) = y_0(x), \\ &y_t - (a(z_x))_x = f \quad \text{in } Q \\ &z = 0 \quad \text{in } \Sigma; \quad z(x,T) = y(x,T) \quad \text{in } \Omega \end{aligned} \tag{3.7}$$

where $N_K(y(t)) \subset H^{-1}(\Omega)$ is the normal cone to $K = \{y \in H_0^1(\Omega); |y_x(x)| \leq 2\delta\}$ at $y(t) \in H_0^1(\Omega)$. By using a penalty function we may replace the first equation in (3.7) by:

$$z_t - a(y_x)_x + 2\varepsilon(y - y^0) + \lambda^{-1}\eta(y_x) = f, \tag{3.8}$$

where $\eta(r) = 0$ if $|r| < 2\delta$, $\eta(r) = r - 2\delta$ if $r > 2\delta$ and $\eta(r) = r + 2\delta$ if $r < -2\delta$.

As regards the minimization problem (3.5) we shall write it as the constrained optimal control problem in $L^2(Q)$,

$$\begin{aligned}\inf\Big\{ \int_Q (\varphi(y_x) + (\varphi + \varphi_0)^*\Big(\int_0^x y_t d\xi - F + v(t))\Big)\, dx\, dt, \varphi'' = \psi \text{ in } \tau;\\ \varphi \in H_0^1(\tau), \psi \in L^2(\tau), \alpha \leq \psi \leq \beta \text{ a.e. in } \tau; v \in L^2(0,T)\Big\}.\end{aligned} \tag{3.9}$$

Here y is fixed, $|y_x| \leq 2\delta$ a.e. in Q. This is a bilinear optimal control problem which can be written as two successive minimization problems. For a fixed φ te optimality conditions for v reads

$$\int_Q \partial(\varphi + \varphi_0)^*\Big(\int_0^x y_t d\xi - f + v\Big)\, dx\, dt = 0.$$

for a fixed v we set $g = \int_0^x y_t d\xi - f + v$ and problem (3.9) reduces to

$$\begin{aligned}\inf\Big\{ \int_Q (\varphi(y_x) + (\varphi + \varphi_0)^*(g))\, dx\, dt; \varphi'' = \psi \text{ in } \tau,\\ \psi \in L^2(\tau), \varphi \in H_0^1(\tau), \alpha \leq \psi \leq \beta \text{ a.e.}\Big\}\end{aligned}$$

or equivalently

$$\inf\{\phi_0(\varphi); \varphi'' = \psi \text{ in } \tau; \psi \in L^2(\tau), \varphi \in H_0^1(\tau), \alpha \leq \psi \leq \beta \text{ a.e. in } \tau\}, \tag{3.10}$$

where

$$\phi_0(\varphi) = \int_Q (\varphi(y_x) + (\varphi + \varphi_0)^*)\, dx.$$

We may write (3.10) as

$$\inf\{\phi(\psi); \psi \in L^2(\tau)\}, \tag{3.11}$$

where $\phi\colon L^2(\tau) \to \bar{\mathbb{R}} =]-\infty, +\infty]$ is given by

$$\phi(\psi) = \begin{cases} \phi_0(\varphi) & \text{if } \varphi'' = \psi \text{ in } \tau; \varphi \in H_0^1(\tau), \alpha \leq \psi \leq \beta \text{ a.e. in } \tau \\ +\infty & \text{otherwise.} \end{cases}$$

Moreover as seen in [2] its subdifferential is given by

$$\partial\phi(\psi) = N_U(\psi) - p, \tag{3.12}$$

where $N_U(\psi)$ is the normal cone to $U = \{\psi \in L^2(\tau); \alpha \le \psi \le \beta \text{ a.e. in } \tau\}$ and $p \in H_0^1(\tau)$ is the solution to

$$\begin{aligned} -p'' &= \mu_1 + \mu_2 && \text{in } \tau \\ p &= 0 && \text{in } \partial\tau, \end{aligned} \tag{3.13}$$

μ_1, $\mu_2 \in H^{-1}(\tau)$ being defined by

$$\begin{aligned} \mu_1(k) &= \int_Q k(y_x)\,dx\,dt \quad \forall k \in H_0^1(\tau) \\ \mu_2(k) &= -\int_Q k(j' + N_\tau)^{-1}(g(x,t))\,dx\,dt \quad \forall k \in H_0^1(\tau), \end{aligned} \tag{3.14}$$

where

$$(j' + N_\tau)^{-1}g = \begin{cases} (j')^{-1}(g) & \text{if } |(j')^{-1}(g)| < 2\delta \\ 2\delta & \text{if } (j')^{-1}(g) \ge 2\delta \\ -2\delta & \text{if } (j')^{-1}(g) \le -2\delta \end{cases}$$

and $j' = \partial j$, $j = \varphi + \varphi_0$. Hence problem (3.11), i.e., $0 \in \partial\phi(\psi)$ is equivalent to

$$\psi = \begin{cases} \alpha & \text{if } p \le 0 \\ \beta & \text{if } p \ge 0 \\ \in (\alpha, \beta) & \text{if } p = 0, \end{cases} \tag{3.15}$$

where p satisfies (3.13), (3.14).

Numerical results will be given in a forthcoming paper.

REFERENCES

1. V. Barbu, Analysis and Control of Nonlinear Infinite Dimensional Sysems, Academic Press, New York, 1993.
2. V. Barbu and K. Kunisch, Identification of nonlinear elliptic equations, *SIAM J. Control & Optimiz.*, to appear.
3. V. Barbu, P. Neittaanmäki and A. Niemistö, A penalty method for the identification of nonlinear elliptic differential operators, *Numerical Functional Analysis and Optimization*, to appear.
4. G. Chavent and P. Lemonnier, Identification de la nonlinearité d'une equation parabolique quasilineaire, *Appl. Math. Optimiz.* vol. 1 (1974) 121–162.
5. C. Kravaris and J. H. Seinfeld, Identification of parameters in distributed parameter systems by regularization, *SIAM J. Control & Optimiz.*, 23 (1985), 217–241.

4 Optimal Control and Calculus of Variations in L^∞

E. N. Barron University of Chicago, Chicago, Illinois

ABSTRACT. Recent results on the optimal control problem with L^∞ cost functional are surveyed. The calculus of variations problem in L^∞ is then considered and we prove an existence theorem and derive the Euler equation.

1. INTRODUCTION

This paper is a survey of L^∞ optimal control and presents some results on the calculus of variations problem in L^∞, (cf. [1], [10]). The motivation for considering problems using the L^∞ norm is the realization that minimization of L^2 functionals is frequently inadequate for many problems of practical significance. For example, in designing an optimal flight path for an aircraft, does one really want to minimize the total energy of forces acting on the craft, or does one want to minimize the pointwise stresses on the craft. When one is designing a structural column pointwise stresses should be minimized, not the total energy in the column. When one is designing an abort plan for a passenger plane, isn't the minimum distance to the ground of interest? Etc., etc.. The reader can supply many more examples. The use of the L^∞ norm puts us into the realm of nonsmooth problems.

2. THE OPTIMAL CONTROL PROBLEM

Consider the controlled system of ordinary differential equations in R^n:

$$d\xi(\tau)/d\tau = f(\tau, \xi(\tau), \zeta(\tau)), \qquad 0 \le t < \tau \le T \tag{2.1}$$

$$\xi(t) = x \in R^n. \tag{2.2}$$

Given a compact control set $Z \subset R^m$, the control functions $\zeta(\cdot)$ are chosen from the class of functions

$$\mathcal{Z}[t, T] = \{\zeta : [t, T] \to Z \mid \zeta \text{ is Lebesgue measureable}\}.$$

We will assume for simplicity that

(A)$f : [0, T] \times R^n \times Z \to R^n$ is jointly continuous and is Lipschitz in x. That is, there is a generic constant K such that

$$|f(t, x, z) - f(t, x', z)| \le K|x - x'|, \quad \forall x, x' \in R^n, \tag{2.3}$$

In addition, $|f(t, x, z)| \le K(1 + |x|)$.

We are also given a function $h : [0, T] \times R^n \times Z \to R^n$ assumed to be jointly continuous and also uniformly Lipschitz in x.

Research of the author was supported in part by grants DMS-9109267 and DMS-9300805 from the National Science Foundation.

The goal is to choose a control function $\zeta \in \mathcal{Z}$ which minimizes the sup norm of h. We approach this problem by studying the value function $V : [0,T] \times R^n \to R^1$:

$$V(t,x) = \inf_{\zeta \in \mathcal{Z}} \operatorname*{ess\,sup}_{t \le r \le T} h(r, \xi(r), \zeta(r)) \tag{2.4}$$

It was proved in [5] (using uniqueness of viscosity solutions) that the $\operatorname{ess\,sup}_{t \le r \le T}$ in (2.4) can be replaced by $\sup_{t \le r \le T}$. Intuitively, jumps of ζ on time intervals of Lebesgue measure zero are not necessary since they cannot change the dynamics. Here is a simple direct proof.

Lemma 2.1.

$$V(t,x) = \inf_{\zeta \in \mathcal{Z}} \operatorname*{ess\,sup}_{t \le r \le T} h(r, \xi(r), \zeta(r)) = \inf_{\zeta \in \mathcal{Z}} \sup_{t \le r \le T} h(r, \xi(r), \zeta(r)) \equiv W(t,x)$$

Proof. Clearly, $V \le W$. Now, given $\varepsilon > 0$, there is a control $\zeta^* \in \mathcal{Z}[t,T]$, and a measureable set $E^* \subset [t,T]$ with $\lambda([t,T] - E^*) = 0$ (λ=Lebesgue measure), so that, using the definition of the essential supremum,

$$\begin{aligned} V &\ge \operatorname*{ess\,sup}_{t \le r \le T} h(r, \xi^*(r), \zeta^*(r)) - \varepsilon \\ &= \inf\{\sup_{r \in E} h(r, \xi^*(r), \zeta^*(r)); E \subset [t,T], \lambda(E^c) = 0\} - \varepsilon \\ &\ge \sup_{r \in E^*} h(r, \xi^*(r), \zeta^*(r)) - 2\varepsilon \end{aligned}$$

Define $\zeta^{**}(r) = \zeta^*(r)$ if $r \in E^*$ and

$$\zeta^{**}(r) \in \arg\min_{z \in Z} h(r, \xi^*(r), z),$$

if $r \in [t,T] - E^*$. Since $\lambda([t,T] - E^*) = 0$, the trajectory associated with ζ^{**} coincides with the trajectory ξ^*. Thus,

$$\begin{aligned} V &\ge \sup_{r \in E^*} h(r, \xi^*(r), \zeta^*(r)) - 2\varepsilon \\ &\ge \sup_{r \in [t,T]} h(r, \xi^*(r), \zeta^{**}(r)) - 2\varepsilon \\ &\ge W - 2\varepsilon. \quad \square \end{aligned}$$

One of the main results of [5] is that the function V is the unique continuous viscosity solution of the problem

$$\max\{V_t + \min_{\{z \in Z | h(t,x,z) \le V(t,x)\}} D_x V \cdot f(t,x,z) \,,\, \min_{z \in Z} h(t,x,z) - V(t,x)\} = 0 \tag{2.5}$$

$$V(T,x) = \min_{z \in Z} h(T,x,z). \tag{2.6}$$

Define the hamiltonian function

$$H(t,x,r,p) = \min_{\{z \in Z | h(t,x,z) \le r\}} p \cdot f(t,x,z) \tag{2.7}$$

for $r \in R^1$ and $p \in R^n$.

Remark 2.1. Observe that the minimum is over the set $\{z \in Z | h(t,x,z) \le r\}$. If this set is empty, H is defined as $+\infty$. When written as in (2.5) we are guaranteed that the set is nonempty.

The L^∞ control problem presents us with an important class of problems for which the hamiltonian is not necessarily continuous. In fact, we have that (see [5])

$$H^*(t,x,r,p) = H(t,x,r-0,p)$$

and

$$H_*(t,x,r,p) = H(t,x,r+0,p)$$

are the upper and lower semicontinuous envelopes of H, respectively.

The value function V satisfies the equation (2.5) in the viscosity sense. That, is

Definition 2.1. V is a viscosity solution of (2.5) if

(i) V is a viscosity subsolution, i.e., for any (t_0, x_0) for which $V - \varphi$ has a maximum, for a smooth function φ, it follows that

$$\max\{\varphi_t + H^*(t_0,x_0,V(t_0,x_0),D_x\varphi(t_0,x_0)\ ,\ \min_{z\in Z} h(t_0,x_0,z) - V(t_0,x_0)\} \geq 0$$

and

(ii) V is a viscosity supersolution, i.e., for any (t_0, x_0) for which $V - \varphi$ has a minimum, for a smooth function φ, it follows that

$$\max\{\varphi_t + H_*(t_0,x_0,V(t_0,x_0),D_x\varphi(t_0,x_0)\ ,\ \min_{z\in Z} h(t_0,x_0,z) - V(t_0,x_0)\} \leq 0.$$

Finally, we may consider slightly more general problems in which we also incur a terminal cost. The appropriate value function is then

$$V(t,x) = \inf_{\zeta\in\mathcal{Z}} \max\{g(\xi(T)),\ \operatorname*{ess\,sup}_{t\leq r\leq T} h(r,\xi(r),\zeta(r))\}.$$

Notice that we use the max operation rather than addition. This is due to the fact that, in some sense, addition is not compatible with the supremum. In this case, the terminal condition becomes

$$V(T,x) = \max\{g(x),\ \min_{z\in Z} h(T,x,z)\}. \tag{2.8}$$

The problem considered in this section has dealt with the free endpoint problem. If we want to consider the fixed endpoint problem, say $\xi(T) = x_T \in R^n$, we may do so by choosing

$$g(x) = 0,\ \text{if } x = x_T,\quad g(x) = +\infty,\ \text{if } x \neq x_T.$$

In this case, the corresponding value function will not necessarily be continuous, but only lower semicontinuous (l.s.c.). For a l.s.c. function we must modify the definition of viscosity solution because it is clear that the definition given above for subsolution may be vacuous.

Definition 2.2. A l.s.c. function V is a l.s.c. viscosity solution of (2.5) if for any (t_0, x_0) for which $V - \varphi$ has a minimum, for a smooth function φ, it follows that

$$\max\{\varphi_t + H_*(t_0,x_0,V(t_0,x_0),D_x\varphi(t_0,x_0)\ ,\ \min_{z\in Z} h(t_0,x_0,z) - V(t_0,x_0)\} \leq 0.$$

and

$$\max\{\varphi_t + H^*(t_0,x_0,V(t_0,x_0),D_x\varphi(t_0,x_0)\ ,\ \min_{z\in Z} h(t_0,x_0,z) - V(t_0,x_0)\} \geq 0.$$

Observe what our definition says when $H^* = H_*$.

An open problem is to establish that the value function is the unique, l.s.c., viscosity solution of (2.5). Since it may even be possible that V is not l.s.c., one may have to expand the statement to the l.s.c. envelope of V. In general, the value function which is l.s.c. is *not* the original value function, but the *relaxed* value function. This leads us to our next topic.

3. Relaxed L^∞ Control

In this section the relaxation of an L^∞ control problem will be given. As we shall see, it does not consist of simply convexifying the trajectories and the cost as is done in the classical problem.

Let $M(Z)$ denote the space of bounded measures on Z. Viewing $M(Z)$ as the dual space of $C(Z)$ = continuous functions on Z, we endow $M(Z)$ with the weak star topology of $C(Z)^*$. The space of relaxed controls (also known as Young measures in a calculus of variations context) is given by

$$\widehat{\mathcal{Z}}[t,T] = \{\mu \in L^\infty([t,T]; M(Z)) \mid \mu(\tau) \text{ is a probability measure a.e. } \tau \in [t,T]\}.$$

Let $\mathcal{M}(Z)$ be the set of probability measures on Z. Then, we may write that $\widehat{\mathcal{Z}}[t,T] = L^\infty([t,T]; \mathcal{M}(Z))$, the space of essentially bounded, Lebesgue measureable maps $\mu : [t,T] \rightarrow \mathcal{M}(Z)$. For any relaxed control $\mu \in \widehat{\mathcal{Z}}[t,T]$ there is a relaxed trajectory given by

$$\widehat{\xi}(\tau) = x + \int_t^\tau \int_Z f(s, \widehat{\xi}(s), z)\, \mu(s, dz) ds. \tag{3.1}$$

For any $\mu \in \mathcal{M}(Z)$ Define the functions

$$\widehat{f}(t,x,\mu) = \int_Z f(t,x,z)\, \mu(dz), \tag{3.2}$$

and

$$\widehat{h}(t,x,\mu) = \operatorname*{ess\,sup}_{z \in Z} h(t,x,z) \equiv ||h(t,x,z)||_{L^\infty(Z;\mu)}, \tag{3.3}$$

where $L^\infty(Z;\mu)$ is the space of essentially bounded, with respect to the measure $\mu \in \mathcal{M}(Z)$, real valued functions on Z. The relaxed L^∞ problem involves the functions $\widehat{f}$ and $\widehat{h}$. The relaxed dynamical vector field is the usual $\widehat{f}$ but the classical relaxation of the cost functional uses $\int_Z h(t,x,z)\mu(dz)$ rather than $\widehat{h}$. Replacing h by $\int_Z h(t,x,z)\, d\mu$ is *not* the correct relaxation. The reader may analyze the following simple example.

Example. On the time interval $[-1,0]$, take $f(t,x,z) = z^2 - 10$, $Z = [0,10]$ and $h(t,x,z) = |x| + z$. Take the initial conditions $x = 0$, $t = -1$. If we use the relaxation $\int_Z h(t,x,z)\, d\mu$, and denote the associated relaxed value by $\widehat{W}(-1,0)$, we will conclude that $\widehat{W}(-1,0) < V(-1,0)$ even though the function f is quadratic and h is linear in z.

This example shows that replacing h by $\int_Z h(t,x,z)\, d\mu$ lowers the cost. This is a Lavrentiev effect appearing in simple problems where it should not appear. The correct relaxation of the L^∞ problem is given next.

Definition 3.1. The relaxed value function associated with the L^∞ problem is

$$\widehat{V}(t,x) = \inf_{\mu(\cdot)\in\widehat{\mathcal{Z}}[t,T]} ||\widehat{h}(s,\widehat{\xi}(s),\mu(s))||_{L^\infty[t,T]}. \tag{3.4}$$

This is the appropriate relaxation in the sense that (i) an optimal relaxed control will exist, and (ii) the relaxed value is the same as the original value function. The usual way of proving (i) is to use convexity and weak lower semicontinuity. Unfortunately, simple examples show that $\mu \mapsto \widehat{h}(t,x,\mu)$ is not convex. But it is true that it is *quasiconvex*. A quasiconvex function, is a function, say $g : X \to R^1$, X a convex set, satisfying

$$g(\lambda x + (1-\lambda)y) \le \max\{g(x), g(y)\}, \qquad 0 \le \lambda \le 1,\ \ x,y \in X.$$

Equivalently, g is quasiconvex if the level set of g, $\{x \in X \mid g(x) \le r\}$ is convex for all $r \in R^1$.

Then, we prove that for each (t,x) fixed, the mapping,

$$\mu \in \mathcal{M}(Z) \mapsto \widehat{h}(t,x,\mu)$$

is a quasiconvex function. In fact,

$$\widehat{h}(t,x,\lambda\mu_1 + (1-\lambda)\mu_2) = \max\{\widehat{h}(t,x,\mu_1), \widehat{h}(t,x,\mu_2)\} \tag{3.5}$$

for $0 < \lambda < 1$.

Next one can show that for each (t,x) fixed,

$$\mu \in \mathcal{M}(Z) \mapsto \widehat{h}(t,x,\mu)$$

is weakly sequentially lower semicontinuous.

We are now ready to state the two main theorems in this section ([3]) which shows that our relaxation is reasonable.

Theorem 3.1. *For each fixed $(t,x) \in [0,T] \times R^n$, there exists an optimal relaxed control $\mu^*(\cdot) \in \widehat{\mathcal{Z}}[t,T]$.*

Finally to establish that the relaxed value and original value coincide, define the relaxed hamiltonian

$$\widehat{H}(t,x,r,p) = \min_{\{\mu\in\mathcal{M}(Z) \mid \widehat{h}(t,x,\mu)\le r\}} p\cdot\widehat{f}(t,x,\mu) \tag{3.6}$$

As usual, $\widehat{H} = +\infty$ if the minimum is over an empty set.

Theorem 3.2. $V(t,x) \equiv \widehat{V}(t,x)$ *on* $[0,T] \times R^n$.

Proof. The relaxed value function is the unique continuous viscosity solution of

$$\max\{\widehat{V}_t + \widehat{H}(t,x,\widehat{V},D_x\widehat{V}),\ \min_{\mu\in\mathcal{M}(Z)} \widehat{h}(t,x,\mu) - \widehat{V}\} = 0, \tag{3.7}$$

and satisfies the terminal condition

$$\widehat{V}(T,x) = \min_{\mu\in\mathcal{M}(Z)} \widehat{h}(T,x,\mu). \tag{3.8}$$

The proof that $V = \widehat{V}$ consists in proving that

$$\min_{\mu\in\mathcal{M}(Z)} \widehat{h}(t,x,\mu) = \min_{z\in Z} h(t,x,z). \tag{3.9}$$

and that $\widehat{H}(t,x,r,p) = H(t,x,r,p)$, i.e.,

$$\min_{\{\mu \in \mathcal{M}(Z) \mid \widehat{h}(t,x,\mu) \le r\}} p \cdot \widehat{f}(t,x,\mu) = \min_{\{z \in Z \mid h(t,x,z) \le r\}} p \cdot f(t,x,z). \tag{3.10}$$

We conclude that $\widehat{V}$ and V both satisfy the same Bellman equation (2.5) (in the viscosity sense) and the terminal condition (2.6). Uniqueness of viscosity solutions then yields that $\widehat{V}$ and V are the same function. □

Full details of the results of this section are in [3]

Remark 3.1. Stochastic problems involving the L^∞ norm were considered in two papers. The paper [4] considers controlled diffusions when the cost also involves the running maximum of the state. This is a special case of the paper [2] in which the full stochastic control problem with L^∞ cost is studied.

4. Calculus of Variations in L^∞.

In this section we will consider the L^∞ calculus of variations problem with ordinary differential equations first studied in [1]. This is a variant of the simple optimal control problem considered above with control set $Z = R^n$ and $f(t,x,z) = z$. The primary difficulty now is the noncompactness of the control set.

The calculus of variations problem in R^n with cost functional $I[u] = ||Du||_{L^\infty(\Omega)}$ was considered in [10].

We consider the problem of minimizing over the class $AC[a,b]$ of absolutely continuous functions, $\tau \mapsto \xi(\tau) \in R^n$, on the interval $[a,b]$, the functional

$$I[\xi] \equiv ||h(s,\xi(s),\xi'(s))||_{L^\infty[a,b]}.$$

The first result from [3] shows that quasiconvexity of h in z is a necessary condition for weak lower semicontinuity of L^∞ functionals in the calculus of variations. The analogous result for classical variational problems is that *convexity* is necessary for weak lower semicontinuity of integral functionals ([8], p. 104).

Theorem 4.1. *If $I[\xi]$ is weakly lower semicontinuous for each fixed $a < b$, then $h(t,x,z)$ is quasiconvex in z for each $(t,x) \in [a,b] \times R^n$.*

To keep the discussion simple we will assume the strong coercivity condition

$$h(t,x,z) \ge \alpha|z|^q + \beta, \quad \exists \alpha > 0, \beta \in R^1, \ q \ge 1. \tag{4.1}$$

Theorem 4.2. *Assume, in addition to (A), that $h(\cdot,\cdot,z)$ is quasiconvex, and that* (4.1) *holds. Then, for each fixed initial position $y \in R^n$, there exists an optimal trajectory.*

Proof. For any trajectory,

$$I[\xi] \ge \alpha||(\xi')^q||_\infty + \beta \ge \beta,$$

so $I[\cdot]$ is bounded below. Set $\mu = \inf_\xi I[\xi]$. Let $\{\xi_k\}$ be a minimizing sequence, $I[\xi_k] \to \mu$. Without loss of generality we may assume that $I[\xi_k] \le I[\xi_1] \equiv K$. Then, by (4.1),

$$\alpha||(\xi_k')^q||_{L^\infty} + \beta \le I[\xi_k] \le K$$

which implies that $||\xi_k'||_{L^\infty} \leq K'$. We also have that, for any $\tau \in [a,b]$,

$$|\xi_k(\tau) - y| \leq \int_a^\tau \left| \frac{d\xi_k(r)}{dr} \right| dr \leq K'(b-a).$$

Therefore, $||\xi_k||_{L^\infty} \leq K''$. Thus, the minimizing sequence is uniformly bounded and equicontinuous. So, there is a subsequence, and an absolutely continuous trajectory, ξ^*, with $\xi^*(a) = y$ and $\xi_k \to \xi^*$ uniformly, and $\xi_k' \rightharpoonup (\xi^*)'$ weak-* in L^∞ as $k \to \infty$. We also know that $||(\xi^*)'|| \leq K'$.

We claim that ξ^* is an optimal trajectory. To prove this claim, since $\xi_k' \rightharpoonup (\xi^*)'$, by Mazur's lemma, for every j there exists an integer n_j, a set of integers $i = 1, 2, \ldots, k$, a set of nonnegative numbers $\alpha_{1j}, \ldots, \alpha_{kj}$, with $\sum_{i=1}^k \alpha_{ij} = 1$, such that, $n_{j+1} > n_j + k$, and, if we define

$$y_j(\tau) = \sum_{i=1}^k \alpha_{ij} \frac{d\xi_{n_j+i}}{d\tau}$$

then $y_j(\tau) \to (\xi^*)'$ for a.e. $\tau \in [t_0, T]$. (We do not distinguish between convergence on a subsequence.) Now define

$$\lambda_j(\tau) = \max_{1 \leq i \leq k} |h(\tau, \xi_{n_j+i}(\tau), (\xi_{n_j+i})'(\tau))|,$$

and

$$\lambda(\tau) \equiv \liminf_{j \to \infty} \lambda_j(\tau).$$

By lower semicontinuity,

$$||\lambda(\tau)||_{L^\infty[a,b]} \leq \liminf_{j \to \infty} \max_{1 \leq i \leq k} ||h(\tau, \xi_{n_j+i}(\tau), \xi_{n_j+i}'(\tau)||_{L^\infty[a,b]} = \mu.$$

Set

$$\sigma_j(\tau) = \sum_{i=1}^k \alpha_{ij} \frac{d\xi_{n_j+i}}{d\tau}, \qquad \vartheta_j(\tau) = \max_{1 \leq i \leq k} |h(\tau, \xi^*(\tau), \xi_{n_j+i}'(\tau))|.$$

Then, since $\xi_k \to \xi^*$ uniformly in τ, $\vartheta_j(\tau) \to \lambda(\tau)$, and $\sigma_j(\tau) \to (\xi^*)'(\tau)$ for a.e. $\tau \in [a,b]$. Define the set valued map

$$\mathcal{L}(t, x, a) = \{z \in R^n \mid a \geq h(t, x, z)\}.$$

Therefore, we have shown that

$$\frac{d\xi_{n_j+i}}{d\tau} \in \mathcal{L}(\tau, \xi^*(\tau), \vartheta_j(\tau)) \text{ a.e. } \tau \in [a,b].$$

Since $h(\cdot, \cdot, z)$ is quasiconvex, for τ fixed, $\mathcal{L}(\tau, \xi^*(\tau), \vartheta_j(\tau))$ is a convex set, and so $\sigma_j(\tau) \in \mathcal{L}(\tau, \xi^*(\tau), \vartheta_j(\tau))$. By continuity, since $\sigma_j \to (\xi^*)'$ and $\vartheta_j \to \lambda$, $(\xi^*)'(\tau) \in \mathcal{L}(\tau, \xi^*(\tau), \lambda(\tau))$, for a.e. $\tau \in [a,b]$. Hence $\lambda(\tau) \geq h(\tau, \xi^*(\tau), (\xi^*)'(\tau))$, a.e. $\tau \in [a,b]$. But $\mu \geq ||\lambda(\tau)||_{L^\infty[a,b]}$, and we conclude that ξ^* is optimal. □

5. The Euler Equation for L^∞ problems

In this section we will derive the Euler equation for an *absolute* minimizer of the variational problem. This will be done under the assumption that the function h is twice continuously differentiable. For simplicity, we will assume that we are in one dimension. The extension to R^n is easily supplied by the reader.

We will consider the fixed endpoint problem:

$$\text{minimize } I_{a,b}[\xi] = ||h(s,\xi(s),\xi'(s))||_{L^\infty[a,b]},\ \xi \in AC[a,b], \xi(a) = x_0, \xi(b) = x_1.$$

Definition 5.1. A trajectory ξ^* is an absolute minimizer of $I_{a,b}[\xi]$ if, for any subinterval $[\alpha,\beta] \subset [a,b]$, ξ^* is a minimizer of $I_{\alpha,\beta}[\xi]$ where $\xi(\alpha) = \xi^*(\alpha)$ and $\xi(\beta) = \xi^*(\beta)$. That is, ξ^* minimizes I on every subinterval.

Theorem 5.1. *If $\xi^* \in C^2[a,b]$ is an absolute minimizer, then*

$$\left(\frac{d}{dt}(h(t,\xi^*(t),(\xi^*)'(t))\right)\cdot h_z(t,\xi^*(t),(\xi^*)'(t)) = 0, \tag{5.1}$$

Proof. The proof, essentially due to Jensen, is based on the idea of a similar proof for a special case in [10]. For simplicity, let us take $h = h(x,z)$ to be independent of t. Also, we will write ξ for ξ^*.

Fix a point $s \in (a,b)$ and choose $\varepsilon > 0$ small enough so that $(s-\varepsilon, s+\varepsilon) \subset (a,b)$. We will show that (5.1) holds at s. Let $\delta \in R^1$ be any real number and consider the trajectory variation

$$y(\tau) = \xi(\tau) + (\delta/2)\varepsilon^2 - (\delta/2)(\tau - s)^2.$$

Notice that $y(s\pm\varepsilon) = \xi(s\pm\varepsilon)$.

Expand ξ in a Taylor's series about s:

$$\xi(\tau) = \xi(s) + \xi'(s)(\tau - s) + \frac{1}{2}\xi''(s)(\tau-s)^2 + o(|\tau - s|^2).$$

Now expand h in a Taylor's series about $(\xi(s),\xi'(s))$ and collect terms to get:

$$h(\xi(\tau),\xi'(\tau)) = h(\xi(s),\xi'(s)) + (\tau - s)\left(h_x\cdot\xi'(s) + h_z\cdot\xi''(s)\right) + o(|\tau - s|).$$

It is now easy to see that

$$\max_{s-\varepsilon\le\tau\le s+\varepsilon} h(\xi(\tau),\xi'(\tau)) = h(\xi(s),\xi'(s)) + \varepsilon\left|h_x\cdot\xi'(s) + h_z\cdot\xi''(s)\right| + o(\varepsilon).$$

Similar calculations yield

$$\max_{s-\varepsilon\le\tau\le s+\varepsilon} h(y(\tau),y'(\tau)) = h(\xi(s) + \frac{\delta}{2}\varepsilon^2, \xi'(s)) + \varepsilon\left|h_x\cdot\xi'(s) + h_z\cdot(\xi''(s) - \delta)\right| + o(\varepsilon).$$

Since ξ is absolutely minimizing,

$$\max_{s-\varepsilon\le\tau\le s+\varepsilon} h(\xi(\tau),\xi'(\tau)) \le \max_{s-\varepsilon\le\tau\le s+\varepsilon} h(y(\tau),y'(\tau))$$

and writing down the details of this, dividing by ε and letting $\varepsilon\to 0$, yields

$$|h_x\cdot\xi'(s) + h_z\cdot\xi''(s)| \le |h_x\cdot\xi'(s) + h_z\cdot(\xi''(s) - \delta)| \equiv F(\delta). \tag{5.2}$$

This defines the function $F : R^1 \to [0,\infty)$. Now, if

$$0 = h_x\cdot\xi'(s) + h_z\cdot\xi''(s) = \frac{d}{ds}h(\xi(s),\xi'(s)) \tag{5.3}$$

then, (5.2) does not yield any information. But then (5.3) says that (5.1) holds. On the other hand, if (5.3) does not hold, then (5.2) says that $F(\delta)$ achieves a minimum at $\delta = 0$. Consequently,

$$\frac{dF(0)}{d\delta} = \operatorname{sgn}\left(h_x \cdot \xi'(s) + h_z \cdot \xi''(s)\right) h_z(\xi(s), \xi'(s)) = 0$$

and this says again that (5.1) must hold. □

It is also possible to derive the Euler equation by using the Pontryagin principle derived in [6].

REFERENCES

1. G. Aronsson, *Minimization problems for the functional* $\sup_x F(x, f(x), f'(x))$, Arkiv For Matematik, 6(1964), pp. 33–53.
2. G.Barles, C. Daher, and M. Romano, *Optimal control on the L^∞ norm of a diffusion*, to appear in SIAM J. Control and Optimization.
3. E.N.Barron and R.Jensen, *Relaxed Minimax Control*, submitted.
4. E.N.Barron, *The Bellman equation for the running max of a diffusion and applications to look back options*, to appear in Applicable Analysis.
5. E.N.Barron and H.Ishii, *The Bellman equation for minimizing the maximum cost*, Nonlinear Anal., TMA, 13(1989), pp. 1067–1090.
6. E.N.Barron, *The Pontryagin maximum principle for minimax problems of optimal control*, Nonlinear Anal., TMA, 15(1990), pp. 1155–1165.
7. L. D. Berkovitz, *Optimal Control Theory*, Springer-Verlag, New York, 1974.
8. L. Cesari, *Optimization Theory and Application*, Springer-Verlag, New York, 1983.
9. M. G. Crandall, L. C. Evans, and P.-L. Lions, *Some properties of viscosity solutions of Hamilton Jacobi equations*, Trans. Amer. Math. Soc., 282(1984), pp. 487–502.
10. R. Jensen, *Uniqueness of Lipschitz Extensions: Minimizing the Sup Norm of the Gradient*, Arch. Rat. Mech. Anal., to appear.

5 Optimal Control of Some Partial Differential Equations with Two Point Boundary Conditions

Zhixiong Cai Barton College, Wilson, North Carolina

Nicolae H. Pavel[1] Ohio University, Athens, Ohio

Shih-liang Wen Ohio University, Athens, Ohio

1 INTRODUCTION

This paper deals with optimal control problems for partial differential equations of parabolic and hyperbolic type with boundary conditions (in the time variable) of periodic and antiperoidic type.

Our results here extend those in [4] on optimal control of the heat and wave equations. Moreover, our technique here is based on "tangent cones" in the sense of Bouligand and Clarke and on some smooth approximations of such cones, e.g., the approximation K_λ (given by (2.33)) of K_β (given by (2.23)). The technique in [4] is mainly based on appropriate "penalization" of cost functional. Some open cases, e.g. the case $|a+b|=ab+1$ in (3.4) are delicate and these will be carried out elsewhere.

2 TWO POINT NONLINEAR BOUNDARY CONDITIONS OF ANTIPERIODIC TYPE

In this section we are concerned with necessary and sometimes sufficient conditions for a pair (y^*,u^*) to be the optimal pair of the problem (OP).

$$\text{(OP) Minimize } L(y,u); \qquad L(y,u) = \int_0^T \Big(g(y(t))+h(u(t))\Big)dt+\ell(y(0),y(T))$$

over all pairs $(y,u) \in M$, where $T>0$,

$$M = M_K = \Big\{(y,u);\ u \in L^2(0,T;U);\ y(t)=e^{tA}y(0)+\int_0^t e^{(t-s)A}Bu(s)ds \quad 0\le t\le T,\ (y(0),y(T)) \in K\Big\} \tag{2.1}$$

[1] The author was supported by NSF under Grant #DMS 9111794

U and H are real Hilbert spaces and $K \subset H \times H$. The basic hypotheses are the following:
(I1) $K \subset H \times H$ is a nonempty weakly-strongly closed subset. Moreover, for every $u \in L^2(0,T;U)$, there is a unique $y_0 \in H$ such that $y(t)$ defined as in (2.1) with $y(0) = y_0$, satisfies $(y(0),y(T)) \in K$.

In other words, we assume that for every $u \in L^2(0,T;U)$, there is a unique function $y = \tilde{P}u$ given by

$$y(t) = (\tilde{P}u)(t) = e^{tA}y(0) + \int_0^t e^{(t-s)A}Bu(s)ds, \quad 0 \le t \le T \tag{2.2}$$

such that $(y(0),y(T)) \in K$, i.e., the "equation"

$$\left(y(0),\ e^{TA}y(0) + \int_0^T e^{(T-s)A}Bu(s)ds\right) \in K, \quad \text{with } u \in L^2(0,T;U) \tag{2.3}$$

has a unique solution $y(0) = y_u \in H$.

(I2) If $\left(y_n^0,\ e^{TA}y_n^0 + z_n\right) \in K$, $n = 1,2,\ldots$ with $\{z_n\}$ bounded in H, then $\{y_n^0\}$ is also bounded in H.

($\mathfrak{I}$1) The C_0-semigroup e^{tA} generated by the maximal dissipative operator A in H is compact for $t > 0$.

($\mathfrak{I}$2) $g:H \to R$ and $h:U \to]-\infty,+\infty]$ are lower semicontinuous convex functionals, with $g(x) < +\infty$ for all $x \in H$, $h(u) \ge \omega(\|u\|^2+1)$, for all $u \in U$ (for some $\omega > 0$).

($\mathfrak{I}$3) $\ell:H \times H \to R$ is a Frechet differentiable convex functional, everywhere defined.

($\mathfrak{I}$4) $B \in L(U,H)$

Remark 2.1. Hypotheses ($\mathfrak{I}$4) can be weakened to

$$(\mathfrak{I}4)' \quad B^* \in (D(A^*);U); \quad \int_0^T \|B^*e^{-tA^*}z\|_U^2\,dt \le C\|z\|^2, \quad \forall z \in H, \text{ for some } C > 0,$$

see e.g. [4]. The Frechet derivative (gradient) of g at x in the direction of v is denoted by $\dot{g}(x)(v)$.

THEOREM 2.1 Hypotheses (I) and ($\mathfrak{I}$) above imply the existence of optimal pairs (y^*,u^*), i.e.

$$\text{Min}\left\{L(y,u);\ (y,u) \in M\right\} = L(y^*,u^*) \tag{2.4}$$

for some $(y^*,u^*) \in M$.

Proof: Part of the proof is routine so we only sketch its main steps. First we show that $\text{Inf}\{L(y,u);\ (y,u) \in M\} = d > -\infty$.

Let $(y_n,u_n) \in M$ be such that $L(y_n,u_n) \to d$ as $n \to \infty$. It follows from ($\mathfrak{I}$2) that u_n

is bounded in $L^2(0,T;U)$, and from (I2) that $y_n(0)$ is bounded in H. Say $u_n \rightharpoonup u^*$ in $L^2(0,T;U)$ and $y_n \rightharpoonup \alpha$ in H. It follows that $y_n \to y^*$ in $L^2(0,T;H)$ where $\rightharpoonup$ ($\to$) denotes weak (strong) convergence respectively,

$$y^*(t) = e^{tA}\alpha + \int_0^t e^{(t-s)A}Bu^*(s)ds, \quad 0 \le t \le T \tag{2.5}$$

and $y_n(T) \to y^*(T)$. One observes that M is strongly-weakly closed so $(y^*,u^*) \in M$ and (2.4) holds.

In what follows we consider the usual inner product on $H \times H$, i.e.

$$<(y_1,y_2),(z_1,z_2)> = <y_1,z_1>_H + <y_2,z_2>_H \tag{2.5)$'$ (}$$

with $<y_1,z_1>_H = <y_1,z_1>$, the inner product of H. We will give, via Bouligand contingent cones [8, p111], necessary conditions for (y^*,u^*) to be an optimal pair. Recall that the tangent cone (Bouligand contingent cone) TM(Y) to M at $Y \in M$ is defined by

$$TM(Y) = \{V \in \tilde{H};\ Y + \lambda V + \lambda R(\lambda) \in M,\ \forall\lambda > 0, \text{ for some } R(\lambda) \to 0 \text{ as } \lambda \downarrow 0\} \tag{2.6}$$

The normal cone NM(Y) to M at $Y \in M$ is defined by

$$NM(Y) = \{Z \in \tilde{H}',\ <Z,V> \le 0,\ \forall V \in TM(Y)\} \tag{2.7}$$

where $\tilde{H}$ is a normed space containing M, $\tilde{H}'$ is the dual of $\tilde{H}$ and $<Z,V>$ is the value of Z at V. Of course, similarly one defines $TK(y_1,y_2)$ and $NK(y_1,y_2)$ with $y_i \in H$, $i = 1,2$. It is easy to check the following

LEMMA 2.1 $TM(y^*,u^*) \subset \{(v,w);\ v \in C([0,T];H),\ w \in L^2(0,T;U),$

$$v(t) = e^{tA}v(0) + \int_0^t e^{(t-s)A}Bw(s)ds, \quad (v(0),v(T)) \in TK(y^*(0),y^*(T))\}$$

i.e., $v' = Av + Bw$ whenever v is a.e. differentiable on [0,T].

The converse inclusion fails, unless K is "sufficiently smooth".
Set

$$p(t) = e^{(T-t)A^*}p(T) - \int_t^T e^{(t-s)A^*}\dot{g}(y^*(s))ds, \quad 0 \le t \le T \tag{2.8}$$

i.e. p is the mild solution of

$$p' = -A^*p + \dot{g}(y^*(t)), \quad \text{a.e. } t \in (0,T) \tag{2.9}$$

and A^* is the adjoint of A.

Of course, in (2.8) we assume that the gradient $\dot{g}(y)$ of g exists and $s \to \dot{g}(y^*(s))$ is in $L^2(0,T;H)$ or only in $L^1(0,T;H)$.

A very general result, containing essentially most of previous results on these problems, is given by Lemma 2.2 below, which has a "geometric formulation". It gives a general principle (a necessary condition) for (y*,u*) to be an optimal pair for (OP). It involves the tangent (and normal) cone to M at (y*,u*).

LEMMA 2.2 Assume that conditions (I) and ($\mathfrak{J}$) holds and that g, h and ℓ are Frechet differentiable. Let (y*,u*) be an optimal pair of (OP). Then for every p given by (2.8) and every $(v,w) \in TM(y^*,u^*)$ we have

$$\int_0^T \langle \dot{h}(u^*(t)) - B^*p(t), w(t)\rangle dt \geq \langle p(0), -p(T)) - \dot{\ell}((y^*(0),y^*(T)),(v(0),v(T))\rangle \tag{2.10}$$

Proof: Condition $\operatorname{Inf}\{L(y,u);\ (y,u) \in M\} = L(y^*,u^*)$ implies that

$$\dot{L}(y^*,u^*))(v,w) \geq 0, \quad \text{for all } (v,w) \in TM(y^*,u^*) \tag{2.11}$$

$$\int_0^T \big(\langle \dot{g}(y^*(t)), v(t)\rangle + \langle \dot{h}(u^*(t), w(t)\rangle\big)dt + \langle \dot{\ell}(y^*(0),y^*(T)),(v(0),v(T))\rangle \geq 0 \tag{2.12}$$

for all $(v,w) \in TM(y^*,u^*)$.

Substituting v given in Lemma 2.1 into (2.12), interchanging the order of integration according to Fubini's theorem, and observing that

$$\langle p(T),v(T)\rangle - \langle p(0),v(0)\rangle = -\langle p(0),-p(T)),(v(0),v(T))\rangle \tag{2.13}$$

with

$$p(0) = e^{TA^*}p(T) - \int_0^T e^{sA^*}\dot{g}(y^*(s))ds, \tag{2.14}$$

one concludes that (2.12) implies (2.10).

LEMMA 2.3 In addition to the hypotheses of Lemma 2.2 assume that for every $\alpha,\beta,\gamma \in H$, the "equation"

$$(e^{TA^*}p_T + \alpha, -p_T) \in (\beta,\gamma) - NK(y^*(0),y^*(T)) \tag{2.15}$$

has a unique solution $p_T \in H$. Then the function p defined by (2.8) with $p(T) = p_T$ given by (2.15) satisfies

$$B^*p(t) = \dot{h}(u^*(t)), \quad \text{a.e. on } [0,T] \tag{2.16}$$

Proof: It follows from (2.10) that

$$\int_0^T <\dot{h}(u^*(t)) - B^*p(t), w(t)> dt \geq 0, \quad \forall w \in L^2(0,T;U) \tag{2.17}$$

which yields (2.16). In order to get (2.17), one chooses $p_T = p(T)$ such that

$$(p(0),-p(T)) - \dot{\ell}(y^*(0),y^*(T)) \in -NK(y^*(0),y^*(T)) \tag{2.18}$$

Remark 2.1 The conclusion of this section is essentially this: Under hypotheses (I1)-($\Im$4), the necessary condition for (y^*,u^*) to be an optimal pair of (OP) are given by $(y^*,u^*) \in M$, $(p(0),p(T))$ satisfies (2.18) and $B^*p(t) = \dot{h}(u^*(t))$. Or more general, one can expect $B^*p(t) \in \partial h(u^*(t))$, a.e on $[0,T]$, where ∂h is the subdifferential of h. If NK is a linear subspace of H, for example, then the above necessary conditions for optimality are sufficient, too.

Some examples of K, including periodic and antiperiodic type boundary conditions are given below

$$K = \{(y_1,y_2) \in H \times H;\ y_2 + Fy_1 = \alpha\} \tag{2.19}$$

where $\alpha \in H$ and $F: H \to H$ is a Frechèt differentiable, possibly nonlinear operator with $\dot{F}(y)$ surjective for $y \in H$.
Then, the tangent cone (space) to K at $(y_1,y_2) \in K$ satisfies

$$TK(y_1,y_2) \subseteq \{(v_1,v_2) \in H \times H;\ v_2 + \dot{F}(y_1)v_1 = 0\} \tag{2.20}$$

If $\dot{F}(y)$ is surjective for $y \in H$, then the inclusion (2.20) is an equality. Moreover, the normal cone (space) is given by

$$NK(y_1,y_2) = \{(\eta_1,\eta_2) \in H \times H;\ \eta_1 - (\dot{F}(y_1))^*\eta_2 = 0,\ (y_1,y_2) \in K\} \tag{2.21}$$

In this case (of (2.19)). condition (2.15) is satisfied if $e^{TA^*} + (\dot{F}(y^*(0)))^*$ is invertible. We actually assume that

$$e^{TA^*} + (\dot{F}(y^*(0)))^* \text{ is invertible with bounded inverse.} \tag{2.22}$$

We are now in a position to study the optimal problem (OP) in (2.1) with $K = K_\beta$, $H = L^2(\Omega)$, and $A = \Delta$ where

$$K_\beta = \{(y_1,y_2) \in L^2(\Omega) \times L^2(\Omega);\ y_2(x) + \beta(y_1(x)) \ni 0, \text{ a.e. } x \in \Omega\} \tag{2.23}$$

with $\beta(r) = ar$ if $r > 0$, $\beta(0) = (-\infty,0]$ and $\beta(r) = \phi$ (empty set) if $r \leq 0$. Here a is an arbitrary positive number. Precisely, we will give necessary conditions for a pair (y^*,u^*) to be an optimal pair of the problem:

$$\text{Minimize } \left\{ L(y,u) = \int_0^T \int_\Omega \left(G(y(t,x)) + \tilde{H}(u(t,x)) \right\} dxdt \right\} \tag{2.24}$$

over all $(y,u) \in C([0,T];L^2(\Omega)) \times L^2(0,T;L^2(\Omega))$ subject to $(y,u) \in M_{K_\beta} \equiv M_\beta$, where $T > 0$ and

$$M_\beta = \Big\{ (y,u) \text{ as in (2.24); } y \text{ is a mild solution of}$$
$$y_t = \Delta y + u \quad \text{a.e. in } (0,T) \times \Omega$$
$$y(t,x) = 0 \quad \text{a.e. on } (0,T) \times \partial\Omega$$
$$(y(0,\cdot),\ y(T,\cdot)) \in K_\beta \Big\} \tag{2.25}$$

Here Δ denotes the Laplace operator with $D(\Delta) = H_0^1(\Omega) \cap H^2(\Omega)$ and Ω is a bounded domain of R^n ($n = 1,2,\ldots$) with smooth boundary $\partial\Omega = \Gamma$.
Therefore,

$$y(t) = e^{t\Delta}y(0) + \int_0^t e^{(t-s)\Delta}u(s)ds, \quad 0 \le t \le T \tag{2.26}$$

with $y(t) = y(t,\cdot)$, i.e. $(y(t))(x) = y(t,x)$ a.e. in $x \in \Omega$.
Clearly, $(y(0), y(T)) \in K_\beta$ is equivalent to

$$\begin{aligned} y(0,x) &\ge 0, && \text{a.e. in } \Omega \\ y(T,x) + ay(0,x) &= 0, && \text{a.e. in } \Omega^+ = \{x \in \Omega,\ y(0,x) > 0\} \\ y(T,x) &\ge 0, && \text{a.e. in } \Omega^0 = \{x \in \Omega,\ y(0,x) = 0\} \end{aligned} \tag{2.27}$$

Note that we can replace zero in the right-hand side of the equalities and inequalities in (2.27) by a function $\psi \in L^2(\Omega)$. We prefer zero for the simplicity of writing.

In order to give necessary conditions for optimality we need to introduce some smooth approximations of β. A first attempt would be the Yosida approximation β_λ^ν of β (as β is maximal monotone in R). This is given by $\beta_\lambda^\nu(r) = \frac{ar}{\lambda + 1}$, if $r \ge 0$, and $\beta_\lambda^\nu(r) = \frac{ar}{\lambda}$, if $r < 0$ with $\lambda > 0$, which is not of class $C^1(R)$, as required in (2.20).

The following approximation β_λ (with $\lambda > 0$) of β is of class $C^1(R)$:

$$\beta_\lambda(r) = \begin{cases} ar, & \text{if } r \ge 0 \\ -\dfrac{a}{2\lambda^2}r^2 + ar, & \text{if } -\lambda \le r \le 0 \\ a(1 + \dfrac{1}{\lambda})r + \dfrac{1}{2}a, & \text{if } r \le -\lambda \end{cases} \tag{2.28}$$

Its inverse function β_λ^{-1} is given by

$$\beta_\lambda^{-1} = \begin{cases} \frac{1}{a} r, & \text{if } r \geq 0 \\ \frac{\lambda}{a}(ar - (a^2\lambda^2 - 2a\lambda)^{1/2}), & \text{if } -a\lambda - \frac{a}{2} \leq r \leq 0 \\ \frac{\lambda}{a(1+\lambda)}(r - \frac{a}{2}), & r \leq -a\lambda - \frac{a}{2} \end{cases} \tag{2.28)'$$

Therefore, there is a constant $c = c(a) > 0$ (independently of $\lambda \leq 1$) such that

$$\begin{aligned} |\beta_\lambda^{-1}(r)| &\leq c(|r| + 1) \quad \text{if } |r| > 1 \\ |\beta_\lambda^{-1}(r)| &\leq c \quad\quad \text{if } |r| \leq 1, \ \forall\lambda \leq 1 \end{aligned} \tag{2.29}$$

Moreover, if r_λ is a bounded sequence with $r_\lambda \leq 0$, then $\beta_\lambda^{-1}(r_\lambda) \to 0$, as $\lambda \downarrow 0$.

Denote by β_λ the L^2-realization of β_λ, i.e.

$$(\beta_\lambda y)(x) = \beta_\lambda(y(x)), \text{ a.e. in } \Omega, \ y \in L^2(\Omega) \tag{2.30}$$

and set $b_\lambda(r) = \beta_\lambda(r) - ar$. It follows that

$$b_\lambda = \beta_\lambda - aI, \ e^{T\Delta} + \beta_\lambda = e^{T\Delta} + b_\lambda + aI \tag{2.30)'$$

So $e^{T\Delta} + \beta_\lambda$ is invertible in $L^2(\Omega)$ with Lipschitz continuous inverse $(e^{T\Delta} + \tilde{\beta}_\lambda)^{-1}$ of Lipschitz constant $1/a$.

The derivative $\dot{b}_\lambda$ of b_λ is given by

$$\dot{b}_\lambda(r) = \begin{cases} 0 & \text{if } r \geq 0 \\ -\frac{a}{\lambda^2} r & \text{if } -\lambda \leq r \leq 0 \\ \frac{a}{\lambda} & \text{if } r \leq -\lambda \end{cases} \tag{2.31}$$

and we have

$$\dot{\beta}_\lambda(y) = \dot{b}_\lambda(y) + aI, \quad e^{T\Delta} + \dot{\beta}_\lambda(y) = e^{T\Delta} + \dot{b}_\lambda(y) + aI \tag{2.32}$$

So $e^{T\Delta} + \dot{\beta}_\lambda(y)$ is also invertible in $L^2(\Omega)$ with $\frac{1}{a}$ - Lipschitz continuous inverse for every $y \in L^2(\Omega)$. We will approximate M_β by $M_{K_{\beta_\lambda}} \equiv M_\lambda$ given below

$$
\begin{aligned}
K_\lambda \equiv K_{\beta_\lambda} &= \left\{(y_1^\lambda, y_2^\lambda) \in L^2(\Omega)\times L^2(\Omega);\ y_2^\lambda(x) + \beta_\lambda(y_1^\lambda(x)) \ni 0,\ \text{a.e. } x \in \Omega\right\} \\
M_\lambda &= \left\{(y^\lambda, u^\lambda) \text{ as in (2.26) with } (y^\lambda(0,\cdot),\ y^\lambda(T,\cdot)) \in K_{\beta_\lambda}\right\}
\end{aligned}
\tag{2.33}
$$

On the basis of (2.19)-(2.22) it follows

$$
\begin{aligned}
TK_\lambda(y_1,y_2) &\subseteq \left\{(v_1,v_2) \in L^2(\Omega)\times L^2(\Omega);\ v_2(x) + \dot\beta_\lambda(y_1(x))v_1(x) = 0,\ \text{a.e. in } \Omega\right\} \\
NK_\lambda(y_1,y_2) &= \left\{(\eta_1,\eta_2) \in L^2(\Omega) \times L^2(\Omega);\ \eta_1(x) - \dot\beta_\lambda(y_1(x))\eta_2(x) = 0,\ \text{a.e. in } \Omega\right\}
\end{aligned}
\tag{2.34}
$$

with $(y_1,y_2) \in K_\lambda$.

The following result will play a crucial role on the sequel.

LEMMA 2.4 Let $(y_1^\lambda, y_2^\lambda) \in K_\lambda$ be such that $y_1^\lambda \rightharpoonup y_1$ and $y_2^\lambda \to y_2$ in $L^2(\Omega)$, as $\lambda\downarrow 0$. Then $(y_1,y_2) \in K_\beta$ and $y_1^\lambda \to y_1$ in $L^2(\Omega)$ as $\lambda\downarrow 0$.

Proof: We may assume, relabeling if necessary, that

$$
y_2^\lambda(x) \to y_2(x),\quad |y_2^\lambda(x)| \le h(x);\quad \text{a.e. in } \Omega \tag{2.35}
$$

with $h \in L^2(\Omega)$ (see [7]). By the definition of K_λ, $y_1^\lambda(x) = \beta_\lambda^{-1}(-y_2^\lambda(x))$. This and (2.28)′ imply that

$$
\lim_{\lambda\downarrow 0} y_1^\lambda(x) \equiv y_0(x) = \begin{cases} -\frac{1}{a}y_2(x) & \text{in } \{x \in \Omega;\ y_2(x) < 0\} \\ 0 & \text{in } \{x \in \Omega,\ y_2(x) \ge 0\} \end{cases} \tag{2.36}
$$

Moreover, (2.29) in conjunction with (2.36) lead us to the conclusion

$$
|y_1^\lambda(x)| \le h(x);\quad \text{a.e. in } \Omega,\quad \forall\lambda \le 1 \tag{2.37}
$$

with $h_1 \in L^2(\Omega)$. Therefore, $y_1^\lambda \to y_0$ in $L^2(\Omega)$, so $y_0 = y_1$. Hence, from (2.36) it follows that $(y_1,y_2) \in K_\beta$.

Finally, in the study of Problem (2.24), the following facts on convex integrands are needed (cf. [3], p61).

Let $G: R \to]-\infty,+\infty]$ be a lower-semicontinuous convex proper function (integrand). It induces a functional $g_G = g: L^2(\Omega) \to]-\infty, +\infty]$ defined by

$$g_G(y) = \begin{cases} \int_\Omega G(y(x))dx, & \text{if } x \to G(y(x)) \text{ is in } L^1(\Omega) \\ +\infty, & \text{otherwise} \end{cases} \tag{2.38}$$

i.e. the effective domain of D(g) of g is $D(g) = \{y \in L^2(\Omega);\ x \to G(y(x)) \in L^1(\Omega)\}$. Moreover, $w \in \partial g(y)$ if and only if $w(x) \in \partial G(y(x))$ a.e. in Ω, where ∂ stands for the subdifferential operator.

It is now clear that Problem (2.24) can be reduced to (OP) as described by (2.1) with $H = L^2(\Omega)$, $g = g_G$ and $h = h_{\bar{H}}$. In order to state some necessary conditions for optimality, we must specify the hypotheses on G and $\bar{H}$, or directly, on $g = g_G$ and $h = h_{\bar{H}}$.

(H1) Suppose that the functional $g = g_G$ and $h = h_{\bar{H}}$ from $L^2(\Omega)$ into $(-\infty, +\infty]$, defined by G and $\bar{H}$ via (2.38) are lower-semicontinuous convex proper functions with

$$g(y) < \infty,\ h(y) \geq \omega\, |y|^2_{L^2(\Omega)},\quad \forall y \in L^2(\Omega) \tag{2.39}$$

for some $\omega > 0$.

In the above hypotheses (as in (H1)) one can prove the existence of optimal pairs (y^*, u^*) for Problem (2.24). The main steps of the proof are outlined in the proof of Theorem 2.1. Necessary conditions for (y^*, u^*) are given below.

THEOREM 2.2 Let (y^*, u^*) be an optimal pair for Problem (2.2). Then there are $p \in C([0,T]; L^2(\Omega))$ and $q \in L^2([0,T] \times \Omega)$ such that (in mild sense)

$$p_t + \nabla p = q,\ q(t,x) \in \partial G(y^*(t,x)),\quad \text{a.e. in } (0,T) \times \Omega \tag{2.40}$$

$$p(t,x) = 0,\quad \text{a.e. on } (0,T) \times \Gamma \tag{2.41}$$

$$p(0,x) + ap(T,x) = 0,\quad \text{a.e. in } \{x \in \Omega;\ y^*(0,x) > 0\} \tag{2.42}$$

$$p(0,x)y^*(T,x) = 0,\quad \text{a.e. in } \{x \in \Omega;\ y^*(0,x) = 0\} \tag{2.43}$$

$$p(t,x) \in \partial \bar{H}(u^*(t,x)),\quad \text{a.e. in } (0,T) \times \Omega \tag{2.44}$$

Remark 2.2 Conditions (2.42) and (2.43), in conjunction with $(y^*(0), y^*(T)) \in K_\beta$, i.e. (2.27), are equivalent to the geometric condition that $(p(0,x), -p(T,x))$ is perpendicular to $(y^*(0,x), y^*(T,x))$ for almost all $x \in \Omega$, i.e.

$$p(0,x)y^*(0,x) - p(T,x)y^*(T,x) = 0,\quad \text{a.e. in } \Omega \tag{2.45}$$

Here p is a mild solution of (2.40)-(2.42), in the sense of (2.8) with q is place of $\dot{g}(y^*)$.

Proof of Theorem 2.2: First of all, the functional L in (2.24) can be written (on the basis of (2.38)) in the form

$$L(y,u) = \int_0^T \big(g(y(t)) + h(u(t))\big)\,dt. \tag{2.46}$$

Let (y^*,u^*) be an optimal pair of Problem (2.24) and let us introduce the following penalization L_1 of L

$$L_1(y,u) = L(y,u)+\frac{1}{2}|u-u^*|^2_{L^2(0,T,L^2(\Omega))}+\frac{1}{2}|y(0)-y(0)^*|_{L^2(\Omega)} \tag{2.47}$$

Denote by $(y_\lambda^*,u_\lambda^*) \in M_\lambda$ an optimal pair of the problem

$$\mathrm{Inf}\big\{L_1(y,u);\ (y,u) \in M_\lambda\big\} = L_1(y_\lambda^*,u_\lambda^*) \tag{2.48}$$

The existence of $(y_\lambda^*,u_\lambda^*)$ is guaranteed by the smoothness of M_λ given in (2.33). Indeed, in this case, due to (2.32) we conclude that (2.22) and accordingly (2.15) is satisfied and Theorem 2.1 applies to (2.48). If g and h are not Frechet differentiable, one uses their Frechet differentiable approximations $g_\mu$and h_μ respectively and then one passes to limit with $\mu \downarrow 0$.

In conclusion, making use of Lemma 2.2 and 2.3 we can claim that $(y_\lambda^*,u_\lambda^*)$ and p_λ satisfy:

$$\left.\begin{aligned}
&y_\lambda^*(t) = e^{t\Delta}y_\lambda^*(0) + \int_0^t e^{(t-s)\Delta}u_\lambda^*(s)ds, \quad 0 \le t \le T \\
&\big(y_\lambda^*(0),\ y_\lambda^*(T)\big)\in K_\lambda,\ \text{i.e. } y_\lambda^*(0)=(e^{T\Delta}+\tilde{\beta}_\lambda)^{-1}\Big(-\int_0^T e^{(T-s)\Delta}u_\lambda^*(s)ds\Big) \\
&p_\lambda(t) = e^{(T-t)\Delta}p_\lambda(T) - \int_t^T e^{(s-t)\Delta}q_\lambda(s)ds, \quad 0 \le t \le T
\end{aligned}\right\} \tag{2.49}$$

with $q_\lambda(t,x) \in \partial G(y_\lambda^*(t,x))$ a.e. in Ω.

$$(p_\lambda(0),-p_\lambda(T)) - (y_\lambda^*(0)-y^*(0),0) \in NK_\lambda(y_\lambda^*(0),y_\lambda^*(T)) \tag{2.50}$$

Note that (2.50) follows from (2.18) with $K = K_\lambda$ (so $NK_\lambda = -NK_\lambda$) and $\ell(y(0),y(T)) = \frac{1}{2}|y(0) - y^*(0)|^2$, so

$$\dot{\ell}(y_\lambda^*(0),y_\lambda^*(T)) = (y_\lambda^*(0) - y^*(0),0). \tag{2.51}$$

Therefore, (2.50) has actually the form

$$p_\lambda(0,x)-(y_\lambda^*(0,x)-y^*(0,x))+\dot{\beta}_\lambda(y_\lambda^*(0,x))p_\lambda(T,x)=0, \text{ a.e. in } \Omega \tag{2.52}$$

Finally, in this case (2.16) becomes

$$p_\lambda(t,x) \in \partial\bar{H}(u^*_\lambda(t,x))+u^*_\lambda(t,x)-u^*(t,x), \text{ a.e.in } [0,T] \times \Omega. \tag{2.53}$$

We now need to deal with the boundedness of y^*_λ and u^*_λ. The key step is the boundedness of u^*_λ. Indeed, $(0,0) \in M_\lambda$, so $L_1(y^*_\lambda,u^*_\lambda) \le L_1(0,0)$, which implies the boundedness of u^*_λ in $L^2(0,T;L^2(\Omega))$. Next it follows the boundedness of $y^*_\lambda(0)$ in $L^2(\Omega)$. Indeed, this is a consequence of (2.49). Say (relabeling if necessary) that $y^*_\lambda(0) \rightharpoonup v$ and $u^*_\lambda \rightharpoonup u$ as $\lambda \downarrow 0$, weakly in $L^2(\Omega)$ and $L^2(0,T;L^2(\Omega))$ respectively. Thus, the compactness of $e^{t\Delta}$ for $t > 0$ and (2.49) imply that $y^*_\lambda(t)$ is strongly convergent, say to $\bar{y}(t)$ in $L^2(\Omega)$, for every $0 < t \le T$ as $\lambda \downarrow 0$. On the basis of Lemma 2.4 it follows that $y^*_\lambda(0)$ is actually strongly convergent to v and $(\bar{y}(T),v) = (\bar{y}(T),\bar{y}(0)) \in K_\beta$. Therefore, $(\bar{y},\bar{u}) \in M_\beta$ and $y^*_\lambda \to \bar{y}$ as $\lambda \downarrow 0$, strongly in $L^2(0,T;L^2(\Omega))$. Actually $y^*_\lambda \to \bar{y}$ in $C([0,T];L^2(\Omega))$.

Let us observe that, according to (2.49), for $u = u^*$, there is a unique $y = y_\lambda$ given by (2.49) with u^* in place of u^*_λ such that $(y_\lambda,u^*) \in M_\lambda$. We have obviously

$$L(y^*_\lambda,u^*_\lambda) \le L_1(y^*_\lambda,u^*_\lambda) \le L_1(y_\lambda,u^*) = L(y_\lambda,u^*) + \frac{1}{2}|y_\lambda(0)-y^*(0)|^2_{L^2(\Omega)} \tag{2.54}$$

Moreover, arguing as in the case of $(\bar{y},\bar{u})$, with u^* in place of $\bar{u}$ it follows that $\lim_{\lambda\downarrow 0} y_\lambda = y^*$ in $C([0,T];L^2(\Omega))$, i.e. $y_\lambda(0) \to y^*(0)$ as $\lambda \downarrow 0$ in $L^2(\Omega)$.

It is now clear that (2.54) and the optimality of (y^*,u^*) yield

$$\begin{aligned} L(y^*,u^*) \le L(\bar{y},\bar{u}) \le \underline{\lim}_{\lambda\downarrow 0} L(y^*_\lambda,u^*_\lambda) \le \overline{\lim}_{\lambda\downarrow 0} L_1(y^*_\lambda,u^*_\lambda) = L(y^*,u^*) \\ \lim_{\lambda\downarrow 0} L(y^*_\lambda,u^*_\lambda) = \lim_{\lambda\downarrow 0} L_1(y^*_\lambda,u^*_\lambda) = L(y^*,u^*). \end{aligned} \tag{2.55}$$

This and (2.47) show that $u^*_\lambda \to u^*$ in $L^2(0,T;L^2(\Omega))$ and $y^*_\lambda \to y^*$ in $C([0,T];L^2(\Omega))$. It remains to deal with the convergence of p_λ.

Substituting $p_\lambda(0)$ given in (2.49) into (2.52), one finds

$$y^*_\lambda(T,x)p_\lambda(T,x) = -\frac{\beta_\lambda(r_\lambda(x))}{\dot{\beta}_\lambda(r_\lambda(x))}(r_\lambda(x) - y^*(0,x) - p_\lambda(0,x)) \tag{2.56}$$

In view of $q_\lambda(t) \in \partial g(y^*_\lambda(t))$ by (2.49) and (H1) - (2.39) it follows that q_λ is bounded in $L^2(0,T;L^2(\Omega))$ and therefore (2.56) implies the boundedness of $p_\lambda(T)$ in $L^2(\Omega)$. Say

$q_\lambda \rightharpoonup q$ in $L^2(0,T;L^2(\Omega))$ and $p_\lambda(T) \rightharpoonup p_T$ as $\lambda \downarrow 0$. Then $p_\lambda(t) \to p(t)$ in $L^2(\Omega)$ as $\lambda \downarrow 0$ for every $t \in [0,T]$, with

$$p(t) = e^{(T-t)\Delta}p_T - \int_t^T e^{(s-t)\Delta}q(s)ds, \quad 0 \le t \le T. \tag{2.57}$$

Passing to the limit as $\lambda \downarrow 0$ in the forth formula of (2.49), one obtains (2.40). Taking into account that $\dot{\beta}_\lambda(r) = a$ for $r > 0$, we see that (2.52) implies (2.42).

Next, let us prove (2.43). We have $y^*_\lambda(T) + \tilde{\beta}_\lambda(y^*_\lambda(0)) = 0$, as $y^*_\lambda(0), y^*_\lambda(T)) \in M_\lambda$. This and (2.52) imply that

$$y^*_\lambda(T,x)p_\lambda(T,x) = -\frac{\beta_\lambda(r_\lambda(x))}{\dot{\beta}_\lambda(r_\lambda(x))}(r_\lambda(x) - y^*(0,x) - p_\lambda(0,x)) \tag{2.58}$$

with $r_\lambda(x) = y^*_\lambda(0,x) \to y^*(0,x)$.

We now observe that

$$\left|\frac{\beta_\lambda(r)}{\dot{\beta}_\lambda(r)}\right| \le 2|r| + \frac{1}{2}\lambda, \quad \forall r \in R, \ \lambda > 0 \tag{2.59}$$

Therefore, if $r = r_\lambda(x) = y^*_\lambda(0,x) \to 0$, as $\lambda \downarrow 0$, then (2.58) implies (2.43). Clearly, for almost all $x \in \{z \in , y^*(0,x) = 0\}$ we have $y^*_\lambda(0,x) \to 0$ and $p_\lambda(0,x) \to p(0,x)$ as $\lambda \downarrow 0$. As far as (2.44) is concerned, it follows from (2.53). Indeed, (2.53) can be written as

$$p_\lambda(t) \in \partial H(u^*_\lambda(t)) + u^*_\lambda(t) - u^*(t), \quad \text{a.e. in } [0,T] \tag{2.60}$$

with $h = h_{\bar{H}}$ defined as in (2.38), and $u^*_\lambda \to u^*$ in $L^2(0,T;L^2(\Omega))$, $p_\lambda \to p$ in $L^2(0,T;L^2(\Omega))$. As ∂H is maximal monotone in $L^2(\Omega)$, one can pass to the limit in (2.60), which yields (2.44). This completes the proof.

3 OPTIMAL CONTROL PROBLEMS FOR WAVE EQUATION

3.1 The Case of Two Point Boundary Conditions of Antiperiodic Type.

In this subsection we first prove the existence and uniqueness of the mild , or generalized solution y, $y \in C([0,T];L^2(\Omega))$, of the problem:

$$y_{tt}(t,x) = \Delta y(t,x) + u(t,x), \quad \text{a.e. in } (0,T) \times \Omega \equiv Q \tag{3.1}$$

$$y(t,x) = 0, \quad \text{a.e. on } (0,T) \times \partial\Omega \equiv \Sigma \tag{3.2}$$

$$ay(0,x)+y(T,x)=f(x), \ by_t(0,x)+y_t(T,x)=g_1(x), \ \text{a.e. in } \Omega \tag{3.3}$$

where $f,g \in L^2(\Omega)$, $u \in L^2((0,T);L^2(\Omega))$, $T > 0$ and

$$|a + b| < ab + 1, \qquad a,b \in R \tag{3.4}$$

We will show that "$|a+b| = ab+1$" is an open critical case. A natural and very general control problem one can associate with (3.1)-(3.3) is the following.

(P) Minimize $L(y,u)$;

$$L(y,U) = \int_0^T \int_Q \tilde{G}(y(t,x),y_t(t,x),u(t,x))dxdt + \tilde{\ell}(y(0),y(T),y_t(0),y_t(T)) \tag{3.5}$$

over all pairs $(y,u) \in C([0,T];L^2(\Omega))\times L^2((0,T);L^2(\Omega))$ subject to (3.1)-(3.3).
We will study (P) in some particular cases of $\tilde{G}$ and $\tilde{\ell}$ to be given in the sequel. First let us define the notion of "mild solution" of (3.1)-(3.3). This will be carried out via C_0-semigroups in the Hilbert space $H=H(\Omega)=H_0^1(\Omega)\times L^2(\Omega)$ endowed with the inner product

$$<\begin{pmatrix}x\\u\end{pmatrix},\begin{pmatrix}y\\v\end{pmatrix}> = <x,y>_{H_0^1(\Omega)} + <u,v>_{L^2(\Omega)};\ x,y \in H_0^1(\Omega),\ u,v \in L^2(\Omega) \tag{3.6}$$

where $<x,y>_{H_0^1(\Omega)} = <\nabla x,\nabla y>_{L^2(\Omega)}$ with $\nabla y = \text{grad } y$.

For simplicity, set $<u,v>_{L^2(\Omega)} = <u,v>$. Denote by $\{e_k\}_1^\infty$ a Hilbertian (orthonormal) basis of $L^2(\Omega)$ of eigenfunctions e_k of Laplace operator Δ with $D(\Delta) = H_0^1(\Omega \cap H^2(\Omega)$ i.e.

$$\Delta e_k = -\lambda_k^2 e_k,\ k = 1,2,..,\ 0<\lambda_k<\lambda_{k+1},\ \lambda_k \to +\infty \text{ as } k \to +\infty \tag{3.7}$$

Then $\{\tilde{e}_k = (1/\lambda_k)e_k\}$ is a Hilbertian basis of $H_0^1(\Omega)$. We refer to Brezis [7,pp87-100] on this subject. By standard arguments, the problem (3.1)-(3.3) can be reduced to

$$V'(t) = AV(T) + Bu(t),\ 0 \le t \le T \tag{3.8}$$
$$V(T) + CV(0) = \Phi \tag{3.9}$$

with $V(t) = \begin{pmatrix}y(t)\\y'(t)\end{pmatrix}$, $y(t) \in H_0^1(\Omega)$; $(y(t))(x) = y(t,x)$ a.e. in Ω.

$$\Phi = \begin{pmatrix}f\\g\end{pmatrix} \in H(\Omega),\ U(t) = \begin{pmatrix}0\\u(t)\end{pmatrix} = Bu(t),\ A = \begin{pmatrix}0 & I\\ \Delta & 0\end{pmatrix},\ B = \begin{pmatrix}0\\1\end{pmatrix},\ C = \begin{pmatrix}aI & 0\\0 & bI\end{pmatrix}. \tag{3.10}$$

It is known that A is maximal monotone, and dissipative, in $H(\Omega) = H$. Denote by e^{At} $S(t)$ the C_0-group generated by A in H. One can check that

$$\begin{pmatrix}\tilde{y}(t)\\ \tilde{v}(t)\end{pmatrix} = S(t)\begin{pmatrix}y_0\\v_0\end{pmatrix} = \begin{cases}\sum_1^\infty(<y_0, e_k> \cos t\lambda_k + \lambda_k^{-1}<v_0,e_k> \sin t\lambda_k)e_k\\ \sum_1^\infty(<v_0,e_k> \cos t\lambda_k - \lambda_k<y_0,e_k> \sin t\lambda_k)e_k\end{cases} \quad t \in R \tag{3.11}$$

for every $y_0 \in H_0^1(\Omega)$ and $v_0 \in L^2(\Omega)$ with $\tilde{v}(t) = \tilde{y}'(t)$.

Of course, by a mild solution V of (3.8) with $V(0) = \begin{pmatrix} y_0 \\ v_0 \end{pmatrix}$ we mean the continuous function $V:[0,T] \to H(\Omega)$ given by

$$V(t) = \begin{pmatrix} y(t) \\ v(t) \end{pmatrix} = e^{tA}\begin{pmatrix} y_0 \\ v_0 \end{pmatrix} + \int_0^t e^{(t-s)A}Bu(s)\, ds, \quad 0 \le t \le T \tag{3.12}$$

By a mild (generalized) solution y of (3.1)+(3.2) corresponding to the initial condition $y(0)=y_0$ and $y_t(0)=v_0$, we mean the first component y of V given by (3.12) (and (3.11)).

Clearly, in view of (3.12), the problem (3.8)+(3.9) leads to the invertibility of $e^{AT} + C$. The result is given by

LEMMA 3.1 If $|a + b| < ab + 1$, then for every $T \in R$, $(C + e^{TA})^{-1} \in L(H(\Omega))$. Here L(H) denotes the space of all linear bounded operators from H into itself.

Proof: On the basis of Banach's theorem on the invertibility of linear bounded operators it is necessary and sufficient to prove that for every $\begin{pmatrix} \eta_1 \\ \eta_2 \end{pmatrix} \in H(\Omega)$, the equation

$$(C + e^{AT})\begin{pmatrix} y_0 \\ v_0 \end{pmatrix} = \begin{pmatrix} \eta_1 \\ \eta_2 \end{pmatrix} \tag{3.13}$$

has a unique solution $\begin{pmatrix} y_0 \\ v_0 \end{pmatrix} \in H(\Omega)$, or equivalently, that the Fourier coefficients $<y_0,\tilde{e}_k>_{H_0^1}$ and $<v_0,e_k>$ of y_0 and v_0 respectively are uniquely determined by η_1 and η_2 and they are in ℓ^2.

It is straightforward that (3.13) implies the system below with $<y_0,e_k>$ and $<v_0,e_k>$,k = 1,2,... as unknowns

$$\begin{cases} (a + \cos T\lambda_k)<y_0,e_k> + \lambda_k^{-1}\sin T\lambda_k <v_0,e_k> = <\eta_1,e_k> \\ -\lambda_k \sin T\lambda_k <y_0,e_k> + (b + \cos T\lambda_k)<v_0,e_k> = <\eta_2,e_k> \end{cases} \tag{3.14}$$

The determinant D_k of (3.14) is

$$D_k = ab + 1 + (a + b) \cos T\lambda_k \ge ab + 1 - |a + b| > 0 \tag{3.15}$$

for every k = 1,2,...Moreover, one can check that $\sum_1^\infty |<y_0,\tilde{e}_k>_{H_0^1}|^2$ and $\sum_1^\infty |<v_0,e_k>|^2$ are convergent. One uses $<y_0,\tilde{e}_k>_{H_0^1} = \lambda_k <y_0,e_k>_{L^2(\Omega)}$, k = 1,2,.... This completes the proof.

Clearly, Lemma 3.1 yields:

THEOREM 3.1 If $|a + b| < ab + 1$, then for every $u \in L^2(0,T;L^2(\Omega))$ and $\Phi \in H(\Omega)$, the problem (3.8)+(3.9) has a unique mild solution V given by (3.12) with

$$V(0) = \begin{pmatrix} y_0 \\ v_0 \end{pmatrix} = (C + e^{TA})^{-1}\Big(\Phi - \int_0^T e^{(T-s)A}Bu(s)\, ds\Big) \tag{3.16}$$

We are now in a position to define the notion of "mild solution" y of (3.1)-(3.3) as being the first component of V given by (3.12) with $\begin{pmatrix} y_0 \\ v_0 \end{pmatrix}$ given by (3.16). It follows that the mild solution $y \in C(0,T;H_0^1(\Omega))$ is uniquely defined. We go back to the following particular case of (P) defined by (3.5). Namely, we will give necessary and sufficient conditions for the pair (y^*,u^*) to be an optimal pair of the problem

(P_1) Minimize $L_1(y,u)$;

$$L_1(y,u) = \int_0^T\!\!\int_\Omega (G(y(t,x)) + \bar{H}(u(t,x))dxdt + \ell(y(0),y(T)) + \ell_1(y'(0),y'(T)) \tag{3.17}$$

over all (y,u), $u \in L^2(0,T;L^2\Omega))$, y - the mild solution of (3.1)-(3.3). Note that $y'(0)$ is actually $y_t(0,\cdot)$, i.e. $(y'(0))(x) = y_t(0,x)$ a.e. in Ω.

We are going to use the geometric approach as in Section 1, involving normal cones. Introduce as in (2.19) , with F = C

$$K = \{(V_1,V_2) \in H(\Omega) \times H(\Omega);\ V_2 + CV_1 = \phi\} \tag{3.18}$$

and with C as in (3.10), and $V_k = \begin{pmatrix} v_k^1 \\ v_k^1 \end{pmatrix}$, $k = 1,2,\ldots$.

In this case $C = C^* = \dot{C}$, so (2.21) becomes

$$NK = \{(N_1,N_2) \in H(\Omega)\times H(\Omega);\ N_1 - CN_2 = 0,\ (y_1,y_2) \in K\} \tag{3.19}$$

As in Section 2, one reduces the problem (P_1) to an abstract problem in $H(\Omega)$, with

$$L_1(V,U) = \int_0^T \big(g_1(V(t) + h(U(t))\big)dt + \ell_2(V(0),V(T)) \tag{3.20}$$

with g and h defined as in (2.38), $g_1(y,y') = g(y)$ and

$$\ell_2(V(0),V(T)) = \ell(y(0),y(T)) + \ell_1(y'(0),y'(T)), \tag{3.21}$$

where $V(t) = \begin{pmatrix} y(t) \\ y'(t) \end{pmatrix} \in H(\Omega)$ is defined as in (3.12) and ℓ, ℓ_1 are lower semicontinuous convex functions mapping from $H(\Omega) \times h(\Omega)$ into $]-\infty,+\infty]$.

First, we will write formally the optimality conditions for (V^*,U^*), by using some principles from Section 2, mainly (2.18). In this case e^{tA} is not compact, but noncompactness of e^{tA} is compensated by the fact that K, being an affine manifold, is now a smooth subset of $H(\Omega)$. The analogous of (2.9) is now

$$P'(t) = - A^*P(t) + \dot{g}_1(V^*(t)), \quad 0 < t < T \tag{3.22}$$

with $A^* = -A$ (the adjoint of A) and

$$g_1(y,y') \equiv \begin{pmatrix} g(y) \\ 0 \end{pmatrix}, \; \dot{g}_1(V^*(t)) \equiv \begin{pmatrix} \dot{g}(y^*(t)) \\ 0 \end{pmatrix}, \; P(t) = \begin{pmatrix} p(t) \\ q(t) \end{pmatrix}. \tag{3.22'}$$

Actually, P is a mild solution of (3.22), i.e.

$$P(t) = e^{(T-t)A^*}P(T) - \int_t^T e^{(s-t)A^*}\dot{g}_1(V^*(s))ds, \quad 0 < t < T, \tag{3.23}$$

where $\dot{g}_1$ is the subdifferential of g_1, or a selection of $\dot{g}_1$ if $\dot{g}_1(V^*)$ is multivalued. According to the results in Section 2 (see Remark 2.1), the necessary and sufficient condition for optimality of (V^*,u^*) are given by: (V^*,u^*) satisfies (3.12) and $(V^*(0),V^*(T)) \in K$, i.e.

$$V^*(T) + CV^*(0) = \phi. \quad \text{(see (3.16))} \tag{3.24}$$

The function P given by (3.23) satisfies

$$(P(0),-P(T)) - \partial\ell_2(V^*(0),V^*(T)) \in NK \tag{3.25}$$

and

$$B^*P(t) \in \partial h(u^*(t)), \quad \text{a.e. on } (0,T), \tag{3.26}$$

where NK is given by (3.19 so $B^*P(t) = q(t)$ and $B^* = (0,1)$. Let's write these conditions in terms of (y^*,u^*) in some particular cases of (P_1). One takes into account the structure of $V^*(t) = \begin{pmatrix} y^*(t) \\ y^{*\prime}(t) \end{pmatrix}$ and $P(t) = \begin{pmatrix} p(t) \\ q(t) \end{pmatrix}$. Therefore, if $\ell = \ell_1 = 0$, then the conditions (3.24)-(3.26) become respectively

$$P(0) + CP(T) = 0, \tag{3.27}$$

i.e.,

$$p(0) + ap(T) = 0, \quad q(0) + bq(T) = 0, \tag{3.28}$$

$$q(t) \in \partial h(u^*(t)) \quad \text{(i.e. } q(t,x) \in \partial G(y^*(t,x)), \text{ a.e. in } Q) \tag{3.29}$$

where y^* is a mild solution of (3.1)-(3.3) with u^* in place of u. Note also that

$p'(0)(x) \equiv p_t(0,x)$ a.e. in Ω and $\dot{g}_1(V^*(s)) = \begin{pmatrix}\dot{g}(y(s))\\ 0\end{pmatrix}$ in (3.23). For example, if $K = H(\Omega)$, we have obviously $NK = 0$ so (3.25) becomes

$$\big(P(0),-P(T)\big) \in \partial\ell_2\big(V^*(0),V^*(T)\big) \tag{3.30}$$

This situation corresponds to (P_1) given by (3.17) when no other restrictions (constraints) on $(y(0),y(T))$ and $(y'(0),y'(T))$ are imposed, except for

$$(y(0),y(T)) \in D(\ell) \text{ and } (y'(0),y'(T)) \in D(\ell_1) \tag{3.31}$$

It can be proved that (3.30) with

$$\ell_2(y(0),y(T),y'(0),y'(T)) = \ell(y(0),y(T))+\ell_1(y'(0),y'(T)) \tag{3.32}$$

is actually

$$g_1(y,y') \equiv \begin{pmatrix}g(y)\\ 0\end{pmatrix},\ \dot{g}_1(V^*(t)) \equiv \begin{pmatrix}\dot{g}(y^*(t))\\ 0\end{pmatrix},\ P(t) = \begin{pmatrix}p(t)\\ q(t)\end{pmatrix}. \tag{3.33}$$

Summarizing, we are led to

Theorem 3.2. Under hypotheses (2.38)-(2.39) on G and $\bar{H}$ and $|a+b| < ab+1$, $a.b \in \mathbb{R}$, the pair (y^*,u^*) is optimal for the problem: Minimize

$$\int_0^T\!\!\int_\Omega \big(G(y(t,x))+\bar{H}(u(t,x))\big)\,dxdt$$

over all $y \in C([0,T];H_0^1(\Omega))$, $u \in L^2(Q)$ with $y_t \in L^2(0,T;L^2(\Omega))$ subject to (3.1)-(3.3) in mild sense, iff there are $q \in C([0,T];L^2(\Omega))$ with $q_t \in C([0,t];H^{-1}(\Omega))$ and $\eta \in L^2(Q)$ such that q satisfies (in mild sense)

$$\begin{aligned} &q_{tt} = \Delta q - \eta,\quad \eta(t,x) \in \partial G(y^*(t,x)) \text{ in } Q,\ q=0 \text{ on } \Sigma,\\ &q(0,x) + bq(T,x) = 0,\ q_t(0,x) + aq_t(T,x) = 0 \text{ in } \Omega, \end{aligned} \tag{3.34}$$

and

$$q(t,x) \in \partial\bar{H}(u^*(t,x)) \text{ in } Q. \tag{3.35}$$

Proof: Let us observe that (3.22) means

$$\begin{aligned} &p'(t) - q(t) \in \partial g(y^*(t))\\ &q'(t) = \Delta p(t) \end{aligned} \tag{3.36}$$

where $\Delta: H_0^1(\Omega) \to H^{-1}(\Omega)$ is defined by $\langle \Delta u,v\rangle = -\int_\Omega \nabla u\nabla v dx$, $u,v \in H_0^1(\Omega)$ so $\Delta p(t) \in H^{-1}(\Omega)$ (the dual of $H_0^1(\Omega)$). Conversely, $p'(t) - q(t)=(-\Delta)^{-1}\eta(t)$ with an η as in (3.34) (and $p'=p_t$). Indeed, it follows by the definition of Δ and the inner product in $H_0^1(\Omega)$ that $\partial g(y^*(t))=(-\Delta)^{-1}\partial G(y^*(t))$. Eliminating p from (3.36) one obtains (3.34), which completes the proof.

3.2 Optimal Control of the Initial Velocity

In this subsection we are concerned with the following optimal control problem.
(P_2) Minimize the functional $L_2(y,u)$

$$L_2(y,u) = \int_0^T\int_\Omega G(y(t,x))dxdt + \int_\Omega h(u(x))dx \tag{3.37}$$

over all pairs (y,u), $u \in L^2(\Omega)$, y - the mild solution of

$$y_{tt}(t,x) = \Delta y(t,x) \quad \text{a.e. in } (0,T) \times \Omega = Q \tag{3.38}$$

$$y(t,x) = 0 \quad \text{a.e. on } (0,T) \times \partial\Omega = \Sigma \tag{3.39}$$

$$y(T,x) + ay(0,x) = z(x), \; y_t(0,x) = u(x), \quad \text{a.e. in } \Omega \tag{3.40}$$

with $z \in H_0^1(\Omega)$ and $|a| > 1$.

We first discuss the existence and uniqueness of the mild solution of (3.38)-(3.40). The mild (generalized solution) of (3.38)-(3.39) corresponding to the initial condition $y(0,x) = y_0(x)$, $y_t(0,x) = u(x)$, a.e. in Ω is by definition the function Y defined by (3.11) with $v_0 = u$, i.e.,

$$Y(t) = S(t)\begin{pmatrix} y_0 \\ u \end{pmatrix} = \begin{pmatrix} y(t) \\ y'(t) \end{pmatrix}, \; t \in R, \; y_0 \in H_0^1(\Omega), \; u \in L^2(\Omega). \tag{3.41}$$

Therefore, given a,z and u as above, we must prove the existence and uniqueness of $y_0 \in H_0^1(\Omega)$ such that y given by

$$y(t) = \sum_1^\infty(<y_0,e_k>\cos t\lambda_k + \lambda_k^{-1}<v_0,e_k>\sin t\lambda_k)e_k \tag{3.42}$$

satisfies the first equality of (3.40), which can be rewritten in $L^2(\Omega)$ as

$$y(T) + ay(0) = z, \;\; y'(0) = u \tag{3.43}$$

As we have mentioned in (3.11), y' is the strong derivative of $t \to y(t)$ in $H_0^1(\Omega)$ norm and

$$y'(t) = \sum_1^\infty(<u,e_k>\cos t\lambda_k - \lambda_k<y_0,e_k>\sin t\lambda_k)e_k. \tag{3.44}$$

Clearly, $y(T) + ay(0) = z$ yields

$$<y_0,e_k>(a + \cos T\lambda_k) = <z,e_k> - \lambda_k^{-1}<u,e_k>\sin T\lambda_k. \tag{3.45}$$

If $|a| > 1$, the element y_0 determined via (3.45) belongs to $H_0^1(\Omega)$ and therefore the mild solution Y given by (3.41), of the problem (3.38)-(3.40) is unique, and depends continuously on z and u. It can be derived from here that the problem

$$P'(t) = -A^*P(t) + \partial g_1(Y^*(t)), \text{ in } (0,T) \tag{3.46}$$

with

$$P(t) = \begin{pmatrix} p(t) \\ q(t) \end{pmatrix} \in H(\Omega), \; Y^* = \begin{pmatrix} y^* \\ y_t^* \end{pmatrix} \in H(\Omega),$$
$$p(0) + a\, p(T) = 0, \; q(T) = 0$$

has also a unique mild solution in $H(\Omega)$, where g_1 is as in (3.20), A^*=-A as in (3.22)′ and (3.23).

It is straightforward that the following result holds:

Theorem 3.3. Let G and h satisfy (2.38)-(2.39). Then (y^*,u^*) is optimal for (P_2) above iff

$$q(0,x) \in -\partial h(u^*(x)) \text{ in } \Omega \tag{3.47}$$

where q is the second component of P in (3.46), i.e., $q \in C([0,T];L^2(\Omega))$, $q \in C([0,T];H^{-1}(\Omega))$

$$q_{tt} = \Delta q - \eta, \quad \eta(t,x) \in \partial G(y^*(t,x)) \text{ in } Q, \tag{3.48}$$
$$q = 0 \text{ on } \Sigma, \tag{3.49}$$

$$q(T,x) = 0, \; q_t(0,x) + aq_t(T,x) = 0 \text{ in } \Omega. \tag{3.50}$$

Proof: The formal equations (3.48)-(3.50) follows as (3.34)-(3.36). Let's motivate (3.47). Say, for simplicity that G and h are of class C^1. Write L_2 in $H(\Omega)$ as

$$L_2(Y,u) = \int_0^T g_1(Y(t))dt + \int_\Omega h(u(x))dx \tag{3.51}$$

with $Y = \begin{pmatrix} y \\ y' \end{pmatrix}$ and g_1 as in (3.20). Then the gradient $\partial L_2(Y^*,u^*)$ in the direction of $V = \begin{pmatrix} v \\ v' \end{pmatrix}$ and $w \in L^2(\Omega)$ is zero (or contains zero, if it is multivalued), i.e.,

$$\partial L_2(Y^*,u^*)(V,w) = \int_0^T < \partial g_1(Y^*),V > dt + \int_\Omega \partial h(u^*)w\, dx = 0 \tag{3.52}$$

for all (V,w) satisfying

$$V' = AV, \quad v(T) + av(0) = 0, \; v'(0) = w. \tag{3.53}$$

Substituting $\partial g_1(Y^*)=P'+A^*P$ into (3.52) and integrating by parts (in view of (3.53)), one obtains (in $L^2(\Omega)$)

$$< q(0) + \partial h(u^*), w > = 0, \quad \forall w \in L^2(\Omega),$$

which yields (3.47). This completes the proof.

Remark 3.1. Problems (3.38)-(3.40) can formally be reduced to a problem of type (3.8)-(3.9) as below

$$Y' = AY \quad \text{in } (0,T), \quad Y(0) + C_1 Y(T) = \phi(u)$$

with $C_1 = \begin{pmatrix} a^{-1} & 0 \\ 0 & 0 \end{pmatrix}$ and $\phi(u) = \begin{pmatrix} a^{-1}z \\ u \end{pmatrix}$.

From (3.45) it follows that $(I + C_1 e^{TA})^{-1} \in L(H(\Omega))$, for $|a| > 1$, therefore,

$$Y(t) = e^{tA}(I + C_1 e^{TA})^{-1}\phi(u),$$

so the problem (P_2) reduces to the minimization of L_2 in (3.51) with respect to $u \in L^2(\Omega)$. Of course, one can consider more general problems of this type with $L_2(y,y'u)$ and less restrictive conditions on G and h. However, the main ideas are pointed out in the frame work of this subsection. The case $|a| \leq 1$ remains also an open case.

REFERENCES

[1] S. Aizicovici and N. H. Pavel, Anti-periodic solutions to a class of nonlinear differential equations in Hilbert space, *J. Funct. Anal.*, 99(1991), 387-408.

[2] A. R. Aftabizadeh, Y. K. Huang and N. H. Pavel, Antiperiodic oscillations of some second order differential equations and optimal control problems, *J. of Computational & Appl. Math.*, (1993, to appear).

[3] V. Barbu, Nonlinear semigroups and differential equations in Banach spaces, Noordhoff Int. Publ., 1976.

[4] V. Barbu and N. H. Pavel, Optimal control problems with two point boundary conditions, *J. Optimiz. Th. Appl.*, (1993, to appear).

[5] V. Barbu and N. H. Pavel, On the invertibility of I±exp(-tA), t>0 with A maximal monotone, *Proc. WCNA.*, Aug.92, Tampa, Florida, (V. Lakshmikantnam Ed.) (to appear).

[6] Z. Cai and N. H. Pavel, Generalized periodic and antiperiodic solutions for the heat equation, *Libertas Mathematica*, Vol 10(1990), 109-121.

[7] H. Brezis, Analyse Fonctionelle, Theorie et Applications, Masson, Paris, New York, 1983.

[8] N. H. Pavel, Differential Equations, Flow-invariance and Applications, Pitman Advanced Publ., Vol 113 (1984).

6 Continuity of a Parametrized Linear Quadratic Optimal Control Problem

Constantin Corduneanu The University of Texas at Arlington, Arlington, Texas

S. Q. Zhu The University of Texas at Arlington, Arlington, Texas

ABSTRACT

In this note we discuss the continuity of the solution of a parametrized linear quadratic optimal control problem in Banach spaces, and our study is carried out in a practical framework, i.e., the system, the cost functional and the control subset are all subjected to perturbations.

Key Words: optimal control, perturbation, approximation, robustness.

1 PROBLEM STATEMENT

Consider the parametrized systems described by the input–output relation

(1.1) $\quad x = T_{\lambda}u,$

where $u \in U$ = a real Banach space, $x \in X$ = a real Banach space, and T_λ is a linear bounded operator mapping U into X. The parameter λ is chosen in a space which could arise from perturbations, approximations, or could describe physical characteristics that are intrinsic to the problem under consideration.

Assume that $C(\lambda)$ is a closed convex subset of U. The optimal control problem considered here is :

$$\inf_{u \in C(\lambda)} \mathcal{K}_\lambda(u), \qquad \mathcal{K}_\lambda(u) := \langle P_\lambda x, x\rangle + \langle Q_\lambda u, u\rangle, \qquad x = T_\lambda u \tag{1.2}$$

where P_λ is a linear bounded operator mapping X into its dual space X^*, Q_λ is a linear bounded operator mapping U into U^* and $\langle .,. \rangle$ is the corresponding dual pairing.

This optimal control problem (for fixed λ) has been studied in [1,2] and [6] for Hilbert spaces and Banach spaces, respectively. And the existence and uniqueness of the solution has been obtained. Moreover, a computing scheme has been proposed in [1].

In this note we shall study the continuity of the optimal solution and the minimal cost with respect to the parameter λ. This issue can be regarded as the robustness (or sensitivity) of the optimal solution with respect to perturbations or uncertainties in the model, or as the validation of an approximating scheme in the sense that the scheme produces approximating solutions. This topic is important in application because perturbations, uncertainties, approximations, etc., are always involved in practical problems.

Precisely, we will investigate the following two problems:

(1.3) Assuming u_λ is the optimal solution to $\inf_{u \in C(\lambda)} \mathcal{K}_\lambda(u)$, does u_λ depend continuously on the parameter λ ?

(1.4) Does the minimal cost $\mathcal{K}_\lambda(u_\lambda)$ depend continuously on the parameter λ ?

Our study will be carried out in two cases, i.e., the case $C(\lambda)$ being free of the parameter λ and the case $C(\lambda)$ depending on λ. This is because in the second case a stronger assumption is needed and no estimates have been obtained. These two cases will be investigated in the next two sections respectively.

2 THE CASE $C(\Lambda) \equiv C$

As stated before, the linear system operator T_λ is assumed to be bounded in U. Usually, one assumes that P_λ is symmetric and nonnegative and Q_λ is symmetric and positive, but we do not assume symmetry here. **Our key assumption is that the composite operator** $R_\lambda := T_\lambda^* P_\lambda T_\lambda + Q_\lambda$ **is positive** (by positive we mean that $\exists\ \gamma_\lambda > 0$ such that

$<R_\lambda u,u> \geq \gamma_\lambda \|u\|^2$ for $\forall\ u \in U$), Obviously, if T_λ, Q_λ, P_λ are bounded with P_λ being nonnegative and Q_λ being positive, this key assumption is satisfied.

Under the key assumption stated above, in [6] it has been proved that there exists a unique solution to $\inf_{u \in C(\lambda)} \mathcal{K}_\lambda(u)$. We need to recall certain techniques from [6] in order to obtain our developments. For simplicity, denote by $B(X_1,X_2)$ the set of all linear bonded operators mapping X_1 into X_2.

THEOREM 1 [6] Assume

1) U,X are two Banach spaces and $T_\lambda \in B(U,X)$, $P_\lambda \in B(X,X^*)$, $Q_\lambda \in B(U,U^*)$;

2) $R_\lambda := T_\lambda^* P_\lambda T_\lambda + Q_\lambda$ is positive.

Then there is a unique solution to $\inf_{u \in C(\lambda)} \mathcal{K}_\lambda(u)$, where $\mathcal{K}_\lambda(.)$ is defined by (1.2).

The proof of Theorem 1 depends on the following technical lemma:

LEMMA 2 [6] Under the assumption of Theorem 1, the following bilinear functional

$$\ll u,v \gg_\lambda := \tfrac{1}{2}[\ <R_\lambda u,v> + <R_\lambda v,u>\ ] = \tfrac{1}{2} <(R_\lambda + R_\lambda^*)u,v> \qquad \forall\ u,v \in U \tag{2.1}$$

defines an inner product on U and the induced norm $|||\cdot|||_\lambda := \ll .,. \gg_\lambda^{1/2}$ is equivalent to $\|.\|$, i.e., there are two positive numbers M_λ, m_λ such that $m_\lambda\ \|\cdot\| \leq |||\cdot|||_\lambda \leq M_\lambda\ \|\cdot\|$.

It is shown in [6] that not every Banach space U could possess a linear bounded positive operator mapping U into U^*. Actually, the existence of such an operator is equivalent to: U is isomorphic to a Hilbert space. Therefore, our control space U is actually a special type of Banach space. But the state (or output) space X is an arbitrary Banach space.

It is clear that under the new norm we have $\mathcal{K}_\lambda(u) = \ll u,u \gg_\lambda$. Therefore, the unique solution is given by

$$u_\lambda = \Pi_\lambda(0), \tag{2.2}$$

where Π_λ is the projection onto the closed convex subset $C(\lambda)$ in the norm $|||\cdot|||_\lambda$, i.e.,

$$\Pi_\lambda v = \text{the solution of } \inf_{w \in C(\lambda)} |||v - w|||_\lambda. \tag{2.3}$$

Thus, the problem (1.3), i.e., the continuity of u_λ as a function of λ, is reduced to the continuity of $\Pi_\lambda(0)$ as a function of λ. We shall prove that $\Pi_\lambda u$ is a continuous function of λ for every $u \in U$.

THEOREM 3 Assume that the conditions in Theorem 1 hold and that $C(\lambda) \equiv C$. If the operator R_λ is continuous with respect to λ, then for each $u \in U$, the projection $\Pi_\lambda u$ is continuous with respect to λ.

Note $<R_\lambda u,u> = <R_{\lambda_0}u,u> + <(R_\lambda - R_{\lambda_0})u,u> \geq (\gamma_{\lambda_0} - \|R_\lambda - R_{\lambda_0}\|)\|u\|^2$. If R_λ is continuous in λ we may naturally assume that its positive index γ_λ is also continuous in γ_λ. For similar reasons we may assume that both m_λ and M_λ in Lemma 2 are continuous in λ provided R_λ is.

PROOF of Theorem 3 Assume $\lambda \longrightarrow \lambda_0$, we will prove that $\Pi_\lambda u \longrightarrow \Pi_{\lambda_0} u$. For simplicity, we take $\lambda_0 = 0$. Suppose that $x_\lambda = \Pi_\lambda u$, $x_0 = \Pi_0 u$. The following is a well-known fact (see [4,7] for example):

(2.4) $$w = \Pi_\lambda v \text{ iff } \ll w - v, y - w \gg_\lambda \geq 0, \qquad \text{for } \forall\, y \in C.$$

Since $x_0 \in C(\lambda) \equiv C$, by (2.4), one has

$$\ll x_\lambda - u , x_0 - x_\lambda \gg_\lambda \geq 0$$

i.e.,

(2.5) $$<[R_\lambda + R_\lambda^*](x_\lambda - u) , (x_0 - x_\lambda)> \geq 0.$$

Similarly, one has

$$\ll x_0 - u , x_\lambda - x_0 \gg_0 \geq 0,$$

i.e.,

$$<[R_0 + R_0^*](x_0 - u) , (x_\lambda - x_0)> \geq 0,$$

i.e.,

$$<[R_0 + R_0^*](x_0 - x_\lambda + x_\lambda - u) , (x_\lambda - x_0)> \geq 0,$$

i.e.,

$$<[R_0 + R_0^*](x_0 - x_\lambda) , (x_\lambda - x_0)> + <[R_0 + R_0^*](x_\lambda - u) , (x_\lambda - x_0)> \geq 0$$

i.e.,

(2.6) $$<[R_0 + R_0^*](x_\lambda - u) , (x_\lambda - x_0)> \geq 2 <R_0(x_0 - x_\lambda) , (x_0 - x_\lambda)> \geq 2\gamma_0 \|x_0 - x_\lambda\|^2.$$

The combination of (2.5) and (2.6) yield,

$$<\{[R_0 + R_0^*] - [R_\lambda + R_\lambda^*]\}(x_\lambda - u) , (x_\lambda - x_0)> \geq 2\gamma_0 \|x_0 - x_\lambda\|^2.$$

Hence, one obtains

(2.7) $\quad \|[R_\lambda - R_0]\| \; \|(x_\lambda - u)\| \geq \gamma_0 \; \|x_0 - x_\lambda\|.$

By symmetry, one also has

(2.8) $\quad \|[R_\lambda - R_0]\| \; \|(x_0 - u)\| \geq \gamma_\lambda \; \|x_0 - x_\lambda\|.$

Since γ_λ may be assumed being continuous in λ, (2.8) shows $x_\lambda \longrightarrow x_0$. However, our conclusion can also be drawn from (2.7). According to (2.7), it is sufficient to show that $\{x_\lambda\}$ is bounded. To show this, we need another known fact from [4,7], i.e., $|||\Pi_\lambda w_1 - \Pi_\lambda w_2\|_\lambda \leq |||w_1 - w_2|||_\lambda$ for all w_1, w_2 in U. Thus,

$$|||x_\lambda - \Pi_\lambda(0)|||_\lambda^2 = |||\Pi_\lambda(u) - \Pi_\lambda(0)|||_\lambda^2 \leq |||u - 0|||_\lambda^2 = |||u|||_\lambda^2 = \langle R_\lambda u \, , \, u\rangle \leq \|R_\lambda\| \; \|u\|^2.$$

Therefore, $\{x_\lambda\}$ is bounded. Consequently, x_λ converges to x_0. This completes the proof. ■

As a direct consequence we have

COROLLARY 4 If R_λ is continuous with respect to λ, then the solution u_λ of $\inf_{u \in C} \mathcal{K}_\lambda(u)$, as a function of λ, is continuous.

Note that the following two estimates for the optimal control follow from (2.7) and (2.8), respectively. Substituting $u = 0$, $u_0 = x_0 = \Pi_0(0)$ and $u_\lambda = x_\lambda = \Pi_\lambda(0)$ in (2.7) and (2.8), one has

(2.9) $\quad \|[R_\lambda - R_0]\| \; \|u_\lambda\| \geq \gamma_0 \; \|u_0 - u_\lambda\|,$

(2.10) $\quad \|[R_\lambda - R_0]\| \; \|u_0\| \geq \gamma_\lambda \; \|u_0 - u_\lambda\|.$

Next, we study the continuity of the minimal cost $\mathcal{K}_\lambda(u_\lambda)$ as a function of λ.

THEOREM 5 If R_λ is continuous with respect to λ, then the minimal cost $\mathcal{K}_\lambda(u_\lambda) = \inf_{u \in C} \mathcal{K}_\lambda(u)$, as a function of λ, is continuous.

PROOF Note

$$\mathcal{K}_\lambda(u_\lambda) - \mathcal{K}_0(u_0) = \mathcal{K}_\lambda(u_\lambda) - \mathcal{K}_0(u_\lambda) + \mathcal{K}_0(u_\lambda) - \mathcal{K}_0(u_0)$$

$$= <[R_\lambda - R_0]u_\lambda, u_\lambda> + [\ <R_0 u_\lambda, u_\lambda> - <R_0 u_0, u_0>\ ].$$

The second part converges to zero because $<R_0(.),(.)>$ is a continuous function. And the first part converges to zero because $\{u_\lambda\}$ is bounded and $R_\lambda \longrightarrow R_0$. ■

3. THE CASE C(λ) IS DEPENDING ON λ

In this section, the space U is assumed to be separable and the parameter λ is in a metrizable space.

First, we need to clarify how $C(\lambda)$ depend on λ. It is obvious that certain continuity of $C(\lambda)$ with respect to λ is necessary. For this purpose we introduce the Hausdorf topology on subsets. If $\lambda \longrightarrow \lambda_0$, define

$$\bar{C} := \limsup_{\lambda \to \lambda_0} C(\lambda) := \{\ x \in U : \exists\ \lambda_i \text{ such that } \exists\ x_i \in C(\lambda_i)\ x_i \longrightarrow x\ \}; \tag{3.1}$$

$$\underline{C} := \liminf_{\lambda \to \lambda_0} C(\lambda) := \{\ x \in U : \exists\ x_\lambda \in C(\lambda) \text{ such that } x_\lambda \longrightarrow x\ \}. \tag{3.2}$$

The set $\bar{C}$ and $\underline{C}$ are called the upper limit and lower limit of $C(\lambda)$, respectively, as $\lambda \longrightarrow \lambda_0$. It is clear that one always has $\underline{C} \subseteq \bar{C}$. The sequence $C(\lambda)$ is said to be convergent to $C(\lambda_0)$ if $\underline{C} = \bar{C} = C(\lambda_0)$. As before, denote by u_λ the unique solution to $\inf_{u \in C_\lambda} \mathcal{K}(u)$ with respect to T_λ, P_λ, and Q_λ.

THEOREM 6 Suppose that the conditions in Theorem 1 hold. Moreover, assume

1. $\{C(\lambda)\}$ converges to $C(\lambda_0)$ as λ converges to λ_0;
2. if $x_k \in C(\lambda_k)$ and $\{x_k\}$ converges to x_0 in $*-$ topology, then $x_0 \in C(\lambda_0)$;

Then for every $u \in U$ the sequence $\{\ \Pi_\lambda u\ \}$ converges to $\Pi_{\lambda_0} u$, where

$$\Pi_\lambda u := \text{the solution of } \inf_{v \in C(\lambda)} |||u - v|||_\lambda.$$

Note that as will be seen in the proof, the condition 2 can be replaced by: $C(\lambda)$ is compact for all λ.

PROOF Again, for simplicity we take $\lambda_0 = 0$. Suppose that $x_\lambda = \Pi_\lambda u$, $x_0 = \Pi_0 u$. First, we prove that $\{x_\lambda\}$ is bounded. Denote $w_\lambda = \Pi_\lambda(0)$. Since

$$|||x_\lambda - w_\lambda|||_\lambda \le |||u - 0|||_\lambda \le M \|u\|,$$

it is sufficient to that show $\{w_\lambda\}$ is bounded. Then

$$|||w_\lambda|||_\lambda = \inf_{y\in C_\lambda} |||y|||_\lambda \le M \inf_{y\in C_\lambda} \|y\|.$$

Let $B_\gamma := \{ x \in U : \|x\| \le \gamma \}$ and $B_\gamma \cap C_0$ be non-empty. Take $y_0 \in B_\gamma \cap C_0$, then $\exists\ y_\lambda \in C(\lambda)$ such that $y_\lambda \longrightarrow y_0$ as $\lambda \longrightarrow 0$. Thus

$$|||w_\lambda|||_\lambda \le M \inf_{y\in C_\lambda} \|y\| \le M \|y_\lambda\| \longrightarrow M \|y_0\| \le M\ \gamma.$$

Therefore, $\{w_\lambda\}$ is bounded, and so is $\{x_\lambda\}$.

According to (2.4), one has

$$\langle\langle x_\lambda - u\ ,\ x - x_\lambda \rangle\rangle_\lambda \ge 0 \qquad \forall\ x \in C(\lambda)$$

i.e.,

$$<[R_\lambda + R_\lambda^*](x_\lambda - u)\ ,\ (x - x_\lambda)> \ \ge 0. \qquad \forall\ x \in C(\lambda).$$

Since $x_0 \in C(0) = \underline{C}$, $\exists\ y_\lambda \in C(\lambda)$ such that $y_\lambda \longrightarrow x_0$. Hence,

$$<[R_\lambda + R_\lambda^*](x_\lambda - u)\ ,\ (y_\lambda - x_0 + x_0 - x_\lambda)> \ \ge 0,$$

i.e.,

$$<[R_\lambda + R_\lambda^*](x_\lambda - u)\ ,\ (y_\lambda - x_0)> + <[R_\lambda + R_\lambda^*](x_\lambda - u)\ ,\ x_0 - x_\lambda)> \ \ge 0. \tag{3.3}$$

Similarly, one has

$$\langle\langle x_0 - u\ ,\ x - x_0 \rangle\rangle_0 \ge 0, \qquad \forall\ x \in C(0)$$

i.e.,

$$<[R_0 + R_0^*](x_0 - u)\ ,\ (x - x_0)> \ \ge 0 \qquad \forall\ x \in C(0). \tag{3.4}$$

As proved in the beginning, $\{x_\lambda\}$ is bounded. Since U is separable and $(U, |||\cdot|||_\lambda)$ is a Hilbert space, $\{x_\lambda\}$ has a subsequence $\{x_{\lambda_i}\}$ which converges to x_∞ in *– topology. By assumption, $x_\infty \in C(0)$. In view of (3.4), one has

$$<[R_0+R_0^*](x_0 - u) \, , \, (x_\infty - x_{\lambda_i} + x_{\lambda_i} - x_0)> \geq 0,$$

i.e.,

$$<[R_0+R_0^*](x_0 - u) \, , \, (x_\infty - x_{\lambda_i})> + <[R_0+R_0^*](x_0 - u) \, , \, (x_{\lambda_i} - x_0)> \geq 0. \tag{3.5}$$

(3.3) and (3.5) lead to

$$<[R_{\lambda_i}+R_{\lambda_i}^*](x_{\lambda_i} - u) \, , \, (y_{\lambda_i} - x_0)> + <[R_0+R_0^*](x_0 - u) \, , \, (x_\infty - x_{\lambda_i})>$$

$$+ <\{[R_{\lambda_i}+R_{\lambda_i}^*](x_{\lambda_i} - u) - [R_0+R_0^*](x_0 - u)\} \, , \, x_0 - x_{\lambda_i}> \geq 0,$$

i.e.,

$$<[R_{\lambda_i}+R_{\lambda_i}^*](x_{\lambda_i} - u) \, , \, (y_{\lambda_i} - x_0)> + <[R_0+R_0^*](x_0 - u) \, , \, (x_\infty - x_{\lambda_i})>$$

$$+ <\{[R_{\lambda_i}+R_{\lambda_i}^*] - [R_0+R_0^*]\}(x_{\lambda_i} - u) \, , \, x_0 - x_{\lambda_i}>$$

$$+ <[R_0+R_0^*]\{(x_{\lambda_i} - u) - (x_0 - u)\} \, , \, x_0 - x_{\lambda_i}> \geq 0,$$

i.e.,

$$<[R_{\lambda_i}+R_{\lambda_i}^*](x_{\lambda_i} - u) \, , \, (y_{\lambda_i} - x_0)> + <[R_0+R_0^*](x_0 - u) \, , \, (x_\infty - x_{\lambda_i})> \tag{3.6}$$

$$+ <\{[R_{\lambda_i}+R_{\lambda_i}^*] - [R_0+R_0^*]\}(x_{\lambda_i} - u) \, , \, x_0 - x_{\lambda_i}> \geq 2\gamma_0\|x_0 - x_{\lambda_i}\|^2.$$

Note that the left-hand side of (3.6) converges to zero, thus $\{x_{\lambda_i}\}$ converges to x_0. Since $\{x_{\lambda_i}\}$ is bounded and every weak limit is equal to x_0, one must have x_λ converges to x_0. This completes the proof. ■

COROLLARY 7 Under the same assumption as Theorem 6, the optimal control $\{u_\lambda = \Pi_\lambda(0)\}$ converges to $u_{\lambda_0} = \Pi_{\lambda_0}(0)$ as $\lambda \longrightarrow \lambda_0$.

COROLLARY 8 Under the same assumption as Theorem 6, the optimal cost

$$\mathcal{K}_\lambda(u_\lambda) = \inf_{u \in C(\lambda)} \mathcal{K}_\lambda(u)$$

converges to $\mathcal{K}_{\lambda_0}(u_{\lambda_0})$ as $\lambda \longrightarrow \lambda_0$.

PROOF Same as the proof of Theorem 5. ■

REFERENCES

[1] Corduneanu, C. LQ–optimal control problem for systems with abstract Volterra operators,*Tekhnicheskaya Kibernetika, to appear. Preprint. University of Texas of at Arlington*, 1992.

[2] Corduneanu, C. An abstract LQ–optimal control problem and its applications, *Libertas Mathematica*, Vol. 12, 1992.

[3] Kato, T. Perturbation theory for linear operators, *Springer–Verlag*, *Berlin New York*, 1976.

[4] Kinderlehrer, D. and G. Stampacchia, An introduction to variational inequalities and their applications, *Academic Press*, *New York*, 1980.

[5] Cioranescu, I. Geometry of Banach spaces, duality mappings and nonlinear problems, Kluwer Academic Publishers, Dordrecht/Boston/London 1990.

[6] Zhu, S.Q., Well–posedness of an optimal control problem on Banach spaces, Preprint, 1993.

[7] G.J. Fix and R. Kannan, A random direction algorithm for an intersection problem, *Computing*, 48(1992)381–385.

7 Some Remarks on Ergodic and Periodic Control

Giuseppe Da Prato Scuola Normale Superiore di Pisa, Pisa, Italy

1 Introduction

We are concerned with a dynamical system governed by the equation:

$$y'(t) = Ay(t) + Bu(t) + f(t), \tag{1.1}$$

where $A : D(A) \subset H \to H$ and $B : U \to H$ are linear operators in the Hilbert spaces H and U respectively. H is the space of *states* and U the space of *controls*. The usual regulator problem in infinite horizon consists in minimizing a quadratic functional of the form

$$J(u,x) = \int_0^{+\infty} [|Cy(t)|^2 + |u(t)|^2]dt$$

over all controls $u \in L^2(0,+\infty;U)$, where $C : H \to H$ is the *obsevation* operator and y is the solution of (1.1) such that $y(0) = x$. This problem is meaningful provided a finite cost condition hold, that is if for any $x \in H$ there exists some control u such that the cost $J(u,x)$ is finite. It is easy to realize that if the exterior force $f(t)$ does not belong to $L^2(0,+\infty;H)$ this is not the case in general.

In this paper we shall assume that $f(t)$ is periodic with minimal period 2π. In this case two approches to the problem have been presented: periodic control and ergodic control. Roughly speaking periodic control consists in looking for periodic solutions of (1.1) and in minimizing the cost along a period. Instead in ergodic control one introduces in the cost a discount factor $e^{-\alpha t}$

$$J_\alpha(u,x) = \int_0^{+\infty} e^{-\alpha t}[|Cy(t)|^2 + |u(t)|^2]dt, \ \alpha > 0.$$

In this way, for any $\alpha > 0$ the problem reduces to a classical regulator problem and can be easyly solved. However what is important in this case is to prove an asymptotic expansion for the optimal cost $J^\star_\alpha$ of the form

$$J^\star_\alpha = J_0 + \frac{1}{\alpha}J_1 + o(\frac{1}{\alpha}).$$

In this paper we want to present some new results about the relationship between periodic and ergodic control problem. This will be done in §4, whereas §2 is devoted to notations and hypotheses that will be used in the sequel, and §3 to recalling some results about periodic control.

2 Notations and Hypotheses

We are given two Hilbert spaces H, and U. If Z is any Hilbert space, we denote bt $L^2_\#(Z)$ the set of all $2\pi-$ periodic functions from $\mathbf{R}$ into Z that are Lebesgue measurable and square integrable in $[0, 2\pi]$. Moreover $C_\#(Z)$ is the set of all continuous and $2\pi-$ periodic functions from $\mathbf{R}$ into Z.

We shall assume

Hypothesis 2.1

(i) *A is the infinitesimal generator of a strongly continuous semigroup e^{tA} in H.*

(ii) $B \in \mathcal{L}(U; H)$.

(iii) $f \in L^2_\#(H)$.

Under these hypotheses, for any $x \in H$, there exists a unique mild solution $y(\cdot, x; u)$ of (1.1) such that $y(0) = x$, given by

$$y(t, x; u) = e^{tA}x + \int_0^t e^{(t-s)A}[Bu(s) + f(s)]ds, \; t \geq 0. \tag{2.1}$$

Moreover $y(\cdot) \in C([0, +\infty[; H)$.

Several of the results below on periodic control can be generalized to the case when B is an unbounded operator, as it happens in treating boundary control problems, see [4], [8], and when coefficients A and B, are time-dependent, see [3], [5] and [6].

We are interested in periodic solutions of (1.1). A function y in $C_\#(H)$ is said to be a periodic solution of (1.1) if $y(\cdot) = y(\cdot, y(0); u)$. As well known, equation (1.1) has not periodic solutions in general; a sufficient condition for existence and uniqueness of a periodic solution is that

$$1 \in \rho(e^{2\pi A}), \tag{2.2}$$

where $\rho(e^{2\pi A})$ is the resolvent set of $e^{2\pi A}$. In this case the unique 2π–periodic solution of equation (1.1) is given by the formula

$$\begin{aligned} y(t) &= e^{tA}(I - e^{2\pi A})^{-1} \int_0^{2\pi} e^{(2\pi - s)A}[f(s) + B(s)u(s)]ds \\ &+ \int_0^t e^{(t-s)A}[f(s) + B(s)u(s)]ds. \end{aligned} \tag{2.3}$$

Now let us recall the definition of periodic and ergodic control problem. To this purpose let us introduce some additional notations and assumptions. For any $x \in H$ and any $u \in L^2(0, +\infty, U)$ we denote by $z(\cdot, x; u)$ the mild solution to the problem

$$\begin{cases} z'(t) = Az(t) + Bu(t), \ t \geq 0 \\ z(0) = x, \end{cases}$$

that is

$$z(t) = e^{tA}x + \int_0^t e^{(t-s)A}Bu(s)ds.$$

Now we assume

Hypothesis 2.2

(i) $C \in \mathcal{L}(H)$.

(ii) (A, B) *is I–stabilizable; that is, for any* $x \in H$ *there exists* $u^x \in L^2(0, \infty; U)$ *such that* $z(\cdot, x; u^x) \in L^2(0, \infty; U)$.

(iii) (A, C) *is detectable; that is,*(A^*, C^*) *is I–stabilizable, where* A^* *and* C^* *denote the adjoint of* A *and* C *respectively.*

We now introduce the *periodic control* problem

Minimize the cost

$$K(u, y) = \frac{1}{2\pi} \int_0^{2\pi} [|Cy(s)|^2 + |u(s)|^2]ds \tag{2.4}$$

over all $(u, y) \in L^2_{\#}(U) \times C_{\#}(H)$ such that y is a periodic

solution of (1.1).

If there exists a pair $(u^\star, y^\star) \in L^2_\#(U) \times C_\#(H)$ that realizes the minimum of problem (2.4), we say that $u^\star$ is an *optimal control*, $y^\star$ an *optimal state*, and $K^* = K(u^\star, y^\star)$ the *optimal cost.*

Instead the *ergodic control* problem, see [1], consists in two steps.

Step 1 For any $\alpha > 0$ we solve the optimal control problem

Minimize the cost

$$J_\alpha(u,x) = \int_0^{+\infty} e^{-\alpha t}[|Cy(t,x;u)|^2 + |u(t)|^2]dt, \; x \in H, \tag{2.5}$$

over all $u : [0, +\infty[\to U$, such that $e^{-\alpha t/2}u(\cdot) \in L^2(0,\infty;U)$.

Step 2 We study the behaviour of $J_\alpha(u,x)$, the optimal cost, as $\alpha \to 0$.

3 Periodic control

We consider here problem (2.4). In order to formulate the existence result it is convenient to introduce the Algebraic Riccati equation

$$A^\star P_\infty + P_\infty A - P_\infty BB^\star P_\infty + C^\star C = 0. \tag{3.1}$$

We recall that, under Hypoteses 2.1 and 2.2 there exists a unique nonnegative solution $P_\infty \in \mathcal{L}(H)$, see for instance [2]. This means that

$$2 < P_\infty x, Ax > -|B^\star P_\infty x|^2 + |Cx|^2 = 0,$$

for all $x \in D(A)$.

The following result is proved in [6], for a more general result that can be proved without assuming detectability of (A, C), see [5].

Theorem 3.1 *Assume Hypotheses* 2.1 *and* 2.2. *Then there exists a unique optimal pair* $(u^\star, y^\star)$ *for problem* (2.4) *and the following statements hold.*

(i) $y^\star(\cdot)$ *is the unique periodic solution to the closed loop equation*

$$y'(t) = (A - BB^\star P_\infty)y(t) - BB^\star r^\star(t) + f(t), \; t \in \mathbf{R}, \tag{3.2}$$

where $r^\star(\cdot)$ *is the unique periodic solution of equation*

$$r'(t) + (A - BB^\star P_\infty)^\star r(t) + P_\infty f(t) = 0, \; t \in \mathbf{R}, \tag{3.3}$$

(ii) $u^\star$ is given by

$$u^*(t) = -B^\star(P_\infty y^*(t) + r^*(t)),\ t \in \mathbf{R}, \tag{3.4}$$

(iv) The optimal cost is given by

$$K^* = \frac{1}{2\pi}\int_0^{2\pi} \{< r^*(t), f(t) > -|B^\star r^*(t)|^2\}dt. \tag{3.5}$$

Remark 3.2 Under Hypotheses 2.1-2.2 the closed loop operator $F = A - BB^* P_\infty$ is stable, see [10] and [2]. This implies in particular that $i\mathbf{R}$ belongs to the resolvent set of F, so that equation (3.3) has a unique periodic solution given by

$$r^*(t) = \int_t^\infty e^{(s-t)(A-BB^*P_\infty)^*} P_\infty f(s) ds. \tag{3.6}$$

For the same reason the solution y^* of the closed loop equation is given by

$$y^*(t) = \int_{-\infty}^t e^{(s-t)(A-BB^*P_\infty)}[f(s) - BB^* r^*(s)]ds. \ \blacksquare \tag{3.7}$$

4 Ergodic control

First we fix $\alpha > 0$ and consider problem (2.5). We say that $u_\alpha^\star : [0, +\infty[\to U$ is an *optimal control* if

(i) $e^{-\alpha t/2} u_\alpha^\star \in L^2(0,\infty;U)$

(ii) $J_\alpha(u_\alpha^\star, x) \le J_\alpha(u, x),\ \forall\, u \in L^2(0,\infty;U)$.

If $u_\alpha^\star$ is the optimal control, then we call $y_\alpha^\star = y(\cdot, x; u_\alpha)$ the *optimal state*, $(u_\alpha^\star, y_\alpha^\star)$ the *optimal pair*, and $J_\alpha^*(x) = J_\alpha(u_\alpha^\star, x)$ the *optimal cost* .

It is convenient to transform problem (2.5), by introducing a new state variable $w(\cdot)$ and a new control $v(\cdot)$. We set

$$y(t) = e^{\alpha t/2} w_\alpha(t),\ u(t) = e^{\alpha t/2} v(t),\ f(t) = e^{\alpha t/2} g_\alpha(t).$$

Then equation (1.1) becomes

$$w'_\alpha(t) = \left(A - \frac{\alpha}{2}\right) w_\alpha(t) + Bv(t) + g_\alpha(t). \tag{4.1}$$

Remark that, since $f \in L^2_\#(H)$, then $g_\alpha \in L^2(0,\infty;U)$.

Now consider the problem

Minimize the cost

$$Q(x,u) = \int_0^{\infty} [|Cw_\alpha(t,x;v)|^2 + |v(t)|^2]dt \tag{4.2}$$

over all $v \in L^2(0,\infty;U)$ where $w_\alpha(\cdot,x;v)$ is the solution to (4.1)

such that $w_\alpha(0,x;v) = x$.

Clearly (v_α^*, w_α^*) is an optimal pair for problem (4.2) if and only if (u_α^*, y_α^*) is an optimal pair for problem (2.5). Concerning problem (4.2) we can prove the following result

Proposition 4.1 *Assume Hypotheses 2.1-2.2, and let $x \in H$. Then there exists a unique optimal pair (v_α^*, w_α^*) for problem (4.2). Moreover the following statements hold*

(i) w_α^* is the mild solution to the closed loop equation

$$\begin{cases} w'(t) = [A - \frac{\alpha}{2} - BB^\star P_{\alpha,\infty}]w(t) - BB^\star r_\alpha^*(t) + g_\alpha(t), t \geq 0 \\ w(0) = x, \end{cases} \tag{4.3}$$

where

$$r_\alpha^*(t) = \int_t^{\infty} e^{(s-t)(A-\frac{\alpha}{2}-BB^\star P_{\alpha,\infty})^*} P_{\alpha,\infty} g_\alpha(s)ds, \ t \geq 0, \tag{4.4}$$

and $P_{\alpha,\infty}$ is the solution to the algebraic Riccati equation

$$\left(A^* - \frac{\alpha}{2}\right) P_{\alpha,\infty} + P_{\alpha,\infty}\left(A^* - \frac{\alpha}{2}\right) - P_{\alpha,\infty}BB^*P_{\alpha,\infty} + C^*C = 0. \tag{4.5}$$

(ii) v_α^* is given by the feedback formula

$$v_\alpha^*(t) = -B^\star[P_{\alpha,\infty}\, w_\alpha^*(t) + r_\alpha^*(t)], \ t \geq 0. \tag{4.6}$$

(iii) The optimal cost is given by

$$\begin{aligned} J_\alpha^*(x) &= \ < P_{\alpha,\infty}x, x > +2 < r_\alpha^*(0), x) > \\ &+ \int_0^{\infty} \left\{2 < r_\alpha^*(s), f(s) > -|B^\star r_\alpha^*(s)|^2\right\} ds. \end{aligned} \tag{4.7}$$

Proof — Since (A, B) is I–stabilizable, then it is stabilizable by a feedback. That is, there exists a linear operator $L \in \mathcal{L}(H; U)$ such that $A - BK$ is stable. It follows obviously that $A - \frac{\alpha}{2} - BK$ is also stable, so that $(A - \frac{\alpha}{2}, B)$ is also I–stabilizable. Similarly $(A - \frac{\alpha}{2}, C)$ is detectable, and so equation (4.5) has a unique nonnegative solution. Now the conclusion is straightforward, see for instance [2], vol II, Part III, Chap. I, §5. ■

Now we are going study the behaviour of the optimal cost $J^*_\alpha(x)$, as $\alpha \to 0$. To this purpose we need a preliminary result on the convergence of $P_{\alpha,\infty}x$, $x \in H$, as $\alpha \to 0$. We remark that a more general result on the convergence of the solutions of an algebraic Riccati equation is known, see [7]. We give however a direct proof for our situation for the reader's convenience.

Proposition 4.2 *Assume Hypotheses* 2.1-2.2, *and let* $x \in H$. *Then*

$$\lim_{\alpha \to 0} P_{\alpha,\infty}x = P_\infty x. \tag{4.8}$$

Proof — Let us introduce two auxiliary optimal control problems

Minimize the cost

$$L(x, u) = \int_0^{+\infty} [|Cy(t)|^2 + |u(t)|^2]dt$$

over all $u \in L^2(0, \infty; U)$ subject to the equation (4.9)

$$\begin{cases} y'(t) = Ay(t) + Bu(t) \\ y(0) = x \end{cases}$$

and

Minimize the cost

$$L_\alpha(x, u) = \int_0^{+\infty} [|Cy_\alpha(t)|^2 + |u(t)|^2]dt$$

over all $u \in L^2(0, \infty; U)$ subject to the equation (4.10)

$$\begin{cases} y'_\alpha(t) = A_\alpha y(t) + Bu(t) \\ y_\alpha(0) = x \end{cases}$$

We shall denote by $y(\cdot) = y(\cdot, x; u)$ and $y_\alpha(\cdot) = y_\alpha(\cdot, x; u)$ the solutions of the initial value problem in (4.9) and in (4.10) respectively. We remark that

$$y_\alpha(\cdot, x; u) = e^{-\alpha t/2} y(\cdot, x; e^{\alpha t/2} u). \tag{4.11}$$

Under Hypotheses 2.1-2.2, there exist unique optimal pairs (u^*, y^*) and (u_α^*, y_α^*) for problems (4.9) and (4.10) respectively. Moreover we have

$$L(u^*, x) = < P_\infty x, x >= \int_0^{+\infty} [|Cy^*(t)|^2 + |u^*(t)|^2] dt \tag{4.12}$$

and

$$L_\alpha(u_\alpha^*, x) = < P_{\alpha,\infty} x, x >= \int_0^{+\infty} [|Cy_\alpha^*(t)|^2 + |u_\alpha^*(t)|^2] dt \tag{4.13}$$

Setting in (4.12), $u(t) = e^{-\alpha t/2} u^*$ and recalling (4.11) we find

$$L_\alpha(u_\alpha^*, x) \leq \int_0^{+\infty} [e^{-\alpha t} |Cy^*(t)|^2 + |u^*(t)|^2] dt \leq L(u^*, x). \tag{4.14}$$

It follows

$$\|u_\alpha^*\|^2_{L^2(0,\infty;U)} \leq L(u^*, x).$$

Thus there exists a subsequence $\alpha_k \to 0$, and functions ρ^*, ξ^* such that

$$\begin{aligned} &u_{\alpha_k}^* \rightharpoonup \rho^* \text{ in } L^2(0, \infty; U), \\ &y_{\alpha_k}^* \rightharpoonup \xi^* \text{ in } L^2(0, T; H), \ \forall\, T > 0, \\ &Cy_{\alpha_k}^* \rightharpoonup C\xi^* \text{ in } L^2(0, \infty; H), \end{aligned} \tag{4.15}$$

as $k \to \infty$. In the equation above, we have denoted by $\rightharpoonup$ the weak convergence in the L^2 norms.

Letting k tend to infinity in the equality

$$y_{\alpha_k}^*(t) = e^{t(A - \alpha_k/2)} x + \int_0^t e^{(t-s)(A-\alpha_k/2)} B u_{\alpha_k}^*(s) ds,$$

we find

$$\xi^*(t) = e^{tA} x + \int_0^t e^{(t-s)A} B \rho^*(s) ds.$$

So we have proved that

$$L(u^*, x) \leq \lambda = \text{ weak } \lim_{k \to \infty} L_{\alpha_k}(u_{\alpha_k}^*, x), \tag{4.16}$$

where

$$\lambda = \int_0^\infty [|C\xi(t)|^2 + |\rho(t)|^2]dt.$$

By a standard argument this implies that all limits in (4.15) are strong, and that

$$\lim_{\alpha \to 0} L_\alpha(u_\alpha^*, x) = L(u^*, x).$$

It follows that

$$\lim_{\alpha \to 0} < P_{\alpha,\infty}x, x >=< P_\infty x, x >, \ \forall x \in H.$$

Since $P_{\alpha,\infty} \leq P_\infty$ by (4.14) the conclusion follows by a standard argument. ∎

We now are in position to prove the main result of the paper.

Theorem 4.3 *Assume Hypotheses 2.1-2.2, and let $x \in H$. Then we have*

$$\lim_{\alpha \to 0} \alpha J_\alpha^* = K^*. \tag{4.17}$$

Proof — We first remark that by Proposition 4.2 it follows that

$$\lim_{\alpha \to 0} r_\alpha^*(t) = r(t), \ t \geq 0, \tag{4.18}$$

so that

$$\lim_{\alpha \to 0} 2 < r_\alpha^*(0), x >= 2 < r^*(0), x > . \tag{4.19}$$

We remark that as $\alpha \to 0$ does not exist finite the limit of $J_\alpha^*(x)$, since r^* is not square integrable in $[0, +\infty[$.

We now compute the limit

$$\lim_{\alpha \to 0} \alpha J_\alpha^*(x).$$

To this purpose remark that

$$r_\alpha^*(t) = e^{-\alpha t/2} h_\alpha(t), t \geq 0,$$

where

$$h_\alpha(t) = \int_t^\infty e^{\sigma(A - \frac{\alpha}{2} - BB^* P_{\alpha,\infty})^*} e^{-\alpha\sigma/2} P_{\alpha,\infty} f(t+\sigma) d\sigma,$$

so that h_α is 2π–periodic. We have

$$\int_0^{+\infty} |B^* r_\alpha^*(t)|^2 dt = \int_0^{+\infty} e^{-\alpha t}|B^* h_\alpha(t)|^2 dt$$

$$= \sum_{k=0}^{\infty} \int_0^{2\pi} e^{-\alpha t - 2k\pi\alpha}|B^* h_\alpha(t)|^2 dt$$

$$= \frac{1}{2 - e^{-2\pi\alpha}} \int_0^{2\pi} e^{-\alpha t}|B^* h_\alpha(t)|^2 dt.$$

It follows, recalling (4.18),

$$\lim_{\alpha\to 0} \int_0^{+\infty} |B^* r_\alpha^*(t)|^2 dt = \frac{1}{2\pi} \int_0^{2\pi} |B^* r^*(t)|^2 dt. \tag{4.20}$$

In a similar way one can prove that

$$\lim_{\alpha\to 0} \int_0^{+\infty} < r_\alpha^*(t), f(t) > dt = \frac{1}{2\pi} \int_0^{2\pi} < r^*(t), f(t) > dt. \tag{4.21}$$

Combining (4.18), (4.18) and (4.18), the conclusion follows. ∎

Remark 4.4 From (4.7), (4.20) and (4.17), we find the following expansion for J_α^*.

$$J_\alpha^*(x) = J_0(\alpha, x) + \frac{1}{\alpha} J_1(\alpha, x), \tag{4.22}$$

where

$$\lim_{\alpha\to 0} J_0(\alpha, x) =< P_\infty x, x > +2 < r^*(0), x >, \; x \in H, \tag{4.23}$$

and

$$\lim_{\alpha\to 0} J_1(\alpha, x) = K^*. \; ∎ \tag{4.24}$$

References

[1] BENSOUSSAN A. (1988) PERTURBATION METHODS IN OPTIMAL CONTROL, Wiley/Gauthier–Vilars Series in Modern Applied Mathematics,

[2] BENSOUSSAN A. , DA PRATO G., DELFOUR M. & MITTER S.K.,(to appear) REPRESENTATION AND CONTROL OF INFINITE DIMENSIONAL SYSTEMS,II Volume, Birkhäuser.

[3] DA PRATO G.(1987), *Synthesis of optimal control for an infinite dimensional periodic problem,* SIAM J. Control Optimiz. **25**, 706- 714.

[4] DA PRATO G.(1992), *Some results on boundary periodic control problems.* Lecture notes in pure and applied Mathematics n. 142, Edited by M. C. Joshi, and A.V. Balakrishnan, Dekker, New-York, Basel, 69–86

[5] DA PRATO G.(1993), *Spectral properties of the closed loop operator and periodic control problems,* Journal of Mathematical Systems, Estimation and Control, Vol 3, n. 1, 41–50.

[6] DA PRATO G. & ICHIKAWA A. (1988), *Quadratic control for linear periodic systems.* Appl. Math. Optim. **18**, 39-66.

[7] LASIECKA I. & TRIGGIANI R. (1991) DIFFERENTIAL AND ALGEBRAIC RICCATI EQUATIONS WITH APPLICATIONS TO BOUNDARY /POINT CONTROL PROBLEMS: CONTINUOUS THEORY AND APPROXIMATION THEORY, Lecture Notes in Control and Information Sciences, No. 164, Springer–Verlag, Berlin, Heidelberg, New York.

[8] LUNARDI A. (1991)*Stabilizability of time parabolic equations,* SIAM J. Control and Optimiz., **29**, 4, 810–828.

[9] LUNARDI A. (1991)*Neuman boundary stabilization of structurally damped time periodic wave and plate equations,* Differential equations with applications in biology, physics, and engineeering, J. A. Goldstein , F.Kappel & W. Schappacher Eds., M. Dekker, 241–257.

[10] PRITCHARD A. & ZABCZYK J. (1981) *Stability and stabilizability of infinite dimensional systems*, SIAM Rev. 23 , 25–52.

8 Optimal Boundary Control of Nonlinear Parabolic Equations

H. O. Fattorini University of California at Los Angeles, Los Angeles, California

T. Murphy University of California at Los Angeles, Los Angeles, California

1 Introduction

Let Ω a bounded domain with boundary $\partial\Omega$ in m-dimensional Euclidean space $\mathbb{R}^m$. Consider the semilinear heat equation in Ω,

$$y_t(t,x) = \Delta y(t,x) + f(t, y(t,x)) \qquad (x \in \Omega\,,\ 0 < t \leq T)\,, \tag{1.1}$$

$$y(0,x) = \zeta(x)\,, \qquad (x \in \Omega)\,, \tag{1.2}$$

with control $u(\cdot,\cdot)$ measurable in $[0,T] \times \partial\Omega$ appearing in a nonlinear boundary condition:

$$\partial_\nu y(t,x) = g(t, y(t,x)) + u(t,x) \qquad (x \in \partial\Omega,\ 0 < t \leq T)\,. \tag{1.3}$$

The **optimal control problem** is that of minimizing a **cost functional** $y_0(t,u)$ among all controls u satisfying a constraint

$$u(t,\cdot) \in U = \textbf{control set} \subseteq L^\infty(\partial\Omega) \quad (0 \leq t \leq \bar{t}) \tag{1.4}$$

The work of both authors was supported by the National Science Foundation under grant DMS-9001793.

whose corresponding solutions $y(t,x,u)$ fulfill a **target condition**

$$y(\bar{t},\cdot,u) \in Y = \textbf{target set} \subseteq C(\bar{\Omega}) \tag{1.5}$$

($C(\bar{\Omega})$ the space of all continuous functions in $\bar{\Omega}$ equipped with the supremum norm.) The terminal time $\bar{t}$ may be fixed or free. A physically interesting example is that where $m = 1$, 2 or 3 and the body Ω undergoes Stefan - Boltzmann cooling at the boundary:

$$\partial_\nu y(t,x) = -y(t,x)^4 + u(t,x) \qquad (x \in \partial\Omega,\ 0 < t \le \bar{t}). \tag{1.6}$$

Here the term $u(t,x)$ represents heat applied at $\partial\Omega$. In practice, u satisfies constraints on $\partial\Omega$, for instance

$$0 \le u(t,x) \le 1 \qquad (x \in \partial\Omega, 0 \le t \le \bar{t}). \tag{1.7}$$

Problems of this type were studied in [19], [13], [14] using a theory of nonlinear programming problems in metric spaces developed in [6], [7], [8], [11], [12] for Hilbert spaces and in [15], [9], [10] for Banach spaces. The specific results used are in [10] and generalize corresponding results in [15]. As a very particular case we obtain the following theorems on the time optimal problem for (1.1)-(1.2)-(1.3) with $f = 0$, cooling law (1.6) and control constraint (1.7) under smoothness conditions on coefficients and domain:

EXISTENCE THEOREM 1.1 Assume that a control $u(\cdot,\cdot) \in L^\infty((0,\bar{t}) \times \Omega)$ satisfying (1.7) a.e. and driving an initial state $\zeta(\cdot) \in C(\bar{\Omega})$ to a closed target set $Y \subseteq C(\bar{\Omega})$ exists. Then there exists a time optimal control $\bar{u}(t,x)$.

THEOREM 1.2 (Pontryagin's maximum principle). Let $\bar{u}(\cdot,\cdot) \in L^\infty((0,\bar{t}) \times \Omega)$ be a time optimal control, $y(t,x,\bar{u})$ the solution corresponding to $\bar{u}$. Let the target set $Y \subseteq C(\bar{\Omega})$ be defined by

$$|y(x) - \bar{y}(x)| \le \varepsilon \qquad (x \in \bar{\Omega}), \quad y(x_j) = c_j, \quad (j = 1,\ 2, \ldots, N) \tag{1.8}$$

where $\bar{y}(\cdot) \in C(\bar{\Omega}), \varepsilon > 0$, $x_j \in \bar{\Omega}$ and the c_j are arbitrary constants. Then there exists a finite Borel measure ν in $\bar{\Omega}$, $\nu \neq 0$, such that

$$\int_{\partial\Omega} z(t,x)\bar{u}(t,x)d\sigma = \max_{v \in L^\infty(\partial\Omega), 0 \le v(x) \le 1} \int_{\partial\Omega} z(t,x)v(x)d\sigma$$

almost everywhere in $0 \le t \le \bar{t} =$ optimal time, where $d\sigma$ is the area differential in $\partial\Omega$ and $z(t,x)$ is the solution of the backwards problem

$$z_t(t,x) = -\Delta z(t,x) \qquad (x \in \Omega, 0 \le t < \bar{t}). \tag{1.9}$$

$$z(\bar{t},\cdot) = \nu \qquad (x \in \bar{\Omega}), \tag{1.10}$$

$$z_\nu(t,x) = -4y(t,x,\bar{u})^3 z(t,x), \qquad (x \in \partial\Omega, 0 \le t < \bar{t}). \tag{1.11}$$

For a more general version of Theorem 1.2, see §6, where the boundary condition is (1.3) and cost functionals other than time are used; we limit ourselves to the case $f = 0$ in the first six sections to avoid unessential complications. Finally, see §7 for control exerted through a Dirichlet boundary condition. The time optimal problem for the linear equation with more general target sets than admitted here (point targets) was considered in [4]. For related treatments of other boundary control problems (both linear and nonlinear) see also [1], [2], [5], [7], [17], [21], [22], [24], [25], [26].

2 The Neumann function

A solution of

$$y_t(t,x) = \Delta y(t,x) \qquad ((t,x) \in \Omega \times (\tau, T]), \tag{2.1}$$

$$y(\tau,x) = \zeta(x) \qquad (x \in \Omega), \tag{2.2}$$

$$\partial_\nu y(t,x) = g(t,x) \qquad ((t,x) \in (\tau,T) \times \partial\Omega), \tag{2.3}$$

is called **strong** or **classical** if it is continuous in $[\tau, T] \times \bar{\Omega}$, continuously differentiable in $((\tau, T] \times \bar{\Omega})$ and satisfies (2.1) in $(\tau, T] \times \Omega$, (2.2) in $\bar{\Omega}$ and (2.3) literally in $(\tau, T] \times \partial\Omega$. A solution is **semi-strong** if it is continuous in $[\tau, T] \times \bar{\Omega}$ and satisfies (2.1) in $(\tau, T] \times \Omega$, (2.2) in $\bar{\Omega}$ and (2.3), where $\partial_\nu y(t,x)$ at a boundary point x is understood as the limit of $\partial_\nu y(t,\eta)$, where $\eta \to x$ in K, K any finite closed cone with vertex in x such that $K \subseteq \Omega$ (see [16]).

The **fundamental solution** of the heat equation in $\mathbb{R}^m$ is

$$\Gamma(t,x;\tau,\xi) = \frac{\exp(-|x-\xi|^2/4(t-\tau))}{(4\pi)^{m/2}(t-\tau)^{m/2}},$$

$x, \xi \in \mathbb{R}^m$, $x \neq \xi$, $\tau < t$. The **Neumann function** $N(t,x;\tau,\xi)$ of the heat equation in $[0,T] \times \Omega$ is defined for $t > \tau$, $\xi \in \Omega$ by

$$N(t,x;\tau,\xi) = \Gamma(t,x;\tau,\xi) - V(t,x;\tau,\xi) \qquad ((t,x) \in [\tau,T] \times \Omega)$$

where $V(t,x;\tau,\xi)$ satisfies [16 p. 155]

$$\begin{aligned} &V_t(t,x;\tau,\xi) = \Delta_x V(t,x;\tau,\xi), && ((t,x) \in (\tau,T] \times \Omega)), \\ &V(\tau,x;\tau,\xi) = 0, && (x \in \Omega), \\ &\partial_{\nu,x} V(t,x;\tau,\xi) = \partial_{\nu,x}\Gamma(t,x;\tau,\xi), && ((t,x) \in (\tau,T] \times \partial\Omega). \end{aligned}$$

A domain $\Omega \subseteq \mathbb{R}^m$ is **of class** $C^{(2)}$ if, locally, its boundary $\partial\Omega$ can be represented in the form $x_j = h(x_1, \ldots, x_{j-1}, x_{j+1}, \ldots, x_m)$ with h twice continuously differentiable; if h is continuously differentiable with λ-Hölder continuous derivatives, Ω is **of class** $C^{(1+\lambda)}$. If $\xi \in \Omega$, $\partial_{\nu,x}\Gamma(t,x;\tau,\xi)$ is infinitely differentiable in $[\tau,T] \times \bar{\Omega}$ thus [16, Theorem 2, p.

144] $V(t,x;\tau,\xi)$ exists; if Ω is of class $C^{(2)}$, $V(t,x;\tau,\xi)$, as a function of t,x is analytic in $(\tau,T]\times\Omega$, continuous in $[\tau,T]\times\bar\Omega$ and continuously differentiable in $(\tau,T]\times\bar\Omega$. Moreover,

$$N(t,x;\tau,\xi) > 0, \qquad (t>\tau, x,\xi\in\Omega) \tag{2.4}$$

$$N^-(t,x;\tau,\xi) = N(\tau,\xi,t,x) \qquad (t>\tau, x,\xi\in\Omega)\,. \tag{2.5}$$

where, in the second inequality $N^-(t,x;\tau,\xi)$ is the Neumann function of the backwards heat equation $y_t = -\Delta y$ in $[0,T]\times\Omega$. By the divergence theorem, if $y(t,x)$ is a strong solution of (2.1)-(2.2)-(2.3)

$$y(t,x) = \int_\Omega N(t,x;\tau,\xi)\zeta(\xi)d\xi + \int_{(\tau,t)\times\partial\Omega} N(t,x;r,\xi)g(r,\xi)d\sigma_\xi dr \qquad (t>\tau, x\in\Omega)\,, \tag{2.6}$$

$d\sigma_\xi$ the area differential on $\partial\Omega$. An approximation argument using the continuous dependence properties of the solution of (2.1)-(2.2)-(2.3) on ζ and g [19] shows that (2.6) extends to semi-strong solutions. Formula (2.6) will be used in §3 to **define** weak solutions of (1.1)-(1.2)-(1.3), for which task good estimates for the Neumann function are needed. Since $N(t,x,\tau,\xi)$ inherits the t-singularity of $\Gamma(t,x;\tau,\xi)$ at $(t,x)=(\tau,\xi)$ and this singularity is not integrable (except for $m=1$) we use the properties of the negative exponential to quench it at the cost of introducing an "artificial singularity" in x in the estimations. The final result is

THEOREM 2.1 Let C be a bounded domain of class $C^{(2)}$, $\mu > 1/2$. The Neumann function $N(t,x;\tau,\xi)(\xi\in\partial\Omega)$ can be extended to $t>\tau$, $\xi\in\partial\Omega$, $x\neq\xi$, is continuous there and satisfies

$$|N(t,x,\tau,\xi)| \le \frac{C}{(t-\tau)^\mu |x-\xi|^{m-2\mu}}\,. \tag{2.7}$$

We may take $\mu = 1/2$ if $m=1$. The estimation (2.7) is based on classical techniques in [16]. For full details see [19] or [13, Theorem 2.1].

3 BOUNDARY VALUE PROBLEMS AS INTEGRAL EQUATIONS

We consider (1.1)-(1.2)-(1.3) in a domain of class $C^{(2)}$, with $g(t,y)$ continuous in $[0,T]\times\mathbb{R}$ and control $u(\cdot,\cdot)$ in $L^\infty((0,T)\times\partial\Omega)$. Given $\zeta(\cdot)\in C(\bar\Omega)$ we call $y(t,x)$ a **weak solution** of (1.1)-(1.2)-(1.3) if it is continuous in $[0,T]\times\bar\Omega$ and satisfies

$$y(t,x) = \int_\Omega N(t,x;0,\xi)\zeta(\xi)d\xi + \int_{(0,t)\times\partial\Omega} N(t,x;\tau,\xi)\{g(\tau,y(\tau,\xi))+u(\tau,\xi)\}d\sigma_\xi d\tau\,. \tag{3.1}$$

For $x \in \partial\Omega$, (3.1) is an integral equation for the restriction $\phi(t,x)$ of $y(t,x)$ to $\partial\Omega$. We solve it, and then construct $y(t,x)$ from (3.1).

Consider the two parameter family of operators

$$(\mathbf{N}(t,\tau)\psi)(x) = \int_{\partial\Omega} N(t,x;\tau,\xi)\psi(\xi)d\sigma_\xi \qquad (\tau < t) \tag{3.2}$$

for $\psi \in L^\infty(\partial\Omega)$. By Theorem 2.1 $\mathbf{N}(t,\tau)$ is bounded from $L^\infty(\partial\Omega)$ into $C(\bar{\Omega})$, continuous in $\tau < t$ in the norm of $L(L^\infty(\partial\Omega), C(\bar{\Omega}))$ and

$$\|\mathbf{N}(t,\tau)\|_{L(L^\infty(\partial\Omega),C(\bar{\Omega}))} \le B(t-\tau)^{-\mu} \quad (0 \le \tau < t \le T) \tag{3.3}$$

where μ $(1/2 \le \mu < 1)$ is the parameter in Theorem 2.1. Let $\mathbf{M}(t,\tau) = \Pi \circ \mathbf{N}(t,\tau)$, $\Pi : C(\bar{\Omega}) \to C(\partial\Omega)$ the restriction operator; $\mathbf{M}(t,\tau)$ is continuous in the norm of the space $L(L^\infty(\partial\Omega), C(\partial\Omega))$ and satisfies the companion of (3.3) in that norm. The **solution operator** $\mathbf{S}(t,\tau)$

$$\mathbf{S}(t,\tau)\zeta(x) = \int_\Omega N(t,x;\tau,\xi)\zeta(\xi)d\xi \tag{3.4}$$

is bounded from $C(\bar{\Omega})$ into $C(\bar{\Omega})$; moreover, it is norm differentiable for $\tau < t$ and strongly continuous and uniformly bounded in $\tau \le t$, with $S(\tau,\tau) = I$. Finally, $\mathbf{T}(t,\tau) = \Pi \circ \mathbf{S}(t,\tau)$; $\mathbf{T}(t,\tau)$ is norm differentiable for $\tau < t$ and strongly continuous for $\tau \le t$. All these operators depend on $t-\tau$, although we keep the present notation in view of extension of the treatment to time dependent equations. The function $g(t,y)$ produces an operator $\mathbf{g} : [0,T] \times C(\partial\Omega) \to C(\partial\Omega)$ defined by $\mathbf{g}(t,\boldsymbol{\phi})(x) = g(t,\phi(x))$ which is norm continuous. In this operator jargon, the integral equation becomes

$$\begin{aligned}\boldsymbol{\phi}(t) &= \mathbf{T}(t,0)\zeta + \int_0^t \mathbf{M}(t,\tau)\mathbf{u}(\tau)d\tau + \int_0^t \mathbf{M}(t,\tau)\mathbf{g}(\tau,\boldsymbol{\phi}(\tau))d\tau \\ &= \mathbf{f}(t) + \int_0^t \mathbf{M}(t,\tau)\mathbf{g}(\tau,\boldsymbol{\phi}(\tau))d\tau\,,\end{aligned} \tag{3.5}$$

where $\boldsymbol{\phi}(t)(x) = \phi(t,x)$, $\mathbf{u}(t)(x) = u(t,x)$; setting $\mathbf{y}(t)(\cdot) = y(t,\cdot)$, (3.1) becomes

$$\mathbf{y}(t) = \mathbf{S}(t,0)\zeta + \int_0^t \mathbf{N}(t,\tau)\mathbf{u}(\tau)d\tau + \int_0^t \mathbf{N}(t,\tau)\mathbf{g}(\tau,\boldsymbol{\phi}(\tau))d\tau\,. \tag{3.6}$$

The first integrals on the right sides of (3.5) and (3.6) demand explanation since $\mathbf{u}(\cdot)$ may not be strongly measurable as a $L^\infty(\Omega)$-valued function. We use that $L^\infty((0,T)\times\Omega)$ can be identified (algebraically and metrically) with the space $L^\infty_w(0,T;L^\infty(\partial\Omega))$ of all $L^1(\Omega)$-weakly measurable $L^\infty(\partial\Omega)$- essentially bounded functions defined in $0 \le t \le T$ endowed with the essential supremum norm (we check easily that $t \to \|\mathbf{u}(t)\|$ is measurable).

LEMMA 3.1 Let $\mathbf{u}(\cdot) \in L^\infty_w(0, T; L^\infty(\partial\Omega))$. Then

$$\tau \to \mathbf{M}(t,\tau)\mathbf{u}(\tau) \quad (0 \leq \tau < t)$$

is a strongly measurable $C(\bar{\Omega})$-valued function.

PROOF See [13, Lemma 3.1].

THEOREM 3.2 Assume $g(t, y)$ is continuous and satisfies a local Lipschitz condition in y; for every $C > 0$ there exists $K = K(C)$ such that

$$|g(t, y') - g(t, y)| \leq K|y' - y| \quad (0 \leq t \leq T, |y|, |y'| \leq C).$$

Then a unique solution of (3.5) exists in an interval $0 \leq t \leq T_0$, $T_0 \leq T$.

PROOF The assumptions imply that the map

$$\mathbf{L}\boldsymbol{\phi}(t) = \mathbf{f}(t) + \int_0^t \mathbf{M}(t,\tau)\mathbf{g}(\tau, \boldsymbol{\phi}(\tau))d\tau$$

is a contraction in a closed ball $B(\mathbf{f}(\cdot), \rho)$ of $C([0, T_0]; C(\partial\Omega))$ (endowed with the supremum norm) where $\rho > 0$ and T_0 is small enough, thus a unique fixed point $\boldsymbol{\phi} = L\boldsymbol{\phi}$ exists. The estimations are based on the **generalized Gronwall inequality** in [18 p. 188]: for details see [13, Lemma 3.2]. The solution $\boldsymbol{\phi}(t)$ is not necessarily defined in $0 \leq t \leq T$. If it is not, we can define a maximal interval of existence $0 \leq t < T_m$ at the end of which $\boldsymbol{\phi}(t)$ must blow up: $\limsup_{t \to T_m} \|\boldsymbol{\phi}(t)\| = \infty$.

Although the boundary condition in

$$\xi_t(t,x) = \Delta\xi(t,x) \qquad (x \in \Omega,\ 0 < t < \bar{t}) \tag{3.7}$$

$$\xi(0,x) = \zeta(x) \qquad (x \in \Omega) \tag{3.8}$$

$$\partial_\nu \xi(t,x) = a(t,x)\xi(t,x) + \delta(t-s)v(x) \qquad (x \in \partial\Omega,\ 0 < t \leq T). \tag{3.9}$$

($0 < s < T, \zeta(\cdot) \in C(\bar{\Omega})$, $a(\cdot,\cdot) \in C([0,T] \times \partial\Omega)$, $v(\cdot) \in L^\infty(\partial\Omega)$, $\delta(t)$ the Dirac delta) is not covered in the preceding theory, this problem can also be solved by means of integral equations; the solution is singular at $t = s$. Likewise, the backwards initial value (or final value) problem

$$z_t(t,x) = -\Delta z(t,x) - f(t,x) \qquad (x \in \Omega,\ 0 \leq t \leq \bar{t}) \tag{3.10}$$

$$z(\bar{t}, \cdot) = \nu \qquad (x \in \Omega) \tag{3.11}$$

$$\partial_\nu \xi(t,x) = b(t,x)\xi(t,x) + v(t,x) \qquad (x \in \partial\Omega, 0 \leq t \leq \bar{t}) \tag{3.12}$$

with $\nu \in \Sigma(\bar{\Omega})$, $f(\cdot,\cdot) \in C([0,T] \times \bar{\Omega})$, $b(\cdot,\cdot) \in C([0,T] \times \partial\Omega)$ and $v(\cdot,\cdot) \in L^\infty((0,T) \times \partial\Omega)$ can be solved in a similar way. See [19], [13] for details.

4 Existence theory

We denote by $\mathbf{U}_{ad}(0,T;U)$ the **admissible control space** consisting of all $\mathbf{u}(\cdot) \in L^\infty_w(0,T;L^\infty(\partial\Omega))$, satisfying (1.4) a.e. in $0 \le t \le T$; if $\mathbf{u}(\cdot) \in L^\infty_w(0,T;\, L^\infty(\partial\Omega))$, $\boldsymbol{\phi}(t,\mathbf{u})$ is the $C(\partial\Omega)$-valued solution of the integral equation (3.5) corresponding to $\mathbf{u}(\cdot)$. Likewise, $\mathbf{y}(t,\mathbf{u})$ is the $C(\bar\Omega)$-valued solution defined by (3.6). These functions may not be defined in the whole interval $0 \le t \le T$.

Lemma 4.1 The operators $\mathbf{N}(t,\tau) : L^\infty(\partial\Omega) \to C(\bar\Omega)$ and $\mathbf{M}(t,\tau) : L^\infty(\partial\Omega) \to C(\partial\Omega)$ are compact.

Lemma 4.2 The operator

$$(\boldsymbol{\Lambda} u)(t) = \int_0^t \mathbf{N}(t,\tau)\mathbf{u}(\tau)d\tau$$

from $L^\infty_w(0,T;L^\infty(\partial\Omega))$ into $C([0,T];C(\bar\Omega))$ is compact.

For proofs see [13, Section 4]. A cost functional $y_0(t,\mathbf{u})$ is **weakly lower semicontinuous** if $y_0(t,\bar{\mathbf{u}}) \le \liminf_{n\to\infty} y_0(t_n,\mathbf{u}_n)$ for every $\{t_n\} \subset \mathbb{R}$ with $t_n \to \bar t$ and every $\{\mathbf{u}_n\} \subset L^\infty_W(0,T;L^\infty(\partial\Omega))$ with $\mathbf{u}_n \to \bar{\mathbf{u}}$ $L^1(0,T;L^1(\partial\Omega))$-weakly.

Let m be the minimum of the cost functional $y_0(t,\mathbf{u})$ under the control constraint and the target condition. Assuming that $-\infty < m < \infty$, a sequence $\{\mathbf{u}^n\}$, $\mathbf{u}^n \in \mathbf{U}_{ad}(0,t_n;U)$ is called a **minimizing sequence** if

$$\limsup_{n\to\infty} y_0(t_n,\mathbf{u}^n) \le m, \ \lim_{n\to\infty} \operatorname{dist}(\mathbf{y}(t_n,\mathbf{u}^n),Y) \to 0\,.$$

Theorem 4.3 Let Y be closed and $\mathbf{U}_{ad}(0,T;U)$ be $L^1(0,T;L^1(\partial\Omega))$-weakly compact in $L^\infty_W(0,T;L^\infty(\partial\Omega))$. Assume that $y_0(t,\mathbf{u})$ is weakly lower semicontinuous. Let $\{\mathbf{u}_n(\cdot)\}$, $\mathbf{u}_n(\cdot) \in \mathbf{U}_{ad}(0,t_n;U)$ be a minimizing sequence with $t_n \to \bar t$ and $\|\boldsymbol{\phi}(t,\mathbf{u}_n)\|_{C(\partial\Omega)} \le C$ $(0 \le t \le \bar t)$. Then (a subsequence of) $\{\mathbf{u}_n\}$ converges $L^1(0,\bar t;L^1(\partial\Omega))$-weakly to an optimal control $\bar{\mathbf{u}}(\cdot) \in \mathbf{U}_{ad}(0,\bar t;U)$.

The proof is standard; see [13, §4] for details.

In the following result $\mathbf{U}_{ad}(0,T;U)$ is equipped with the **Ekeland distance**

$$d(\mathbf{u},\mathbf{v}) = \operatorname{meas}\{t;\mathbf{u}(t) \ne \mathbf{v}(t)\}\,. \tag{4.1}$$

($\{t;\mathbf{u}(t) \ne \mathbf{v}(t)\}$ is the union of the countable collection of measurable sets $\{t;\langle y_n,\mathbf{u}(t)\rangle \ne \langle y_n,\mathbf{v}(t)\rangle\}$, $\{y_n\}$ a countable dense set in $L^1(\partial\Omega)$; hence the set between curly brackets in (4.1) is measurable.)

Lemma 4.4 Let $\bar{\mathbf{u}}(\cdot) \in \mathbf{U}_{ad}(0,\bar t;U)$ be such that $\boldsymbol{\phi}(t,\bar{\mathbf{u}})$ exists in $0 \le t \le \bar t$. Then there exists $\rho > 0$ such that if $d(\mathbf{u},\bar{\mathbf{u}}) \le \rho$ then $\boldsymbol{\phi}(t,\mathbf{u})$ also exists in $0 \le t \le \bar t$, and

$$\|\boldsymbol{\phi}(t,\mathbf{v}) - \boldsymbol{\phi}(t,\mathbf{u})\| \le Cd(\mathbf{v},\mathbf{u})^{1-\mu} \quad (0 \le t \le \bar t, \mathbf{u},\mathbf{v} \in B(\bar u,\rho))\,. \tag{4.2}$$

For a proof, see [13], Lemma 4.4 and Lemma 4.5.

Using results specific to parabolic equations (the maximum principle) global existence of solutions can be insured a priori; for instance, this is the case if $\zeta(x) \geq 0$ and $g(t,0) = 0$, $g(t,y) < 0$ for $y \neq 0$. See [13, Lemma 4.7].

5 Directional derivatives

Let V a metric space, E a Banach space, $g : V \to E$. An element $\xi \in E$ is a (one sided) **directional derivative** of g at $u \in V$ if and only if there exists $u : [0,\delta] \to V$ with $d(u(h),u) \leq h$ and $g(u(h)) = g(u) + h\xi + o(h)$ as $h \to 0+$. The set of all directional derivatives of g at u is denoted $\partial g(u)$; it is star-shaped and closed. We compute below certain elements of $\partial \mathbf{f}(\mathbf{u})$, where $\mathbf{f} : \mathbf{U}_{ad}(0,\bar{t};U) \to E$ is

$$\mathbf{f}(\mathbf{u}) = \mathbf{y}(\bar{t},\mathbf{u}) \in C(\bar{\Omega}) \tag{5.1}$$

and $\mathbf{u} \in \mathbf{U}_{ad}(0,\bar{t};U)$ is such that $\boldsymbol{\phi}(t,\mathbf{u})$ (thus $\mathbf{y}(t,\mathbf{u})$) exists in $0 \leq t \leq \bar{t}$. The space $\mathbf{U}_{ad}(0,\bar{t};U)$ is equipped with the distance (4.1). Derivatives are constructed by means of McShane's "multispike perturbations" $\mathbf{u}(h) = \mathbf{u}_{\mathbf{s},\mathbf{p},h,\mathbf{v}}(t)$ defined by

$$\mathbf{u}_{\mathbf{s},\mathbf{p},h,\mathbf{v}}(t) = \begin{cases} v_j & (s_j - p_j h \leq t \leq s_j,\ j = 1,2,\ldots,m) \\ \mathbf{u}(t) & \text{elsewhere} \end{cases} \tag{5.2}$$

where $\mathbf{s} = (s_1, s_2, \ldots, s_m)$, $0 < s_1 < s_2 < \cdots < s_m < \bar{t}$, $\mathbf{p}$ a probability vector $(p_1, p_2, \ldots, p_m)$, $p_j \geq 0$, $\Sigma p_j = 1$, and $\mathbf{v} = (v_1, v_2, \ldots, v_m)$, $v_j \in U$.

Theorem 5.1 Assume that $g(t,y)$ is differentiable with respect to y and that $g(t,y)$, $\partial_y g(t,y)$ are continuous in $[0,\bar{t}] \times \mathbb{R}$. Then there exists a set e of full measure in $0 \leq s \leq \bar{t}$ such that if $s_j \in e$ $(j = 1,2,\ldots,m)$ then

$$\boldsymbol{\xi}(t,\mathbf{s},\mathbf{p},\mathbf{u},\mathbf{v}) = \lim_{h \to 0+} \frac{\mathbf{y}(t,\mathbf{u}_{\mathbf{s},\mathbf{p},h,\mathbf{v}}) - \mathbf{y}(t,\mathbf{u})}{h} \tag{5.3}$$

exists in the norm of $C(\bar{\Omega})$, convergence uniform outside of the intervals $|t - s_j| < \varepsilon$ for any $\varepsilon > 0$. We have

$$\boldsymbol{\xi}(t,\mathbf{s},\mathbf{p},\mathbf{u},\mathbf{v}) = \sum_{j=1}^{m} p_j \boldsymbol{\xi}(t,s_j,\mathbf{u},v_j) \tag{5.4}$$

where $\boldsymbol{\xi}(t,s,\mathbf{u},v)(x) = \xi(t,x,s,u,v)$ is the solution of the linear initial value problem

$$\xi_t(t,x,s,u,v) = \Delta_x \xi(t,x,s,u,v) \qquad ((t,x) \in (0,\bar{t}] \times \Omega) \tag{5.5}$$

$$\xi(0,x,s,u,v) = 0 \qquad (x \in \Omega) \tag{5.6}$$

$$\begin{aligned}\partial_\nu \xi(t,x,s,u,v) &= \partial_y g(t,y(t,x,u))\xi(t,x,s,u,v) \\ &\quad + \delta(t-s)(v(x) - u(s,x)) \qquad ((t,x) \in (0,\bar{t}] \times \partial\Omega)\,.\end{aligned} \tag{5.7}$$

Moreover,

$$\|h^{-1}(\mathbf{y}(t, \mathbf{u_{s,p,h,v}}) - \mathbf{y}(t, \mathbf{u})) - \boldsymbol{\xi}(t, \mathbf{s}, \mathbf{p}, \mathbf{u}, \mathbf{v})\|_{C(\bar\Omega)} \le C \sum_{j=1}^{m} \kappa(t, s_j, h, 1 - 2\mu) \tag{5.8}$$

for sufficiently small h, with $\kappa(t, h, s, \beta) = h(t - s - h)$ for $\beta \ge 0$, h the Heaviside function: for $\beta < 0$,

$$\kappa(t, h, s, \beta) = \begin{cases} 0 & (0 \le t \le s) \\ (t - (s - h))^{\beta} & (s - h \le t \le s) \\ (t - s)^{\beta} & (s \le t \le \bar t)\,. \end{cases} \tag{5.9}$$

The proof is in [13, Theorem 5.1]. Directional derivatives of the cost functional are computed below. Let

$$\mathbf{f}_0(\mathbf{u}) = y_0(t, \mathbf{u}) = \int_{(0,t)\times\Omega} f_0(\tau, y(\tau, x, u))dxd\tau + \int_{(0,t)\times\partial\Omega} g_0(\tau, y(\tau, x, u), u(\tau, x))d\sigma d\tau\,. \tag{5.10}$$

THEOREM 5.2 Assume that $f_0(t, y)$ and $g_0(t, y, u)$ are continuous in all variables and continuously differentiable with respect to y. Let $\mathbf{u} \in \mathbf{U}_{ad}(0, \bar t; U)$, $\mathbf{s}, \mathbf{p}, h, \mathbf{v}$ be as in Theorem 5.1. Then there exists a set $e_0 \subseteq e$ of full measure in $0 \le t \le \bar t$ such that if $s_j \in e_0$ then

$$\xi_0(\bar t, \mathbf{s}, \mathbf{p}, \mathbf{u}, \mathbf{v}) = \lim_{h\to 0+} \frac{y_0(\bar t, \mathbf{u_{s,p,h,v}}) - y_0(\bar t, \mathbf{u})}{h} \tag{5.11}$$

exists and equals

$$\xi_0(t, x, \mathbf{s}, \mathbf{p}, \mathbf{u}, \mathbf{v}) = \sum_{j=1}^{m} p_j \xi_0(t, x, s_j, \mathbf{u}, v_j)\,, \tag{5.12}$$

where

$$\begin{aligned} &\xi_0(t, x, s, \mathbf{u}, v) \\ &= \int_{(s,t)\times\Omega} \partial_y f_0(\tau, y(\tau, x, u))\xi(\tau, x, s, u, v)dxd\tau \\ &\quad + \int_{(s,t)\times\partial\Omega} \partial_y g_0(\tau, y(\tau, x, u))\xi(\tau, x, s, u, v)d\sigma d\tau \\ &\quad + \int_{\partial\Omega} g_0(s, y(s, x, u), v(x))d\sigma - \int_{\partial\Omega} g_0(s, y(s, x, u), u(s, x))d\sigma \end{aligned} \tag{5.13}$$

6 THE MAXIMUM PRINCIPLE

We apply the abstract nonlinear programming theory in [10, §2] to the optimal control problem written in the form

$$\text{minimize} \quad \mathbf{f}_0(\mathbf{u}) \quad \text{subject to} \quad \mathbf{f}(\mathbf{u}) \in Y \tag{6.1}$$

where $\mathbf{f}$ (resp. $\mathbf{f}_0$) is defined by (5.1) (resp. (5.19)) in the space $\mathbf{U}_{ad}(0, \bar{t}; U)$ equipped with the distance (4.6), which makes the space complete. The function $\mathbf{f}$ is (Hölder) continuous (Lemma 4.4), and a simple estimation of the integrals defining the cost functional (5.10) gives

LEMMA 6.1 Let $\bar{\mathbf{u}}, \rho, \mu$ be as in Corollary 4.6, and let the assumptions of Theorem 5.2 be fulfilled. Then

$$|y_0(t, \mathbf{v}) - y_0(t, \mathbf{u})| \leq C d(\mathbf{v}, \mathbf{u})^{1-\mu} \qquad (\mathbf{u}, \mathbf{v} \in B(\bar{\mathbf{u}}, \rho)) \, . \tag{6.2}$$

Continuity of $\mathbf{f}$ and $\mathbf{f}_0$ is more than enough to apply Theorem 2.8 in [10], which is based on Ekeland's variational principle. Let $\bar{\mathbf{u}}$ be a solution of (6.1) (that is, an optimal control). [10, Theorem 2.8] provides a sequence $\{\mathbf{u}^n\} \in \mathbf{U}_{ad}(0, \bar{t}; U)$ with $\mathbf{u}_n \to \bar{\mathbf{u}}$ ($d(\mathbf{u}^n, \bar{\mathbf{u}}) \to 0$ as fast as we wish) such that, if $\{\mathbf{D}_n\}$ is a sequence of convex sets with $\mathbf{D}_n \subseteq \partial(\mathbf{f}_0, \mathbf{f})(\mathbf{u}^n)$ there exists a sequence $\{(\mu_n, \nu_n)\} \subseteq \mathbb{R} \times C(\bar{\Omega})^* = \mathbb{R} \times \Sigma(\bar{\Omega})$ such that

$$\mu_n^2 + \|\nu_n\|^2 = 1, \qquad \mu_n \geq 0, \qquad \mu_n \eta^n + \langle \nu_n, \xi^n \rangle \geq -\delta_n$$

for $(\eta^n, \xi^n) \in \mathbf{D}_n$, where $\delta_n \to 0$. If (μ, ν) is the $\mathbb{R} \times C(\bar{\Omega})$-weak limit of (a subsequence of) $\{(\mu_n, \nu_n)\}$ we obtain the Kuhn-Tucker inequality

$$\mu \eta + \langle \nu, \xi \rangle \geq 0 \tag{6.3}$$

for (η, ξ) in the set $\liminf_{n \to \infty} \mathbf{D}_n$ of all limits of sequences $\{(\eta^n, \xi^n)\}$, $(\eta^n, \xi^n) \in \mathbf{D}_n$. Here, $\mathbf{D}_n$ consists of all elements

$$(\xi_0(\bar{t}, \mathbf{s}, \mathbf{p}, \mathbf{u}, \mathbf{v}), \boldsymbol{\xi}(\bar{t}, \mathbf{s}, \mathbf{p}, \mathbf{u}, \mathbf{v})) \, , \tag{6.4}$$

where $\mathbf{u} = \mathbf{u}^n$, $\boldsymbol{\xi}(\bar{t}, \mathbf{s}, \mathbf{p}, \mathbf{u}^n, \mathbf{v})$ (resp. $\xi_0(\bar{t}, \mathbf{s}, \mathbf{p}, \mathbf{u}^n, \mathbf{v})$) given by (5.4) (resp. (5.12)). In order that $\mathbf{D}_n$ be convex we must allow for the possibility that two or more of the s_j in the vector $\mathbf{s}$ coincide.

LEMMA 6.2 Let $\mathbf{u} \in \mathbf{U}_{ad}(0, \bar{t}; U)$. Then there exists a set e of full measure in $0 \leq t \leq \bar{t}$ such that (6.4) belongs to $\partial(\mathbf{f}_0, \mathbf{f})(\mathbf{u})$ for $\mathbf{s} = (s_1, s_2, \ldots, s_m)$, $s \in e$, $0 < s_1 \leq s_2 \leq \cdots \leq s_m < \bar{t}$.

The next task is to compute elements of $\liminf_{n \to \infty} \mathbf{D}_n$:

LEMMA 6.3 Let $\{t_n\}$ be a sequence in $0 \le t \le \bar{t}$ with $t_n \to \bar{t}$, $\{\mathbf{u}^n\}$ a sequence in $\mathbf{U}_{ad}(0, t_n, U)$ such that $\Sigma d_n(\mathbf{u}^n, \bar{\mathbf{u}}) < \infty$. Then there exists a set e of full measure in $0 \le s \le \bar{t}$ such that if $s_j \in e$ and n is large enough then the directional derivatives $\boldsymbol{\xi}(\bar{t}, \mathbf{s}, \mathbf{p}, \mathbf{u}^n, \mathbf{v})$, $\xi_0(\bar{t}, \mathbf{s}, \mathbf{p}, \mathbf{u}^n, \mathbf{v})$ exist and converge to (6.4), that is, (6.4) belongs to $\liminf_{n\to\infty} \mathbf{D}_n$ if $s_j \in e$.

For proofs of Lemma 6.2 and Lemma 6.3 see the corresponding results in [13]. The maximum principle for the general optimal control problem is

THEOREM 6.4 Let $\bar{u}(t,x) \in L^\infty((0,T)\times\partial\Omega) = L^\infty_w(0,T; L^\infty(\partial\Omega))$ be an optimal control. Then there exists a set e of full measure in $0 \le t \le \bar{t}$ and $(\mu, \nu) \in \mathbb{R} \times \Sigma(\bar{\Omega})$, $\mu \ge 0$ such that, if $z(t,x)$ is the solution of the final value problem

$$z_t(t,x) = -\Delta z(t,x) - \mu \partial_y f_0(t, y(t,x,\bar{u})) \qquad ((t,x) \in [0,\bar{t}) \times \Omega) \tag{6.5}$$

$$z(\bar{t}, \cdot) = \nu \qquad (x \in \bar{\Omega}) \tag{6.6}$$

$$\partial_\nu z(t,x) = \partial_y g(t, y(t,x,\bar{u})) z(t,x) - \mu \partial_y g_0(t, y(t,x,\bar{u}) \qquad (t,x) \in [0,\bar{t}) \times \partial\Omega \tag{6.7}$$

then we have

$$\begin{aligned} &\int_{\partial\Omega} \{z(s,x)\bar{u}(s,x) + \mu g_0(s, y(s,x,\bar{u}), \bar{u}(s,x))\} d\sigma \\ &\quad = \min_{v\in U} \int_{\partial\Omega} \{z(s,x)v(x) + \mu g_0(s, y(s,x,\bar{u}), v(x))\} d\sigma \end{aligned} \tag{6.8}$$

a.e. in $0 \le t \le \bar{t}$.

For the proof we use the Kuhn-Tucker inequality (6.3) for the elements (6.4) of $\liminf_{n\to\infty} \mathbf{D}_n$; for single spikes we obtain

$$\mu\xi_0(\bar{t}, x, s, \bar{u}, v) + \langle \nu, \xi(\bar{t}, x, s, \bar{u}, v) \rangle \ge 0$$

so that

$$\begin{aligned} \mu &\int_{(s,\bar{t})\times\Omega} \partial_y f_0(\tau, y(\tau,x,\bar{u})) \xi(\tau, x, s, \bar{u}, v) dx d\tau \\ &+ \mu \int_{(s,\bar{t})\times\partial\Omega} \partial_y g_0(\tau, y(\tau,x,\bar{u})) \xi(\tau, x, s, \bar{u}, v) d\sigma d\tau \\ &+ \mu \int_{\partial\Omega} \{g_0(s, y(t,x,\bar{u}), v(x)) - g_0(s, y(t,x,\bar{u}), \bar{u}(s,x))\} d\sigma \\ &+ \int_{\bar{\Omega}} \xi(\bar{t}, x, s, \bar{u}, v) \nu(dx) \ge 0 \end{aligned}$$

and (6.8) results from an application of the divergence theorem (in a slightly generalized form).

For the time optimal problem, we apply Theorem 2.4 in [10] to the sequence of functions

$\mathbf{f}_n(\mathbf{u}) = \mathbf{y}(t_n, \mathbf{u})$, $\mathbf{f}_n : \mathbf{U}_{ad}(0, t_n; U) \to C(\bar{\Omega})$, where $\{t_n\}$ is a sequence with $t_n < \bar{t} =$ optimal time, $t_n \to \bar{t}$ (see [10] for details). If $\bar{\mathbf{u}}$ is a time optimal control Theorem 2.4 provides a sequence $\{\mathbf{u}^n\}$, $\mathbf{u}^n \in \mathbf{U}_{ad}(0, t_n; U)$ (with $d_n(\mathbf{u}^n, \bar{\mathbf{u}}) \to 0$ as fast as we wish, d_n the distance (4.1)) such that, if $\{\mathbf{D}_n\}$ is a sequence of convex sets with $\mathbf{D}_n \subseteq \partial \mathbf{f}_n(\mathbf{u}^n)$ there exists a sequence $\{\nu_n\} \subseteq C(\bar{\Omega})^* = \Sigma(\bar{\Omega})$ such that

$$\|\nu_n\| = 1, \qquad \langle \nu_n, \xi^n \rangle \geq -\delta_n \tag{6.9}$$

for every $\xi^n \in \mathbf{D}_n$, where $\delta_n \to 0$. Taking limits,

$$\langle \nu, \xi \rangle \geq 0 \tag{6.10}$$

for $\liminf_{n\to\infty} \mathbf{D}_n$. For single spikes,

$$\int_{\bar{\Omega}} \xi(\bar{t}, x, s, \bar{u}, v)\nu(dx) \geq 0\,, \tag{6.11}$$

for all s in a total set e and all $v \in e$; operating as before, we obtain

THEOREM 6.5 Let $\bar{u}(t,x)$ be an optimal control for the time optimal problem. Then there exists $\nu \in \Sigma(\bar{\Omega})$ such that, if $z(t,x)$ is the solution of

$$z_t(t,x) = -\Delta z(t,x) \qquad (t,x) \in [0,\bar{t}) \times \Omega \tag{6.12}$$

$$z(\bar{t}, \cdot) = \nu \qquad (x \in \bar{\Omega}) \tag{6.13}$$

$$\partial_\nu z(t,x) = \partial_y g(t, y(t,x,u))z(t,x) \qquad (t,x) \in [0,\bar{t}) \times \partial\Omega \tag{6.14}$$

then we have

$$\int_{\partial\Omega} z(s,x)\bar{u}(s,x)d\sigma = \min_{v\in U} \int_{\partial\Omega} z(s,x)v(x)d\sigma \tag{6.15}$$

a.e. in $0 \leq t \leq \bar{t}$.

Without special assumptions, the multipliers (μ, ν) in Theorem 6.4 and ν in Theorem 6.5 may be zero, destroying the maximum principle. Conditions for the abstract nonlinear programming problem that prevent this are given in [10, Lemma 2.5]. For the general problem, this result requires a compact set Q such that

$$\Delta = \bigcap_{n\geq 1} (\overline{\text{conv}}(\Delta_n) + Q) \tag{6.16}$$

contains an interior point, $\Delta_n = \Pi(\mathbf{D}_n) - N_Y(\mathbf{y}^n) + Q$, $\{\mathbf{D}_n\}$ the sequence in [10, Theorem 2.8], Π the canonical projection from $\mathbb{R} \times C(\bar{\Omega})$ into $C(\bar{\Omega})$, $\{\mathbf{y}_n\}$ a sequence in the target set Y with $\mathbf{y}^n \to \bar{\mathbf{y}} = \mathbf{y}(\bar{t}, \bar{\mathbf{u}})$ and $N_Y(\bar{\mathbf{y}})$ the tangent cone to Y at $\bar{\mathbf{y}}$. It can be shown that, due to the smoothing properties of the heat equation, $\Pi(\mathbf{D}_n) + Q$ will never satisfy this condition by itself, thus we must rely on a "large" target set Y, such as the one defined by (1.8). In the time optimal case, $\Delta_n = \mathbf{D}_n - N_Y(\mathbf{y}^n) + Q$ and the same comments apply.

7 THE DIRICHLET BOUNDARY CONDITION

The treatment is (up to a point) similar. To fix ideas, we consider

$$y_t(t,x) = \Delta y(t,x) + f(t, y(t,u)) \qquad ((t,x) \in \Omega \times (\tau, T]) \tag{7.1}$$
$$y(\tau, x) = \zeta(x) \qquad (x \in \bar{\Omega}) \tag{7.2}$$
$$y(t,x) = u(t,x) \qquad ((t,x) \in [\tau, T] \times \partial\Omega) \tag{7.3}$$

with $f(t,y)$ continuous. We use the Green function

$$G(t,x;\tau,\xi) = \Gamma(t,x;\tau,\xi) - V(t,x;\tau,\xi),$$

($t > \tau$ and $\xi \in \Omega$) with $\Gamma(t,x;\tau,\xi)$ the fundamental solution of the heat equation (§2) and $V(t,x;\tau,\xi)$ the solution of the initial-boundary value problem

$$V_t(t,x;\eta,\xi) = \Delta_x V(t,x;\tau,\xi) \qquad ((t,x) \in (\tau,T] \times \Omega)$$
$$V(\tau,x;\tau,\xi) = 0 \qquad (x \in \Omega)$$
$$V(t,x;\tau,\xi) = \Gamma(t,x;\tau,\xi) \qquad ((t,x) \in (\tau,T] \times \partial\Omega).$$

If $y(t,x)$ is a classical solution of (7.1)-(7.2)-(7.3), then

$$\begin{aligned} y(t,x) &= \int_\Omega G(t,x;\tau,\xi)\zeta(\xi)d\xi \\ &= \int_{(t,\tau)\times\Omega} G(t,x;\tau,\xi) f(\tau, y(\tau,\xi)) d\xi dt \\ &\quad + \int_{(t,\tau)\times\partial\Omega} \partial_{\nu,\xi} G(t,x;r,\xi) u(r,\xi) d\sigma_\xi dr \end{aligned} \tag{7.4}$$

($t > \tau, x \in \Omega$). It results from this formula with $f = 0$ and the maximum principle that $G(t,x;\tau,\xi) \geq 0$ ($x, \xi \in \Omega$), $\partial_{\nu,\xi} G(t,x;\tau,\xi) \geq 0$ ($x \in \Omega$, $\xi \in \partial\Omega$), and

$$\int_\Omega G(t,x;\tau,\xi)d\xi + \int_{(0,t)\times\partial\Omega} \partial_{\nu,\xi} G(t,x;r,\xi) d\sigma_\xi dr = 1. \tag{7.5}$$

We cast (7.4) in operator form using $\mathbf{G}(t,\tau) : L^\infty(\Omega) \to C(\bar{\Omega})$ given by

$$(\mathbf{G}(t,\tau)\zeta)(x) = \int_\Omega G(t,x;\tau,\xi) y(\xi) d\xi$$

and $\mathbf{N}(t,\tau) : L^\infty(\partial\Omega) \to L^\infty(\Omega)$ given by

$$(\mathbf{N}(t,\tau)u)(x) = \int_{\partial\Omega} \partial_{\nu,\xi} G(t,x;\tau,\xi)u(\xi)d\sigma_\xi \,.$$

It follows from (7.5) that $\mathbf{G}(t,\tau)$ has uniformly bounded $L(L^\infty(\Omega),\, C(\bar{\Omega}))$ norm in $t > \tau$ and that

$$\|\mathbf{N}(t,\tau)\|_{L(L^\infty(\partial\Omega),L^1(\Omega))} \leq \alpha(t-\tau) \tag{7.6}$$

with $\alpha(t)$ continuous and summable in $t > 0$. Equation (7.4) becomes

$$\mathbf{y}(t) = \mathbf{G}(t,0)\zeta + \int_0^t \mathbf{G}(t,\tau)\mathbf{f}(\tau,\mathbf{y}(\tau))d\tau + \int_0^t \mathbf{N}(t,\tau)\mathbf{u}(\tau)d\tau \tag{7.7}$$

where $\mathbf{u}(t)(x) = u(t,x)$, $\mathbf{y}(t)(x) = y(t,x)$. There are two important differences with (3.6): $\mathbf{y}(t)$ does not take values in $C(\bar{\Omega})$ but in $L^\infty(\Omega)$ (or, rather in the space $BC(\Omega)$ of bounded continuous functions in Ω) and the second integral has to be understood pointwise, since the $L(L^\infty(\partial\Omega), L^\infty(\Omega))$-norm of $\mathbf{N}(t,\tau)$ has an nonintegrable singularity for $t = \tau$. Solutions are sought in the space $C_w(0,T;L^\infty(\Omega))$ of all $L^1(\Omega)$-weakly continuous $L^\infty(\Omega)$-valued functions equipped with the supremum norm. The pertinent existence and extension results are

THEOREM 7.1 Assume $f(t,y)$ is continuous and satisfies a local Lipschitz condition in y, that is, for every $C > 0$ there exists $K = K(C)$ such that

$$|f(t,y') - f(t,y)| \leq K|y'-y| \; (0 \leq t \leq T, |y|, |y'| \leq c)\,.$$

Then a unique solution of (7.6) exists in an interval $0 \leq t \leq T_0$, $T_0 \leq T$; a maximal interval of existence can be obtained as in §3. The solution $\boldsymbol{\xi}(t)(x) = \xi(t,x)$ of the linear initial - boundary value problem

$$\xi_t(t,x) = \Delta\xi(t,x) + f(t,x)\xi(t,x) + g(t,x) \qquad (x \in \Omega, 0 < t < \bar{t}) \tag{7.8}$$

$$\xi(0,x) = \nu \qquad (x \in \Omega) \tag{7.9}$$

$$\xi(t,x) = \delta(t-s)v(x) \qquad (x \in \partial\Omega, 0 < t \leq T)\,. \tag{7.10}$$

$(0 < s < \bar{t})$ is defined as well by integral equations, first in $L^1(0,T;L^1(\Omega))$; then, using bounds similar to (2.7) for the Green function we show that $\boldsymbol{\xi}(t)$ is a weakly continuous $L^\infty(\Omega)$-valued function for $t > s$.

The existence theory in §4 extends without major changes.

THEOREM 7.2 Let Y be closed and $\mathbf{U}_{ad}(0,T;U)$ $L^1(0,T;L^1(\partial\Omega))$-weakly compact in $L^\infty_w(0,T;L^\infty(\partial\Omega))$, and assume $y_0(t,\mathbf{u})$ is weakly lower semicontinuous. Let $\{\mathbf{u}_n(\cdot)\}$,

$\mathbf{u}_n(\cdot) \in \mathbf{U}_{ad}(0, t_n; U)$ be a minimizing sequence with $t_n \to \bar{t}$ and $\|\mathbf{y}(\cdot, \mathbf{u}_n)\|_{L^\infty(\Omega)} \leq C$ $(0 \leq t \leq \bar{t})$. Then (a subsequence of) $\{\mathbf{u}_n\}$ converges $L^1(0, \bar{t}; L^1(\partial\Omega))$-weakly to an optimal control $\bar{\mathbf{u}}(\cdot) \in \mathbf{U}_{ad}(0, \bar{t}; U)$.

Using the maximum principle, global existence can be insured a priori; if $f(t, y) \leq 0$, $\zeta(x) \geq 0$, $u(t, x) \geq 0$ the solution of (7.1)-(7.2)-(7.3) exists in $[0, T] \times \Omega$ and

$$0 \leq y(t, x) \leq \max(\|\zeta\|_{L^\infty(\Omega)}, \|u\|_{L^\infty((0,T)\times\partial\Omega)}) \qquad ((t, x) \in [0, T] \times \Omega).$$

LEMMA 7.3 Let f satisfy the assumptions of Theorem 7.1 and let $\mathbf{U}_{ad}(0, T; U)$ be bounded in $L^\infty_w(0, T; L^\infty(\partial\Omega))$ and consist of nonnegative functions. Then there exists a constant C such that

$$\|y(t, u) - y(t, v)\|_{L^1(\Omega)} \leq C \int_{\{t; u(t) \neq v(t)\}} \alpha(\sigma) d\sigma \ (0 \leq t \leq T) \tag{7.11}$$

Let V be a metric space, E a Banach space, $g : V \to E$. The element $\xi \in E$ is a **variation** of g at u if there exist sequences $\{u_k\} \subseteq V$, $\{h_k\} \subset \mathbb{R}_+$ such that $d(u_k, u) \leq h_k \to 0$ and $g(u_k) = g(u) + h_k\xi + o(h_k)$. A directional derivative is a variation but the converse is not true. Variations of the analogue of (5.1) corresponding to the Dirichlet boundary condition are again constructed via the multispike perturbations (5.2).

THEOREM 7.4 Assume that $f(t, y) \leq 0$ and that $f(t, y)$ is differentiable with respect to y and that $f(t, y)$, $\partial_y f(t, y)$ are continuous in $[0, \bar{t}] \times \mathbb{R}$; moreover, assume that controls in $\mathbf{U}_{ad}(0, \bar{t}; U)$ are positive. Then, given $\mathbf{u}(\cdot) \in \mathbf{U}_{ad}(0, \bar{t}; U)$ there exists a sequence $\{h_k\} \subset \mathbb{R}_+$ with $h_k \to 0$ and a set $e \subseteq (0, \bar{t})$ of full measure such that if $s_j \in e$ $(j = 1, 2, \ldots, m)$ then

$$\boldsymbol{\xi}(t, \mathbf{s}, \mathbf{p}, \mathbf{u}, \mathbf{v}) = \lim_{k\to\infty} \frac{\mathbf{y}(t, \mathbf{u}_{\mathbf{s},\mathbf{p},h_k,\mathbf{v}}) - \mathbf{y}(t, \mathbf{u})}{h_k} \tag{7.12}$$

exists in the norm of $C(\bar{\Omega})$, convergence uniform outside of the intervals $|t - s_j| < \varepsilon$ for any $\varepsilon > 0$. We have

$$\boldsymbol{\xi}(t, \mathbf{s}, \mathbf{p}, \mathbf{u}, \mathbf{v}) = \sum_{j=1}^{m} p_j \boldsymbol{\xi}(t, s_j, \mathbf{u}, v_j) \tag{7.13}$$

where $\boldsymbol{\xi}(t, s, \mathbf{u}, v)(x) = \xi(t, x, s, u, v)$ is the solution of the linear initial value problem

$$\xi_t(t, x, s, u, v) = \{\Delta + \partial_y f(t, y(t, x, u))\}\xi(t, x, s, u, v) \qquad ((t, x) \in (0, \bar{t}] \times \Omega) \tag{7.14}$$

$$\xi(0, x, s, u, v) = 0 \qquad (x \in \Omega) \tag{7.15}$$

$$\xi(t, x, s, u, v) = \delta(t - s)(v(x) - u(s, x)) \quad ((t, x) \in (0, \bar{t}] \times \partial\Omega). \tag{7.16}$$

Treatment of cost functionals like (5.10) (with $g(t, y, u) = g(t, u)$) follows the lines of §5 and we omit the companion results, limiting ourselves below to the time optimal problem.

For reasons explained in §8, the basic space in the application of the nonlinear programming theory will not be $L^\infty(\Omega)$ or $BC(\Omega)$ but the space $E^p(\Omega, K) \subseteq L^p(\Omega)$ ($1 \le p < \infty$, $K \subset \Omega$ compact) consisting of all functions in $L^p(\Omega)$ that are (equivalent to a function) continuous in K. The norm of the space is

$$\|f\|_{E^p(\Omega,K)} = \max(\|f(\cdot)\|_{C(K)}, \|f(\cdot)\|_{L^p(\Omega\setminus K)}) . \tag{7.17}$$

The dual space $E^p(\Omega, K)^*$ consists of all pairs (ν, ϕ), where $\nu \in \Sigma(K)$ and $\phi \in L^q(\Omega\setminus K)$ ($1/p + 1/q = 1$). The norm in $E^q(\Omega, K)^*$ is

$$\|(\nu, \phi)\|_{E^p(\Omega,K)^*} = |\nu|(K) + \|\phi\|_{L^q(\Omega\setminus K)} \tag{7.18}$$

and the canonical duality

$$\langle(\nu, \phi), f\rangle = \int_K f(x)\nu(dx) + \int_{\Omega\setminus K} \phi(x)f(x)dx .$$

The theory in [10] only requires lower semicontinuity of the function $\|\mathbf{f}(\mathbf{u}) - \mathbf{y}\|_{E^p(\Omega,K)}$ for $\mathbf{y} \in E^p(\Omega, K)$ fixed, and this is a consequence of Lemma 7.3.

For the time optimal problem we apply Theorem 2.4 in [10] to the sequence of functions $\mathbf{f}_n(\mathbf{u}) = \mathbf{y}(t_n, \mathbf{u})$, $\mathbf{f}_n : U_{ad}(0, t_n : U) \to E^p(\Omega, K)$, where $\{t_n\}$ is a sequence with $t_n < \bar{t} =$ optimal time, $t_n \to \bar{t}$ (see [10] for details). If $\bar{\mathbf{u}}$ is a time optimal control Theorem 2.4 provides a sequence $\{\mathbf{u}^n\}$, $\mathbf{u}^n \in U_{ad}(0, t_n; U)$ (with $d_n(\mathbf{u}^n, \bar{\mathbf{u}}) \to 0$ as fast as we wish, d_n the distance (4.1)) such that, if $\{\mathbf{D}_n\}$ is a sequence of convex sets with $\mathbf{D}_n \subseteq \partial\mathbf{f}_n(\mathbf{u}^n)$ there exists a sequence $\{(\nu_n, \phi)\} \subseteq E^p(\Omega, K)^*$ such that

$$\|(\nu_n, \phi_n)\| = 1 , \quad \langle(\nu_n, \phi_n), \xi^n\rangle \ge -\delta_n \tag{7.19}$$

for every $\xi^n \in \mathbf{D}_n$, where $\delta_n \to 0$. If (ν, ϕ) is the weak limit of (a subsequence of) $\{\nu_n, \phi_n\}$ we have

$$\langle(v, \phi), \xi\rangle \ge 0 . \tag{7.20}$$

for every $\xi \in \liminf \mathbf{D}_n$. Elements of $\liminf \mathbf{D}_n$ are computed just as in §5. Using (7.20) for single spikes,

$$\int_K \xi(\bar{t}, x, s, \bar{u}, v)\nu(dx) + \int_{\Omega\setminus K} \xi(\bar{t}, x, s, \bar{u}, v)\phi(x)dx \ge 0 \tag{7.21}$$

for all s in a total set e and all $v \in U$. Via the divergence theorem we obtain the result below, where $\nu_K \in \Sigma(\bar{\Omega})$ is the measure defined by $\nu_K(dx) = \nu(dx)$ in K, $\nu_K(x) = \phi(x)dx$ in $\Omega\setminus K$.

THEOREM 7.5 Let $\bar{u}(t,x)$ be an optimal control for the time optimal problem. Then there exists $(\nu,\phi) \in E^q(\Omega,K)$ such that, if $z(t,x)$ is the solution of

$$z_t(t,x) = -\Delta z(t,x) - \partial_y f(t,y(t,x,u))z(t,x) \quad (t,x) \in [0,\bar{t}) \times \Omega \tag{7.22}$$

$$z(\bar{t},\cdot) = \nu_K \quad (x \in \bar{\Omega}) \tag{7.23}$$

$$z(t,x) = 0 \quad (t,x) \in [0,\bar{t}) \times \partial\Omega \tag{7.24}$$

then we have

$$\int_{\partial\Omega} \partial_\nu z(s,x)\bar{u}(s,x)d\sigma = \min_{v \in U} \int_{\partial\Omega} \partial_\nu z(s,x)v(x)d\sigma \tag{7.25}$$

8 FINAL COMMENTS

For control on the Dirichlet data, it would seem natural to use the space $BC(\Omega)$ as basic space for application of the nonlinear programming theory; in fact, (7.11) implies that $\|\mathbf{f}(\mathbf{u}) - \mathbf{y}\|_{BC(\Omega)}$ is lower semicontinuous in $\mathbf{u}$, the only thing required in [10]. Moreover, the variations (7.13) are computed in the norm of $BC(\Omega)$. The dual space of $BC(\Omega)$ is the space $\Sigma_{rba}(\Omega)$ of bounded, regular finitely additive measures on Ω (by Alexandroff's theorem, each $\nu \in \Sigma_{rba}(\Omega)$ must be countably additive on each compact set $K \subset \Omega$). We obtain in this way the exact analogue of Theorem 7.5, with $\nu \in \Sigma_{rba}(\Omega)$ instead of $\nu \in E_p(\Omega;K)^*$; if the target set is of the form (1.8), then it is guaranteed that $\nu \neq 0$ in $\Sigma_{rba}(\Omega)$. *However, this does not mean that the adjoint vector is nonzero*; in fact, there exist measures $\nu \in \Sigma_{rba}(\Omega)$, $\nu \neq 0$ such that

$$\int_{\bar{\Omega}} g(x)\nu(dx) = 0 \qquad (g(\cdot) \in C(\bar{\Omega})) \tag{8.1}$$

so that, expressing the function $z(t,x)$ in (7.22)-(7.23)-(7.24) by means of the Green function (of the backwards equation), we obtain $z(t,x) \equiv 0$. A nonzero $\nu \in \Sigma_{rba}(\Omega)$ satisfying (8.1) can be constructed as follows for $m = 2$; take $\Omega =$ open unit circle, $f_0 \in BC(\Omega)$ such that (in polar coordinates) $f(1,\phi) = \phi$. Then $\|f_0 - g\| \geq \pi$ for every $g(\cdot) \in C(\bar{\Omega})$, thus by the Hahn - Banach theorem there exists an element $\nu \in \Sigma_{rba}(\Omega)$ satisfying (8.1) and $\int_\Omega f_0(x)\nu(dx) \neq 0$.

At the moment, we don't know if nontriviality of the adjoint vector obtains in the $BC(\Omega)$ setting.

The present treatment of either the Dirichlet boundary condition or the boundary condition (1.3) does not apply to a point target $Y = \{\bar{y}\}$ (see the comments at the end of §6). There are indications that the point target case for the general control problem as well as for the time optimal problem could be treated in controllability subspaces such as those provided in [20] or in [4]. This has been done in the linear case (using separation theorems for convex sets) in [5]. However, the nonlinear case seems to be open.

REFERENCES

1. V. Barbu, Boundary control problems with nonlinear state equations, SIAM J. Control Optimization 20 (1982) 46-65 .

2. V. Barbu and N. Pavel, Optimal control problems for boundary control systems, to appear.
3. H. O. Fattorini, The time optimal control problem in Banach spaces, Appl. Math. Optimization 1 (1974/75) 163-188.
4. H. O. Fattorini, Boundary control of temperature distributions in a parallelepipedon, SIAM J. Control 13 (1975) 1-13.
5. H. O. Fattorini, The time-optimal problem for boundary control of the heat equation, **Calculus of Variations and Control Theory**, Academic Press, New York (1976) 305-320.
6. H. O. Fattorini, The maximum principle for nonlinear nonconvex systems in infinite dimensional spaces, **Distributed Parameter Systems**, Springer Lecture Notes in Control and Information Sciences 75 (1985) 162-178.
7. H. O. Fattorini, A unified theory of necessary conditions for nonlinear nonconvex control systems, Applied Math. Optim. 15 (1987) 141-185.
8. H. O. Fattorini, Optimal control of nonlinear systems: convergence of suboptimal controls, I, Lecture Notes in Pure and Applied Mathematics vol. 108, Marcel Dekker, New York (1987) 159-199.
9. H. O. Fattorini, Optimal control problems for distributed parameter systems governed by semilinear parabolic equations in L^1 and L^∞ spaces, **Optimal Control of Partial Differential Equations**, Springer Lecture Notes in Control and Information Sciences 149 (1991) 68-80.
10. H. O. Fattorini, Optimal control problems in Banach spaces, to appear in Applied Mathematics and Optimization.
11. H. O. Fattorini and H. Frankowska, Necessary conditions for infinite dimensional control problems, Springer Lecture Notes in Control and Information Sciences 111 (1990) 381-392.
12. H. O. Fattorini and H. Frankowska, Necessary conditions for infinite dimensional control problems, Mathematics of Control, Signals and Systems 4 (1990) 41-67.
13. H. O. Fattorini and T. Murphy, Optimal problems for nonlinear parabolic boundary control problems, to appear in SIAM J. Control & Optimization.
14. H. O. Fattorini and T. Murphy, Optimal problems for nonlinear parabolic boundary control problems: the Dirichlet boundary condition.
15. H. Frankowska, Some inverse mapping theorems, Ann Inst. Henri Poincaré 7 (1990) 183-234.
16. A. Friedman, **Partial Differential Equations of Parabolic Type**, Prentice-Hall, Englewood Cliffs, N.J.
17. K. Glashoff and N. Weck, Boundary control of parabolic differential equations in arbitrary dimension: supremum-norm problems, SIAM J. Control and Optimization 14 (1976) 662-681.
18. D. Henry, **Geometric Theory of Semilinear Parabolic Equations**, Springer, Berlin 1981.
19. T. Murphy, Doctoral Dissertation, University of California at Los Angeles, 1992.
20. D. L. Russell, A unified boundary controllability theory for hyperbolic and parabolic partial differential equations, Studies in Applied Math. 52 (1973) 189-211.
21. E. Sachs, A parabolic control problem with a condition of Stefan-Boltzmann type, Z. Angew. Math. Mech. 58 (1978) 443-449.

22. E. J. P. G. Schmidt, Boundary control for the heat equation with non-linear boundary condition, J. Differential Equations 78 (1989) 89-121.
23. D. Tiba, **Optimal Control of Nonsmooth Distributed Parameter Systems**, Springer Lecture Notes in Mathematics 1459 (1990).
24. F. Tröltzsch, A minimum principle and a generalized bang-bang principle for a distributed optimal control problem with constraints in control and state, Z. Angew. Math. Mech 59 (1979) 737-739.
25. F. Tröltzsch, **Optimality Conditions for Parabolic Control Problems and Applications**, Teubner-Texte zur Mathematik, Band 62, B. G. Teubner Verlagsgesselschaft, Lepzig 1984.
26. L. v. Wolfersdorf, Optimal control for processes governed by mildly nonlinear differential equations of parabolic type I, II, Z. Angew Math. Mech. 56 (1976) 531-538, 57 (1977) 11-17.

9 Numerical Approximations of Solutions to Riccati Equations Arising in Boundary Control Problems for the Wave Equation

Erik Hendrickson University of Virginia, Charlottesville, Virginia

Irena Lasiecka University of Virginia, Charlottesville, Virginia

1 INTRODUCTION

1.1 Description of the problem

This paper considers finite dimensional approximations of Algebraic Riccati equations with unbounded inputs such as those arising in hyperbolic boundary control problems. Approximations of Riccati Equations in the case of bounded control operators have been studied in [G-1], [B-K], [K-K], and references therein. Convergence results, and in some cases rates of convergence, for approximations of Riccati solutions arising in optimization problems with unbounded controls are given in [L-T-1], [L-5], and [L-T] in the case of analytic semigroups and in [L-2], [L-T] in the general hyperbolic case. All these convergence results depend upon the verification, via a suitable numerical scheme, of certain control theoretic properties such as a "uniform cost condition" and a "uniform trace condition" (in the case of unbounded control) to hold uniformly in the parameter of discretization. While these control theoretic properties hold for the original continuous problem, they may fail in a discrete setting. This is particularly true in the case of hyperbolic dynamics and Finite Element Methods (FEM) as observed in numerical experiments in [B-I-W] and [H-1]. As a result, the developed numerical procedures for computations of Riccati solutions do not produce, in such cases, convergent algorithms. To cope with the difficulty, we propose an algorithm which combines a regularization and an approximation (e.g. FEM) procedure. We shall prove that this algorithm produces a convergent approximation to the solution of the Riccati Equation (see Theorem 1.2). Our theoretical results are illustrated by an example of the wave equation with boundary control (see Theorem 3.1).

1.2 Abstract model

Let $\mathcal{A}$ be a positive, self-adjoint, densely-defined operator on a Hilbert space $\mathcal{H}$ with a dense domain $\mathcal{D}(\mathcal{A})$. Define a control operator $\mathcal{B} \in \mathcal{L}(U;[\mathcal{D}(\mathcal{A}^{1/2})]')$, where U represents another Hilbert space and $[\mathcal{D}(\mathcal{A}^{1/2})]'$ is the dual space to $\mathcal{D}(\mathcal{A}^{1/2})$ with respect to the $\mathcal{H}$-inner product. We also introduce a damping operator denoted by $\mathcal{D}$, which is assumed nonnegative, self-adjoint on $\mathcal{H}$ and $\mathcal{D} \in \mathcal{L}(\mathcal{D}(\mathcal{A}^{1/2});\mathcal{H})$. The boundary control model is given by the following equation,

$$w_{tt} + \mathcal{A}w + \mathcal{D}w_t = \mathcal{B}u \quad on \quad [\mathcal{D}(\mathcal{A})]', \tag{1.1}$$

with the following initial data,

$$w(t=0) = w_0 \in \mathcal{D}(\mathcal{A}^{1/2}); \quad w_t(t=0) = w_1 \in \mathcal{H}.$$

and the control function $u \in L_2(0,\infty;U)$.

As we shall see later, this abstract model encompasses many physical examples described by plate and wave equations with boundary controls (see example of a wave equation with boundary control in Sec. 3). Equation (1.1) can be written as a first order equation on a Hilbert space $H \equiv \mathcal{D}(\mathcal{A}^{1/2}) \times \mathcal{H}$. Indeed, let

$$A = \begin{bmatrix} 0 & I \\ -\mathcal{A} & -\mathcal{D} \end{bmatrix} : H \to H; \quad \mathcal{D}(A) = \mathcal{D}(\mathcal{A}) \times \mathcal{D}(\mathcal{A}^{1/2}); \tag{1.2}$$

$$Bg = \begin{bmatrix} 0 \\ \mathcal{B}g \end{bmatrix}. \tag{1.3}$$

With the above notation (1.1) is equivalent to

$$y_t = Ay + Bu \quad on \quad [\mathcal{D}(A^*)]'; \quad y(0) = y_0 \in H, \; where \; y = (w, w_t)^T. \tag{1.4}$$

It is well known that A generates a C_0-semigroup of contractions on H and the control operator B satisfies,

$$A^{-1/2}B \in \mathcal{L}(U;H). \tag{1.5}$$

Remark 1.1 Notice that the control operator B is unbounded (see (1.5) and the damping operator $\mathcal{D}$ is not assumed to be strictly positive (in particular we can take $\mathcal{D} = 0$). These features are representative of elastic systems with very low (if any) inherent internal damping and of the *boundary* nature of the controls employed. On the other hand, these properties, as we shall see later, contribute to major mathematical difficulties and the novelty of the problem under study.

With (1.4) we associate the cost functional

$$J(u,y) = \int_0^\infty [\,\| Ry(t) \|_Z^2 + \| u(t) \|_U^2]\, dt, \tag{1.6}$$

where $R \in \mathcal{L}(H;Z)$ and Z is a Hilbert space. Consider the following control problem,

$$\min_{u \in L_2(0,\infty;U)} J(u,y(u)), \quad \textit{where } y(u) \textit{ satisfies } (1.4). \tag{P}$$

It is well known that the necessary and sufficient condition for the unique solvability of (P) is the following Finite Cost Condition,

$$\forall\ y_0 \in H, \textit{ there exists } u \in L_2(0,\infty;U) \textit{ such that } J(u,y(u)) < \infty. \tag{F.C.C.}$$

We shall formulate another condition which is referred to as Detectability Condition (D.C.),

$$\textit{There exists an operator } K \in \mathcal{L}(Z;H) \textit{ such that} \tag{D.C.}$$

$$A + KR \textit{ is exponentially stable on } H.$$

Assuming that the (F.C.C.) is satisfied, we are interested in feedback control associated with the problem (P). It is also well known that this feedback control can be given via a Riccati operator, provided that there exists a solution to an appropriate Riccati equation. In the case when the control operator B is bounded, the standard existence result can be found for instance in [B-1]. In our unbounded control case, the following recent result of [F-L-T] is pertinent.

Theorem 1.1 [F-L-T]:
In addition to (F.C.C.) assume that for each finite T there exists a constant c_T such that

$$\forall\ y \in \mathcal{D}(A^*), \quad \int_0^T \| B^* e^{A^* t} y \|_U^2 \, dt \le c_T \| y \|_H^2 \tag{H.1}$$

where $(Bu,v)_H = (u,B^*v)_U, \; u \in U, \; v \in \mathcal{D}(B^*) \supset \mathcal{D}(A^*)$.

then there exists $P \in \mathcal{L}(H)$, $P = P^* \ge 0$ to the following Algebraic Riccati Equation

$$(A^*Px,y)_H + (PAx,y)_H + (R^*Rx,y)_H = (B^*Px,B^*Py)_U, \text{ for } \textit{all } x,y \in \mathcal{D}(A) \tag{A.R.E.}$$

and

$$B^*P \in \mathcal{L}(\mathcal{D}(A);U) \cap \mathcal{L}(\mathcal{D}(A_P);U) \textit{ where } A_P \equiv A - BB^*P \textit{ generates a } C_o\textit{semigroup on } H. \tag{1.7}$$

The optimal control u^0 is given in a feedback from as

$$u^0(t) = -B^*Py^0(t) \quad a.e. \textit{ in } t. \tag{1.8}$$

In addition, if the condition (D.C.) holds then the solution P is unique within the class of self-adjoint, positive operators and such that regularity property (1.7) holds. Moreover,

$$\| e^{A_P t} \|_{\mathcal{L}(H)} \le Ce^{-\omega t}, \quad \text{for } \textit{some } \omega > 0. \tag{1.9}$$

Remark 1.2 Since the control operator B is unbounded, the gain operator B^*P may not be defined at all. The regularity property (1.7) from Theorem 1.1 enables us to define the gain operator B^*P as an unbounded but densely defined operator on H. This regularity property, which gives meaning to the nonlinear term in the (A.R.E.), plays a critical role in the derivation of this equation.

In what follows we shall assume standing hypotheses (H.1), (F.C.C.), and (D.C.).

The main goal of this paper is to provide an implementable approximation theory for the solutions to Algebraic Riccati Equations (ARE). Approximation framework for Riccati Equations arising in elastic systems has been provided in the literature only in the special cases when (i) the control operator is bounded ([G-1], [G-2]) or (ii) the structural damping is present and the corresponding semigroup is analytic (see [L-T-1]). Thus the main contribution and the novelty of this paper is that (i) we consider unbounded control operators as they arise in boundary control problems (see Remark 1.1) and (ii) no inherent or structural damping is assumed (i.e the governing semigroup is not analytic).

1.3 Approximation framework

We introduce the following approximating subspaces and operators.

Let h denote the discretization parameter which is assumed to tend to zero, $V_h \subset \mathcal{D}(\mathcal{A}^{1/2})$ be the approximating subspace. On V_h, we shall introduce a "discrete" norm, $\| \; \|_{\hbar}$, and the inner product $(.,.)_{\hbar}$ such that

$$\| y_h \|_{\hbar} = \| y_h \|_{\mathcal{H}}, \quad y_h \in V_h \tag{1.10}$$

$$\| y \|_{\hbar} \leq \| y \|_{\mathcal{H}}, \quad y \in \mathcal{H} \tag{1.11}$$

Let π_h represent the orthogonal projection of $\mathcal{H}$ onto V_h with the property:

$$\exists \; y_h = \pi_h y \in V_h \ni \| \pi_h y - y \|_{\mathcal{H}} \to 0, \; y \in \mathcal{H}$$

Similarly,

$$\| \pi_h y - y \|_{\mathcal{D}(\mathcal{A}^{1/2})} \to 0 \, , \; y \in \mathcal{D}(\mathcal{A}^{1/2}) \, .$$

Assumptions on $\mathcal{A}$, $\mathcal{B}$, and $\mathcal{D}$

Let $\mathcal{A}_h$: $V_h \to V_h$ be a Galerkin approximation of $\mathcal{A}$ such that

$$(i) \quad (\mathcal{A}_h x_h, \xi_h)_{\hbar} = (\mathcal{A} x_h, \xi_h)_{\mathcal{H}}; \quad x_h, \, \xi_h \in V_h. \tag{1.12}$$

$$(ii) \quad \| (\mathcal{A}^{-1} - \mathcal{A}_h^{-1} \pi_h) \, x \|_{\mathcal{D}(\mathcal{A}^{1/2})} \to 0; \quad x \in \mathcal{H}$$

Let $\mathcal{D}_h$: $V_h \to V_h$ be a Galerkin approximation of $\mathcal{D}$, such that

$$(\mathcal{D}_h x_h, \xi_h)_{\hbar} = (\mathcal{D} x_h, \xi_h)_{\mathcal{H}}; \quad x_h, \, \xi_h \in V_h. \tag{1.13}$$

Let $\mathcal{B}_h$: $U \to V_h$ be such that

(i) $\quad \| (\mathcal{B}_h^* \pi_h - \mathcal{B}^*) x \|_U \to 0; \quad x \in \mathcal{D}(\mathcal{A}^{1/2})$. (1.14)

Here $(\mathcal{B}^* v, g)_U = (v, \mathcal{B}g)_{\mathcal{H}}$ and $(\mathcal{B}_h^* v_h, g)_U = (v_h, \mathcal{B}_h g)_{\hbar}$ for $g \in U$, $v \in \mathcal{D}(\mathcal{B}^*) \supset \mathcal{D}(\mathcal{A}^{1/2})$.

(ii) $\quad \| \mathcal{A}^{-1} (\mathcal{B}_h - \mathcal{B}) u \|_{\mathcal{D}(\mathcal{A}^{1/2})} \to 0; \quad u \in U$.

(iii) $\quad \| (\mathcal{A}_h^{-1} - \mathcal{A}^{-1}) \mathcal{B}_h u \|_{\mathcal{D}(\mathcal{A}^{1/2})} \to 0; \quad u \in U$.

Define A_h: $H_h \to H_h$ and B_h: $U \to H_h$, where $H_h \equiv \hat{V}_h \times V_h$ and $\hat{V}_h$ consists of elements in V_h supplied with a norm in $\mathcal{D}(\mathcal{A}^{1/2})$, i.e.,

$$\| y_h \|_{\hat{V}_h}^2 = \| \mathcal{A}^{1/2} y_h \|_{\mathcal{H}}^2 = (\mathcal{A} y_h, y_h)_{\mathcal{H}} = (by\,(1.12)) = (\mathcal{A}_h y_h, y_h)_{\hbar} = \| \mathcal{A}_h^{1/2} y_h \|_{\hbar}^2$$

and

$$A_h \equiv \begin{bmatrix} 0 & \pi_h \\ -\mathcal{A}_h & -\mathcal{D}_h \end{bmatrix},$$

$$B_h g \equiv \begin{bmatrix} 0 \\ \mathcal{B}_h g \end{bmatrix}.$$

It can be easily verified that $B_h^* v^h = \mathcal{B}_h^* v_2^h$ where $v^h = (v_1^h, v_2^h) \subset \hat{V}_h \times V_h \equiv H_h$ and that the above approximation properties postulated in (1.12)-(1.14) are the usual consistency and stability properties satisfied by most of the approximating schemes like Finite Element (FEM) and Finite Difference (FD) methods.

We first consider the following approximation of the Algebraic Riccati Equations: Find $P_h \in \mathcal{L}(H_h)$ such that for all $x_h, y_h \in V_h \times V_h \equiv H_h$,

$$(A_h^* P_h x_h, y_h)_{H_h} + (P_h A_h x_h, y_h)_{H_h} + (R_h^* R_h x_h, y_h)_{H_h} - (B_h^* P_h x_h, B_h^* P_h y_h)_U = 0. \qquad \text{(ARE-h)}$$

In [L-2] it was proved that there exists a unique solution $P_h \in \mathcal{L}(H_h)$ to the (ARE-h) which converges strongly in H to P, the solution to the (ARE), <u>provided</u> that the following <u>stability</u> conditions hold.

<u>Approximate Trace Regularity:</u>

There exists $T > 0$ and the constant $c_T > 0$ (independent of h) such that

$$\int_0^T \| B_h^* e^{A_h^* t} \Pi_h x \|_U^2 dt \le c_T \| x \|_H^2, \qquad \text{(H.1-h)}$$

where,

$$\Pi_h \equiv \begin{bmatrix} \pi_h & 0 \\ 0 & \pi_h \end{bmatrix}$$

and,

Discrete Finite Cost Condition:
For each $y_0 \in H$ there exists $u \in L_2(0,\infty;U)$ and a constant α (independent of h) such that

$$J(u, y_h(u)) \leq \alpha \| y_0 \|_H^2 , \qquad \text{(F.C.C.-h)}$$

where $y_h(u)$ satisfies

$$\dot{y}_h(t) = A_h y_h(t) + B_h u(t) , \; y_h(0) = \Pi_h y(0).$$

Remark 1.3 Notice that the (F.C.C-h) is a necessary condition for the convergence of the Riccati operator P_h (a stronger version of this condition was also assumed in [G-1] where the case of B-bounded was treated).

Though stability conditions (H.1-h) and, in particular (F.C.C-h), are natural from the point of view of approximations ((F.C.C-h) has been assumed in all works (see [G-1], [G-2] and references therein) dealing with approximations to Riccati Equations), their requirements are not satisfied by several numerical schemes of practical interest, such as the Finite Element methods. In fact, it has recently been verified that for two-dimensional models of wave or plate equations these conditions are violated (see [B-I-W]). In order to cope with the difficultly, we propose an approach which is based on a combination of a certain regularization procedure applied to the original problem and an approximation algorithm applied to the regularized system. The advantage of this algorithm is that it does not require conditions (H.1-h) and (F.C.C-h). Precise statements of these results are given in the next section.

1.4 Statement of the Main Results

Let $\varepsilon = (\varepsilon_1, \varepsilon_2)$ denote the parameter of regularization, where $|\varepsilon_i| \leq 1$ and ε_i are supposed to tend to zero. We consider the following regularized operator

$$A_{h\varepsilon} \equiv A_h - \varepsilon_1 \begin{bmatrix} 0 & 0 \\ 0 & \pi_h \end{bmatrix} - \varepsilon_2 B_h B_h^* \quad on \; [\mathcal{D}(A)]' . \qquad (1.15)$$

With (1.15) we associate a discrete Algebraic Riccati Equation, for $x_h, y_h \in H_h$,

$$(A_{h\varepsilon}^* P_{h\varepsilon} x_h, y_h)_{H_h} + (P_{h\varepsilon} A_{h\varepsilon} x_h, y_h)_{H_h} + (R_h^* R_h x_h, y_h)_{H_h} = (B_h^* P_{h\varepsilon} x_h, B_h^* P_{h\varepsilon} y_h)_U . \qquad \text{(A.R.E.-}\varepsilon\text{,h)}$$

Our main abstract result is the following theorem.

Theorem 1.2

Assume the continuous hypotheses (H.1) and (F.C.C). Moreover, assume the approximation properties (1.12)-(1.14). Then, there exists a unique solution $P_{h\varepsilon} \in \mathcal{L}(H_h)$ of the discrete Algebraic Riccati Equation (ARE-ε,h) and

$$\lim_{\varepsilon_1 \to 0} \lim_{\varepsilon_2 \to 0} \lim_{h \to 0} \| P_{h\varepsilon} \Pi_h x - P x \|_H = 0, \quad x \in H, \qquad (1.16)$$

$$\lim_{\varepsilon_1\to 0}\lim_{\varepsilon_2\to 0}\lim_{h\to 0} \| u^0_{h\varepsilon} - u^0 \|_{L_2(0,\infty;U)} = 0, \qquad (1.17)$$

$$\lim_{\varepsilon_1\to 0}\lim_{\varepsilon_2\to 0}\lim_{h\to 0} \| y^0_{h\varepsilon} - y^0 \|_{L_2(0,\infty;H)} = 0, \qquad (1.18)$$

$$\lim_{\varepsilon_1\to 0}\lim_{\varepsilon_2\to 0}\lim_{h\to 0} J(u^0_{h\varepsilon}, y^0_{h\varepsilon}) = J(u^0, y^0),$$

where $y^0_{h\varepsilon}$ and $u^0_{h\varepsilon}$ satisfies

$$\dot{y}^0_{h\varepsilon}(t) = (A_{h\varepsilon} - B_h B^*_h P_{h\varepsilon})\, y^0_{h\varepsilon}(t);\ \ y_{h\varepsilon}(0) = \Pi_h y_0\,,$$

$$u^0_{h\varepsilon}(t) = -\,B^*_h P_{h\varepsilon} y^0_{h\varepsilon}(t), \qquad (1.19)$$

and there exist $\omega > 0$ and $C > 0$ independent of h such that for all $\varepsilon > 0$ there exists $h_0 > 0$,

$$\| y_{h\varepsilon}(t) \|_{\mathcal{L}(H)} \le Ce^{-\omega t},\ h < h_0. \quad ■ \qquad (1.20)$$

In the special case when the control operator B is bounded, there is no need for the second regularization with respect to the parameter ε_2.

Corollary Assume that $B \in \mathcal{L}(U;H)$ and that the (F.C.C.) together with the approximation properties (1.12) are satisfied. Then all the statements of the Main Theorem hold true with $\varepsilon_2 = 0$ and $\mathcal{B}_h = \pi_h \mathcal{B}$. ■

Remark 1.4 The main result in (1.16) states the convergence properties of the Riccati operators. The discrete control feedback given by (1.19) provides a near optimal performance of the system (see (1.18)). Moreover, the corresponding discrete control system is uniformly exponentially stable (see (1.18), where the constants C, $\omega > 0$ are independent on the parameter of discretization). The second section is devoted to the proof of Theorem 1.2. The last section provides an example of the theory applied to the wave equation with boundary control.

2 PROOF OF THEOREM 1.2

2.1 Regularized problem: convergence in ε.

We introduce the following *regularized* Riccati Equation,

$$(A^*_\varepsilon P_\varepsilon x, y)_H + (P_\varepsilon A_\varepsilon x, y)_H + (R^* R x, y)_H = (B^* P_\varepsilon x, B^* P_\varepsilon y)_H \qquad \text{(A.R.E.-}\varepsilon)$$

for all $x,\ y \in \mathcal{D}(A_\varepsilon)$, where $A_\varepsilon : H \to H$ is given by

$$A_\varepsilon = A \;-\; \varepsilon_1 \begin{bmatrix} 0 & 0 \\ 0 & I \end{bmatrix} \;-\; \varepsilon_2 BB^*. \qquad (2.1)$$

Lemma 2.1

(i) $A_\varepsilon : H \to H$ generates a C_o semigroup of contractions.

(ii) $\int_0^\infty \| B^* e^{A_\varepsilon^* t} x \|_U^2 \, dt \le \frac{C}{\varepsilon_2} \|x\|_H^2$.

(iii) $\| e^{A_\varepsilon t} \|_{\mathcal{L}(H)} \le C e^{-\omega t}$, where ω depends only on ε_1.

Proof of Lemma 2.1

The generation statement of part (i) follows from the Lummer Phillips Theorem (see proof of Theorem 2.2 in [L-3] and also [H-1]).

Proof of parts (ii) and (iii)

Step 1: Let $A_{\varepsilon_1} \equiv A - \varepsilon_1 \begin{bmatrix} 0 & 0 \\ 0 & I \end{bmatrix}$. A standard perturbation result in [P-1] implies that A_{ε_1} generates a C_o semigroup, with $\mathcal{D}(A_{\varepsilon_1}) = \mathcal{D}(A)$. We shall show

$$\| e^{A_{\varepsilon_1} t} \|_{\mathcal{L}(H)} \le M e^{-\omega t} \quad \textit{where } \omega \textit{ depends on } \varepsilon_1 . \tag{2.2}$$

To assert (2.2) we shall use a multipliers method. Let $z(t) = e^{A_{\varepsilon_1} t} x$, $x \in H$. For $x \in \mathcal{D}(A_{\varepsilon_1}) = \mathcal{D}(A)$, we have $z_t \in C[0,T;H] = C[0,T;\mathcal{D}(\mathcal{A}^{1/2}) \times \mathcal{H}]$. Hence with $z \equiv (w, w_t)$, $w \in C[0,T;\mathcal{D}(\mathcal{A})]$, $w_t \in C[0,T;\mathcal{D}(\mathcal{A}^{1/2})]$, and $w_{tt} \in C[0,T;\mathcal{H}]$, we write

$$w_{tt} + \mathcal{A}w + \mathcal{D}w_t + \varepsilon_1 w_t = 0 \quad in\ \mathcal{H}, \quad w(0) = x_0 \in \mathcal{D}(\mathcal{A}), \quad w_t(0) = x_1 \in \mathcal{D}(\mathcal{A}^{1/2}) . \tag{2.3}$$

Denote $E(t) \equiv \|w_t(t)\|_{\mathcal{H}}^2 + \|\mathcal{A}^{1/2} w(t)\|_{\mathcal{H}}^2$. Multiplying equation (2.2) by w_t and integrating by parts (which is justified in view of the regularity of w stated above) yields

$$2\varepsilon_1 \int_0^T \|w_t(t)\|_{\mathcal{H}}^2 \, dt + 2\int_0^T (\mathcal{D}w_t, w_t)_{\mathcal{H}} + E(T) = E(0). \tag{2.4}$$

On the other hand, multiplying (2.2) by w and integrating by parts gives

$$\int_0^T \|w_t(t)\|_{\mathcal{H}}^2 \, dt - \int_0^T \|\mathcal{A}^{1/2} w(t)\|_{\mathcal{H}}^2 \, dt \le C\, E(0), \tag{2.5}$$

where we have used

$$2\int_0^T (\mathcal{D}w_t, w)_{\mathcal{H}} \, dt = (\mathcal{D}w(T), w(T))_{\mathcal{H}} - (\mathcal{D}w(0), w(0))_{\mathcal{H}} \le C\, [\ \|\mathcal{A}^{1/2} w(T)\|_{\mathcal{H}}^2 + \|\mathcal{A}^{1/2} w(0)\|_{\mathcal{H}}^2].$$

Multiplying equation (2.5) by ε_1 and combining with (2.4) gives

$$\varepsilon_1 \int_0^\infty (\|w_t(t)\|_{\mathcal{H}}^2 + \|\mathcal{A}^{1/2} w(t)\|_{\mathcal{H}}^2) \, dt \le C\, E(0).$$

Datko's result (see [P-1]) yields the desired conclusion in (2.2).

Step 2: proof of part (ii)

Notice first that

$$\mathcal{D}(A_\varepsilon^*) = \mathcal{D}(A_\varepsilon) \subset \mathcal{D}(B^*) \tag{2.6}$$

and

$$(Au,v)_H \le C \, \| u \|_{\mathcal{D}(A_\varepsilon)} \, \| v \|_{\mathcal{D}(A_\varepsilon)} \quad \text{for } u,v \in \mathcal{D}(A_\varepsilon). \tag{2.7}$$

Indeed, $\mathcal{D}(A_\varepsilon) = \mathcal{D}(A_\varepsilon^*)$ consists of elements $z \equiv (z_1, z_2)$ such that

$$z_2 \in \mathcal{D}(\mathcal{A}^{1/2}). \tag{2.8}$$

On the other hand, $B^* z = \mathcal{B}^* z_2 \in U$, where we have used (2.8) and the regularity of $\mathcal{B}$. As for (2.7), we have

$$(Au,v)_H = (u_2, v_1)_{\mathcal{D}(\mathcal{A}^{1/2})} - (\mathcal{A}u_1 + \mathcal{D}u_2, v_2)_{\mathcal{H}}$$

$$= (\mathcal{A}^{1/2} u_2, \mathcal{A}^{1/2} v_1)_{\mathcal{H}} - (\mathcal{A}^{1/2} u_1 + \mathcal{A}^{-1/2} \mathcal{D} u_2, \mathcal{A}^{1/2} v_2)_{\mathcal{H}}. \tag{2.9}$$

Since $\mathcal{A}^{1/2} u_2$, $\mathcal{A}^{1/2} v_1$, $\mathcal{A}^{1/2} u_1$, $\mathcal{A}^{1/2} v_2$ are in $\mathcal{H}$, (2.7) follows from (2.9). Next, denote $v_\varepsilon(t) \equiv e^{A_\varepsilon^* t} x$, where $x \in \mathcal{D}(A_\varepsilon)$. Hence,

$$\dot v_\varepsilon(t) = (A_{\varepsilon_1}^* - \varepsilon_2 B B^*) v_\varepsilon(t) \quad \textit{in } H. \tag{2.10}$$

Multiplying equation (2.10) by v_ε, integrating by parts, and noticing (after using (2.7)) that $(A_{\varepsilon_1}^* x, x)_H \le 0$ for $x \in \mathcal{D}(A_\varepsilon)$ we obtain

$$\| v_\varepsilon(T) \|_H^2 + \varepsilon_2 \int_0^T \| B^* v_\varepsilon(t) \|_U^2 \, dt \le \| x \|_H^2 \, ; \; x \in \mathcal{D}(A_\varepsilon), \tag{2.11}$$

where we have used the regularity result (2.6). The inequality (2.11) followed by the usual density argument yields part (ii).

Step 3: proof of part (iii)

Let $v_\varepsilon(t) = e^{A_\varepsilon^* t} x$. By the variation of parameters formula

$$v_\varepsilon(t) = e^{A_{\varepsilon_1}^* t} x - \varepsilon_2 \int_0^t e^{A_{\varepsilon_1}^* (t-s)} B B^* v_\varepsilon(s) \, ds = e^{A_{\varepsilon_1}^* t} x - \varepsilon_2 L(B^* v_\varepsilon)(t).$$

where $(Lu)(t) = \int_0^t e^{A_{\varepsilon_1}^* (t-s)} Bu(s) ds$. By virtue of hypothesis (H.1) and the exponential stability of $A_{\varepsilon_1}^*$ (see (2.2)) we have (see [L-4])

$$L \in \mathcal{L}(L_2(0,\infty;U); L_2(0,\infty;H)). \tag{2.12}$$

Hence,

$$\int_0^\infty \|v_\varepsilon(t)\|_H^2\, dt \le C_{\varepsilon_1}\ [\|x\|_H^2 + \varepsilon_2^2 \|L\|^2_{(L_2(0,\infty;U);\, L_2(0,\infty;H))} \int_0^\infty \|B^* v_\varepsilon(t)\|_U^2\, dt]. \quad \text{(2.13)}$$

From (2.13) and from part (ii), we have

$\int_0^\infty \|v_\varepsilon(t)\|_H^2\, dt \le C_{\varepsilon_1} \|x\|_H^2$ *which yields part (iii) by Datko's result.* ■

From Lemma 2.1 it follows that operator A_ε complies with all the assumptions of Theorem 1.1 (applied with A replaced by A_ε). Hence, by the result of Theorem 1.1,

Corollary 2.1
For each value of $\varepsilon > 0$, there exists a unique solution $P_\varepsilon \in \mathcal{L}(H)$ to the Algebraic Riccati Equation (ARE-ε). Moreover,

$B^* P_\varepsilon \in \mathcal{L}(\mathcal{D}(A_\varepsilon);U) \cap \mathcal{L}(\mathcal{D}(A_{P,\varepsilon};U)$

where $A_{P,\varepsilon} \equiv A_\varepsilon - BB^* P_\varepsilon$ generates an exponentially stable semigroup.

The main aim here is to establish convergence of P_ε to P. This is a technical result and for a rather lengthy proof we refer to [H-1].

Theorem 2.1 [H-1]
Under the assumptions of Theorem 1.1, we have

$$\lim_{\varepsilon_1 \to 0} \lim_{\varepsilon_2 \to 0} P_\varepsilon = P \quad \textit{strongly in } H. \tag{2.14}$$

Moreover, for each initial data $y_0 \in H$ we have

$$\lim_{\varepsilon_1 \to 0} \lim_{\varepsilon_2 \to 0} u_\varepsilon^0 = u^0 \quad \textit{in } L_2(0,\infty;U),$$

$$\lim_{\varepsilon_1 \to 0} \lim_{\varepsilon_2 \to 0} y_\varepsilon^0 = y^0 \quad \textit{in } L_2(0,\infty;H),$$

$$\lim_{\varepsilon_1 \to 0} \lim_{\varepsilon_2 \to 0} J_\varepsilon(u_\varepsilon^0, y_\varepsilon^0) = J(u^0, y^0),$$

where $u_\varepsilon^0 = -B^* P_\varepsilon y_\varepsilon^0(t)$, and $y_\varepsilon^0(t)$ is given by

$$y_\varepsilon^0(t) = e^{A_{P,\varepsilon} t} y_0. \tag{2.15}$$

2.2 Approximated problem: Convergence in h

We shall next examine the question of convergence of $P_{h\varepsilon}$ with respect to h, the parameter of discretization. The main result in this direction is the following.

Theorem 2.2
Assume the continuous hypotheses (H.1), (F.C.C), and (D.C.). Moreover, assume the approximation properties (1.12)-(1.14) are satisfied. Then for each value of ε and h there exists a unique solution $P_{h\varepsilon}$ of the Riccati Equation (ARE-εh) and moreover,

$$\lim_{h\to 0} \| P_{h\varepsilon}\Pi_h x - P_\varepsilon x \|_H = 0; \quad x \in H. \tag{2.16}$$

$$\lim_{h\to 0} \| u^0_{h\varepsilon} - u^0_\varepsilon \|_{L_2(0,\infty;U)} = 0; \quad y_0 \in H. \tag{2.17}$$

$$\lim_{h\to 0} \| y^0_{h\varepsilon} - y^0_\varepsilon \|_{L_2(0,\infty;H)} + \| y^0_{h\varepsilon} - y^0_\varepsilon \|_{C(0,\infty;H)} = 0; \quad y_0 \in H. \tag{2.18}$$

$$J(u^0_{h\varepsilon}, y^0_{h\varepsilon}) \to J(u^0_\varepsilon, y^0_\varepsilon). \tag{2.19}$$

$\| y^0_{h\varepsilon}(t) \|_{\mathcal{L}(H)} \le Ce^{-\omega t}$, where ω, C depend only on $\varepsilon > 0$. Recall P_ε is the unique solution of the regularized Riccati Equation (ARE-ε).

Proof of Theorem 2.2

The proof of Theorem 2.2 is based on the application of the convergence Theorems 1 and 2 in [L-2]. Indeed, the statements of Theorem 2.2 follow from Theorems 1 and 2 in [L-2] provided the following conditions hold (see p. 320-321 in [L-2]).

Approximate Trace Regularity:
There exists a constant $C_{T,\varepsilon}$ such that

$$\int_0^T \| B^*_h e^{A^*_{h\varepsilon} t} \Pi_h x \|^2_U \, dt \le C_{T,\varepsilon} \| x \|^2_H. \tag{2.20}$$

Discrete Finite Cost Condition:
There exist constants $C_\varepsilon, \omega_\varepsilon > 0$ (independent of h) such that

$$\| e^{A_{h\varepsilon} t} x_h \|_{H_h} \le C_\varepsilon e^{-\omega_\varepsilon t} \| x_h \|_{H_h}; \quad x_h \in H_h \equiv \hat{V}_h \times V_h. \tag{2.21}$$

As $h \to 0$, the following convergence hold:

$$\| [(A_{h\varepsilon})^{-1}\Pi_h - (A)^{-1}_\varepsilon] x \|_H \to 0, \quad x \in H. \tag{2.22}$$

$$\| (A)^{-1}_\varepsilon (B_h - B) u \|_H \to 0, \quad u \in U; \tag{2.23-i}$$

$$\| [(A_{h\varepsilon})^{-1} - (A)^{-1}_\varepsilon] B_h u \|_H \to 0, \quad u \in U; \tag{2.23-ii}$$

$$\| (B^*_h \Pi_h - B^*)(A^*_\varepsilon)^{-1} x \|_U \to 0, \quad x \in H; \tag{2.24-i}$$

$$\| B^*_h [(A^*_{h\varepsilon})^{-1}\Pi_h - \Pi_h (A^*_\varepsilon)^{-1}] x \|_U \to 0, \quad x \in H. \tag{2.24-ii}$$

Remark 2.1 Notice that the condition (F.C.C-h) in [L-2] is a direct consequence of (2.21).

We shall show that conditions (2.21)-(2.24) are satisfied.

Conditions (2.20) and (2.21)

Let $z_h(t) = e^{A^*_{h\varepsilon}t}\Pi_h x$. Hence,

$$\dot{z}_{h2} = -z_{h1} \tag{2.25}$$

$$\ddot{z}_{h1} + \mathcal{A}_h z_{h1} + \mathcal{D}_h \dot{z}_{h1} + \varepsilon_1 \dot{z}_{h1} + \varepsilon_2 \mathcal{B}_h \mathcal{B}_h^* \dot{z}_{h1} = 0.$$

Taking the inner product (in V_h) with $\dot{z}_{h1}$, recalling (1.12), and integrating from 0 to T yields:

$$\frac{1}{2}E_h(T) + \int_0^T (\mathcal{D}_h \dot{z}_{h1}(t), \dot{z}_{h1}(t))_{\mathcal{H}}\, dt + \varepsilon_1 \int_0^T \| \dot{z}_{h1}(t) \|_{\mathcal{H}}^2\, dt + \varepsilon_2 \int_0^T \| \mathcal{B}_h^* \dot{z}_{h1}(t) \|_U^2 dt = \frac{1}{2}E_h(0), \tag{2.26}$$

where by (1.12i) and (1.10),

$$E_h(t) = \| \dot{z}_{h1}(t) \|_{\mathcal{H}}^2 + \| \mathcal{A}^{1/2} z_{h1}(t) \|_{\mathcal{H}}^2 = \| z_h(t) \|_{H_h}^2 = \| z_h(t) \|_H^2 .$$

Next, taking the inner product (in V_h) with $z_{h1}(t)$ and integrating by parts the result produces

$$\int_0^T \| \dot{z}_{h1}(t) \|_{\mathcal{H}}^2 - \int_0^T \| \mathcal{A}^{1/2} z_{h1}(t) \|_{\mathcal{H}}^2 - \frac{1}{2}\varepsilon_2 \| \mathcal{B}_h^* z_{h1}(T) \|_U^2 \le C\, E_h(0), \tag{2.27}$$

where we have used the regularity of $\mathcal{B}_h^*$ from (1.14), together with (2.26). Combining (2.26) with (2.27) yields, in particular,

$$\frac{\varepsilon_1}{2} \int_0^T \| \dot{z}_{h1}(t) \|_{\mathcal{H}}^2 + \frac{\varepsilon_1}{2} \int_0^T \| \mathcal{A}^{1/2} z_{h1}(t) \|_{\mathcal{H}}^2 + \varepsilon_2 \int_0^T \| \mathcal{B}_h^* \dot{z}_{h1}(t) \|_U^2 dt \le C\, E_h(0). \tag{2.28}$$

Thus,

$$\int_0^\infty \| B_h^* z_h(t) \|_U^2 dt = \int_0^\infty \| \mathcal{B}_h^* \dot{z}_{h1}(t) \|_U^2 dt \le \frac{C}{\varepsilon_2}\, E_h(0) = \frac{C}{\varepsilon_2} \| x \|_H^2 , \tag{2.29}$$

which proves (2.20), and

$$\varepsilon_1 \int_0^T E_h(t)\, dt \le C\, E_h(0), \tag{2.30}$$

Standard semigroup theory yields (2.21).

Proof of (2.22)

Straightforward computations yield

$$A^{-1} = A_{\varepsilon_1}^{-1} = \begin{bmatrix} -\mathcal{A}^{-1}(\mathcal{D} + \varepsilon_1 I) & -\mathcal{A}^{-1} \\ I & 0 \end{bmatrix}; \quad (A^*)^{-1} = \begin{bmatrix} -\mathcal{A}^{-1}(\mathcal{D} + \varepsilon_1 I) & \mathcal{A}^{-1} \\ I & 0 \end{bmatrix}; \tag{2.31}$$

$$A_\varepsilon^{-1} = \begin{bmatrix} -\mathcal{A}^{-1}(\mathcal{D}+\varepsilon_1 I+\varepsilon_2\mathcal{B}\mathcal{B}^*) & -\mathcal{A}^{-1} \\ I & 0 \end{bmatrix}; \quad (A_\varepsilon^*)^{-1} = \begin{bmatrix} -\mathcal{A}^{-1}(\mathcal{D}+\varepsilon_1 I+\varepsilon_2\mathcal{B}\mathcal{B}^*) & \mathcal{A}^{-1} \\ I & 0 \end{bmatrix}. \tag{2.32}$$

By direct computations, using (2.32) and its discrete counterpart, with $x = (x_1, x_2)$,

$$\|(A_{h\varepsilon}^{-1}\Pi_h - A_\varepsilon^{-1})x\|_H \le \|(\mathcal{A}_h^{-1}(\mathcal{D}_h\pi_h + \varepsilon_1\pi_h + \varepsilon_2\mathcal{B}_h\mathcal{B}_h^*\pi_h)x_1$$

$$-\mathcal{A}^{-1}(\mathcal{D}+\varepsilon_1 I+\varepsilon_2\mathcal{B}\mathcal{B}^*)x_1\|_{\mathcal{D}(\mathcal{A}^{1/2})} + \|(\mathcal{A}_h^{-1}\pi_h - \mathcal{A}^{-1})x_2\|_{\mathcal{D}(\mathcal{A}^{1/2})}.$$

$$\le C[\|(\mathcal{A}_h^{-1}\pi_h - \mathcal{A}^{-1})x_1\|_{\mathcal{D}(\mathcal{A}^{1/2})} + \|(\mathcal{A}_h^{-1}\pi_h - \mathcal{A}^{-1})x_2\|_{\mathcal{D}(\mathcal{A}^{1/2})} +$$

$$\|(\mathcal{A}_h^{-1}\mathcal{B}_h\mathcal{B}_h^*\pi_h - \mathcal{A}^{-1}\mathcal{B}\mathcal{B}^*)x_1\|_{\mathcal{D}(\mathcal{A}^{1/2})} + \|(\mathcal{A}_h^{-1}\mathcal{D}_h\pi_h - \mathcal{A}^{-1}\mathcal{D})x_1\|_{\mathcal{D}(\mathcal{A}^{1/2})}. \tag{2.33}$$

The first two terms on the right hand side of (2.33) tend to zero by virtue of approximation property (1.12-ii). As for the third term in (2.33) we write

$$\|(\mathcal{A}_h^{-1}\mathcal{B}_h\mathcal{B}_h^*\pi_h - \mathcal{A}^{-1}\mathcal{B}\mathcal{B}^*)x_1\|_{\mathcal{D}(\mathcal{A}^{1/2})} \le \|(\mathcal{A}_h^{-1} - \mathcal{A}^{-1})\mathcal{B}_h\mathcal{B}^*x_1\|_{\mathcal{D}(\mathcal{A}^{1/2})} \tag{2.34}$$

$$+ \|\mathcal{A}_h^{-1}\mathcal{B}_h(\mathcal{B}_h^*\pi_h - \mathcal{B}^*)x_1\|_{\mathcal{D}(\mathcal{A}^{1/2})} + \|\mathcal{A}^{-1}(\mathcal{B}-\mathcal{B}_h)\mathcal{B}^*x_1\|_{\mathcal{D}(\mathcal{A}^{1/2})}.$$

Note that by (1.14-iii) we obtain

$$\|\mathcal{A}_h^{-1}\mathcal{B}_h\|_{\mathcal{L}(U;\mathcal{D}(\mathcal{A}^{1/2}))} \le C\|\mathcal{A}^{-1}\mathcal{B}_h\|_{\mathcal{L}(U;\mathcal{D}(\mathcal{A}^{1/2}))},$$

and by (1.14-ii),

$$\le C\|\mathcal{A}^{-1}\mathcal{B}\|_{\mathcal{L}(U;\mathcal{D}(\mathcal{A}^{1/2}))} = C\|\mathcal{A}^{-1/2}\mathcal{B}\|_{\mathcal{L}(U,\mathcal{H})} \le M. \tag{2.35}$$

Returning to (2.34) and using (1.14-i)-(1.14-iii), we obtain, as $h \to 0$

$$\|(\mathcal{A}_h^{-1}\mathcal{B}_h\mathcal{B}_h^*\pi_h - \mathcal{A}^{-1}\mathcal{B}\mathcal{B}^*)x_1\|_{\mathcal{D}(\mathcal{A}^{1/2})} \le \|\mathcal{A}_h^{-1}\mathcal{B}_h\|_{\mathcal{L}(U;\mathcal{D}(\mathcal{A}^{1/2}))}\|(\mathcal{B}_h^*\pi_h - \mathcal{B}^*)x_1\|_U \tag{2.36}$$

$$+ \|\mathcal{A}^{-1}(\mathcal{B}-\mathcal{B}_h)\mathcal{B}^*x_1\|_{\mathcal{D}(\mathcal{A}^{1/2})} + \|(\mathcal{A}_h^{-1} - \mathcal{A}^{-1})\mathcal{B}_h\mathcal{B}_h^*x_1\|_{\mathcal{D}(\mathcal{A}^{1/2})} \to 0.$$

where in the last step we have used (1.12-ii) again.

Proof of (2.23-i): Using (1.14-ii) and the representation of A_ε^{-1} in (2.32), we have

$$\|A_\varepsilon^{-1}(B_h - B)u\|_H = \|\mathcal{A}^{-1}(\mathcal{B}_h - \mathcal{B})u\|_{\mathcal{D}(\mathcal{A}^{1/2})} \to 0 \ \text{as} \ h \to 0.$$

Proof of (2.23-ii): By (2.32) and (1.14-iii), we have

$$\|(A_{h\varepsilon}^{-1} - A_\varepsilon^{-1})B_h u\|_H = \|(\mathcal{A}_h^{-1} - \mathcal{A}^{-1})\mathcal{B}_h u\|_{\mathcal{D}(\mathcal{A}^{1/2})} \to 0 \ \text{as} \ h \to 0.$$

Proof of (2.24-i): By (2.32) and (1.14-i), we obtain

$$\| (B_h^* \Pi_h - B^*) A_\varepsilon^{*-1} x \|_U = \| (\mathcal{B}_h^* \pi_h - \mathcal{B}^*) x_1 \|_U \quad \to 0 \ \textit{as}\ h \to 0.$$

Proof of (2.24-ii): By the representation in (2.32), we obtain
$\| B_h^* [(A_{h\varepsilon}^*)^{-1} \Pi_h - \Pi_h (A^{*-1})] x \|_U = 0$ for *all h.*

Hence, we have verified that the approximation hypotheses (2.21)-(2.24) hold. Thus Theorem 2.2 has been proved.

Proof of Theorem 1.2

The statements (1.16)-(1.19) clearly follow by combining the results of Theorem 2.1 with the results from Theorem 2.2. As for (1.20), we simply notice that (1.18) implies that $\forall\, \varepsilon > 0\ \exists\ h_0$ such that for all $h < h_0$

$$\int_0^\infty \| R y_{h\varepsilon}^0 (t) \|_Z^2 \le C \ \| y_0 \|_H^2 , \tag{2.37}$$

where the constant C does not depend neither on ε nor h. Since R is strictly positive,

$$\int_0^\infty \| y_{h\varepsilon}^0 (t) \|_H^2 \le \tilde{C} \| y_0 \|_H^2 ,$$

where the conclusion of (1.20) follows by repeating a standard semigroup argument as in [P-1].

3 WAVE EQUATION WITH BOUNDARY CONTROL

In this section, we illustrate the applicability of the theory within the class of dynamics described by the wave equation with boundary control. Let Ω be an open, bounded domain in R^n, with boundary Γ which is assumed to be sufficiently regular or convex (polyhedral). We consider the following problem:

$$w_{tt} = \Delta w \qquad \textit{in}\ \ (0,\infty) \times \Omega \equiv Q, \tag{3.1a}$$

$$w(0,\cdot) = w_0 \ ; \ w_t(0,\cdot) = w_1 \qquad \textit{in}\ \ \Omega, \tag{3.1b}$$

$$w\,|_\Sigma = u(t) \qquad \textit{in}\ \ (0,\infty) \times \Gamma \equiv \Sigma. \tag{3.1c}$$

With (3.1) we associate the cost functional

$$J(u, w(u)) = \int_0^\infty (\| w(t) \|_{L_2(\Omega)}^2 + \| u(t) \|_{L_2(\Gamma)}^2)\, dt. \tag{3.2}$$

Notice that the uncontrolled system (with $u(t) = 0$) is unstable (there are infinitely many eigenvalues on the imaginary axis). Thus, one of the purposes of introducing a control into the model is to stabilize the overall system (in addition to solving the minimization problem). We are then led to consider the following optimization problem:

$$\textit{Minimize } J(u,w(u)) \text{ for } \textit{all } u \in L_2(0,\infty;L_2(\Gamma)) \text{ and} \qquad \text{(O.P.)}$$

$$w = w(u) \textit{ subject to the dynamics in } (3.1).$$

3.1 Abstract Setting

To put problem (3.1),(3.2) into the abstract model (1.2)-(1.4), we introduce the following spaces,

$$U = L_2(\Gamma), \quad \mathcal{H} = H^{-1}(\Omega) \equiv (H_0^1(\Omega))'. \tag{3.3}$$

and the positive, self-adjoint (on $L_2(\Omega)$) operator

$$\mathcal{A}h = -\Delta h, \quad \text{for } h \in H_2(\Omega) \cap H_0^1(\Omega). \tag{3.4}$$

We shall consider the $H^{-1}(\Omega)$ realization of $\mathcal{A}$ which will be denoted by the same symbol. This is to say that $\mathcal{D}(\mathcal{A}) = \{h \in H_0^1(\Omega);\ -\Delta h \in H^{-1}(\Omega)\} \equiv H_0^1(\Omega)$ and $\mathcal{D}(\mathcal{A}^{1/2}) = L_2(\Omega)$. Let the operator $D\colon\ L_2(\Gamma) \to L_2(\Omega)$(the Dirichlet map) be defined as

$$Dv = y \iff (\Delta y = 0 \ \ in \ \ \Omega,\ \ y\,|_\Gamma = v). \tag{3.5}$$

We introduce the control operator $\mathcal{B}$: $U \to (\mathcal{D}(\mathcal{A}))'$, where

$$\mathcal{B}u \equiv \mathcal{A}Du. \tag{3.6}$$

Since $\mathcal{A}^{-1/2}\mathcal{B} = \mathcal{A}^{1/2}D \in \mathcal{L}(L_2(\Gamma);H^{-1}(\Omega))$, $\mathcal{B}$ is linear and bounded from U into $(\mathcal{D}(\mathcal{A}))'$ (notice that the duality is considered with respect to the $H^{-1}(\Omega)$ inner product). With the above notation, problem (3.1),(3.2) is a special case of the problem (1.1),(1.6), with damping operator $\mathcal{D} = 0$ and $R = \begin{bmatrix} I & 0 \\ 0 & 0 \end{bmatrix}$. We need to verify the continuous hypotheses (F.C.C), (D.C.), and (H.1).

Condition (H.1)

From (1.2), (1.3), and (3.5), we calculate (see [L-T], p.72)

$$B^*\begin{bmatrix} z_1 \\ z_2 \end{bmatrix} = D^* z_2 = -\frac{\partial}{\partial \nu}\mathcal{A}^{-1} z_2 \tag{3.7}$$

and

$$B^* e^{A^* t}\begin{bmatrix} z_1 \\ z_2 \end{bmatrix} = \frac{\partial}{\partial \nu}\phi(t), \quad (z_1,z_2) \in L_2(\Omega) \times H^{-1}(\Omega) \quad .$$

where $\phi(t)$ solves the corresponding homogeneous problem,

$$\phi_{tt} = \Delta\phi \tag{3.8}$$

$$\phi(T,\cdot) = \phi_0 = -\mathcal{A}^{-1} z_2 \in H_0^1(\Omega)$$

$$\phi_t(T,\cdot) = \phi_1 = z_1 \in L_2(\Omega)$$

$$\phi|_\Gamma = 0.$$

Thus, by (3.7), (3.8), an equivalent formulation of assumption (H.1) is the following trace inequality,

$$\int_\Sigma \left|\frac{\partial \phi}{\partial \nu}\right|^2 d\Sigma \le C_T(\|\phi_0\|^2_{H_0^1(\Omega)} + \|\phi_1\|^2_{L_2(\Omega)}) = C_T(\|\mathcal{A}^{-1}z_2\|^2_{H_0^1(\Omega)} + \|z_1\|^2_{L_2(\Omega)}). \quad (3.9)$$

It should be noted that the inequality (3.9) does not follow from *a priori* (optimal) interior regularity, $\phi(t) \in C([0,T];H_0^1(\Omega))$, of the solution to problem (3.8). Instead, (3.9) is an independent trace regularity result and was established in [L-T-4] and [L-L-T].

(F.C.C.) and (D.C.) conditions

Exact controllability of problem (3.1) over a finite time interval $[0,T]$ on the state space $H \equiv L_2(\Omega) \times H^{-1}(\Omega)$ within the class of $L_2([0,T];L_2(\Gamma))$ controls u is a sufficient condition for the (F.C.C.) to hold. For this problem, exact controllability, which is equivalent to a reversed inequality in (3.9), has been established in [L-6]. The (D.C.) follows from a well known fact that the system

$$w_{tt} + w_t = \Delta w \quad on\ Q,$$

$$w(0,\cdot) = w_0,\ w_t(0,\cdot) = w_1 \quad in\ \Omega$$

$$w|_\Gamma = 0 \quad in\ \Sigma,$$

is exponentially stable on $L_2(\Omega) \times H^{-1}(\Omega)$. Thus the $(D.C.)$ holds with $K = \begin{bmatrix} -I & 0 \\ 0 & 0 \end{bmatrix}$.

Since (H.1), (F.C.C.), and (D.C.) are satisfied for our problem, Theorem 1.1 applies and provides for the existence of a unique solution to the (ARE), $P \in \mathcal{L}(L_2(\Omega) \times H^{-1}(\Omega))$, such that $Px \equiv \begin{bmatrix} P_{11}x_1 + P_{12}x_2 \\ P_{21}x_1 + P_{22}x_2 \end{bmatrix} \equiv \begin{bmatrix} p_{1x} \\ p_{2x} \end{bmatrix}$ satisfies

$$\int_\Omega (p_{1x}y_2 - p_{2x}y_1 + x_2p_{1y} - x_1p_{2y} + x_1y_1)\, d\Omega = \int_\Gamma \left(\frac{\partial}{\partial \nu}\mathcal{A}^{-1}p_{2x}\, \frac{\partial}{\partial \nu}\mathcal{A}^{-1}p_{2y}\right) d\Gamma \quad (3.10)$$

for all $x = (x_1,x_2)$, $y = (y_1,y_2)$ in $H_0^1(\Omega) \times L_2(\Omega)$. Moreover, from (1.8) in Theorem 1.1, we obtain that the system (3.1) with feedback control

$$u(t) = \frac{\partial}{\partial \nu}\mathcal{A}^{-1}p_2(w(t),w_t(t)) \quad (3.11)$$

is exponentially stable on $L_2(\Omega) \times H^{-1}(\Omega)$. This is to say that there exists constants $C, \omega > 0$ such that

$$\| w(t) \|_{L_2(\Omega)} + \| w_t(t) \|_{H^{-1}(\Omega)} \le Ce^{-\omega t} [\ \| w_0 \|_{L_2(\Omega)} + \| w_1 \|_{H^{-1}(\Omega)}\ ].$$

Remark 3.1: Initially, the regularity and controllability results needed in the (H.1) and (F.C.C.) had been proven on smooth domains. However, careful analysis of the proof reveals that these results hold as well for either convex domains (see [L-6]) or polyhedral domains up todimension three (see [G-3]). ■

Remark 3.2: Uniform exponential decay rates for the wave equation (3.1) with boundary, velocity feedback

$$u(t) = \frac{\partial}{\partial\nu} \mathcal{A}^{-1} w_t(t) \tag{3.12}$$

have been proven first for convex domains in [L-T-2] and for smooth domains in [L-T-5]. ■

In view of (3.12), we can certainly say that the solution, P, to the Riccati equation, provides another stabilizing feedback for the problem. However, as we shall see later, from a computational point of view, the stabilizing feedback based only on velocity feedback (as in (3.12)) is inadequate, whereas the Riccati-based feedback is much more "robust". Indeed, numerical experiments performed using only a velocity feedback show that the exponential decay of the system is lost for the conformal FEM approximation (see [B-I-W]). This is to say that the margin of stability is not uniformly maintained and in fact decreases as the accuracy of the approximation increases. Instead, as we shall see later, the algorithm in Theorem 1.2 for the approximation of the Riccati-based feedback produces a uniform margin a stability (with respect to the parameter of discretization).

Next, we consider the approximation scheme for the Riccati Equation, (3.10). We introduce a family of finite dimensional subspaces $V_h \subset H_0^1(\Omega)$ with orthogonal projection $Q_h\colon L_2(\Omega) \to V_h$ and having the following properties:

$$\| Q_h z - z \|_{H^{\alpha}(\Omega)} \to 0 \quad as \quad h \to 0, \quad z \in H_0^{\alpha}(\Omega); \ |\alpha| \le 1 \tag{3.13}$$

$$\| \frac{\partial}{\partial\nu} z_h \|_{L_2(\Gamma)} \le Ch^{-1/2} \ \| z_h \|_{H_0^1(\Omega)}$$

$$\| z_h \|_{H_0^{\alpha}(\Omega)} \le Ch^{-s} \ \| z_h \|_{H_0^{\alpha-s}(\Omega)}, \quad 0 \le \alpha \le 1$$

$$\| \frac{\partial}{\partial\nu} (Q_h z - z) \|_{L_2(\Gamma)} \le Ch^{1/2} \ \| z \|_{H^2(\Omega)} .$$

It is well known that the above approximation properties are satisfied for spline approximations defined on a uniform (quasi-uniform) mesh. We define the discrete operators $\mathcal{A}_h \equiv Q_h \mathcal{A} Q_h$, i.e.

$$\int_{\Omega} \mathcal{A}_h \psi_h \phi_h \, d\Omega = \int_{\Omega} \nabla\psi_h \nabla\phi_h \, d\Omega = \int_{\Omega} \mathcal{A}\psi_h \phi_h \, d\Omega, \quad \text{for } \psi_h, \phi_h \in V_h \subset H_0^1(\Omega), \tag{3.14}$$

and $\mathcal{B}_h\colon L_2(\Gamma) \to V_h$, $\mathcal{B}_h \equiv Q_h \mathcal{A} D$ or equivalently,

$$\int_{\Omega} \mathcal{B}_h u \psi_h \, d\Omega = -\int_{\Gamma} u \frac{\partial}{\partial\nu} \psi_h \, d\Gamma. \tag{3.15}$$

By using the well known elliptic error estimates (see [T-1]), we obtain

$$\|(\mathcal{A}_h^{-1}\pi_h - \mathcal{A}^{-1})z\|_{H_0^s(\Omega)} \leq Ch^{2-s} \|z\|_{L_2(\Omega)}, \quad 0 \leq s \leq 1, \tag{3.16}$$

and we infer that approximation property (1.12ii) is satisfied. On V_h we define the following discrete inner product

$$(\psi_h, \phi_h)_{\hbar} = (\mathcal{A}_h^{-1}\psi_h, \phi_h)_{L_2(\Omega)} = \int_\Omega \mathcal{A}_h^{-1}\psi_h \phi_h \, d\Omega.$$

Since

$$\|\mathcal{A}_h^{1/2}\phi_h\|_{L_2(\Omega)} = \|\mathcal{A}^{1/2}\phi_h\|_{L_2(\Omega)} = \|\nabla\phi_h\|_{L_2(\Omega)}, \quad \phi_h \in V_h, \tag{3.17}$$

we can show that

$$\|\phi_h\|_{\hbar} = \|\phi_h\|_{\mathcal{H}} = \|\mathcal{A}^{-1/2}\phi_h\|_{L_2(\Omega)}, \quad \phi_h \in V_h \tag{3.18i}$$

$$\|\phi\|_{\hbar} \leq \|\phi\|_{\mathcal{H}}, \quad \phi \in \mathcal{H} \tag{3.18ii}$$

Indeed, by duality,

$$\|\phi_h\|_{\hbar} = \|\mathcal{A}_h^{-1/2}\phi_h\|_{L_2(\Omega)} = \sup_{\psi_h \in V_h} \frac{\left|\int_\Omega \phi_h\psi_h \, d\Omega\right|}{\|\mathcal{A}_h^{1/2}\psi_h\|_{L_2(\Omega)}}$$

and by (3.17),

$$\sup_{\psi_h \in V_h} \frac{\left|\int_\Omega \phi_h\psi_h \, d\Omega\right|}{\|\mathcal{A}^{1/2}\psi_h\|_{L_2(\Omega)}} \equiv \|\mathcal{A}^{-1/2}\phi_h\|_{L_2(\Omega)} = \|\phi_h\|_{\mathcal{H}}.$$

This proves (3.18i). As for (3.18ii),

$$\|\phi\|_{\hbar} = \|\mathcal{A}_h^{-1/2}Q_h\phi\|_{L_2(\Omega)} = \sup_{\psi \in L_2(\Omega)} \frac{|(\mathcal{A}_h^{-1/2}Q_h\phi, \psi)_\Omega|}{\|\psi\|_{L_2(\Omega)}}$$

$$\leq \sup_{\psi \in L_2(\Omega)} \frac{|(Q_h\phi, \mathcal{A}_h^{-1/2}Q_h\psi)_\Omega|}{\|\psi\|_{L_2(\Omega)}} \leq \sup_{\psi \in L_2(\Omega)} \frac{|(Q_h\phi, \mathcal{A}_h^{-1/2}Q_h\psi)_\Omega|}{\|Q_h\psi\|_{L_2(\Omega)}}$$

Changing variables, $\hat{\psi}_h \equiv \mathcal{A}_h^{-1/2}Q_h\psi$, yields,

$$= \sup_{\hat{\psi}_h} \frac{|(Q_h\phi, \hat{\psi}_h)_\Omega|}{\|\mathcal{A}_h^{1/2}\hat{\psi}_h\|_{L_2(\Omega)}} = \sup_{\hat{\psi}_h} \frac{|(\phi, \hat{\psi}_h)_\Omega|}{\|\mathcal{A}_h^{1/2}\hat{\psi}_h\|_{L_2(\Omega)}}$$

By (3.17),

$$= \sup_{\hat{\psi}_h} \frac{|(\phi, \hat{\psi}_h)_\Omega|}{\|\mathcal{A}^{1/2}\hat{\psi}_h\|_{L_2(\Omega)}} = \sup_{\hat{\psi}_h} \frac{|(\mathcal{A}^{-1/2}\phi, \mathcal{A}^{1/2}\hat{\psi}_h)_\Omega|}{\|\mathcal{A}^{1/2}\hat{\psi}_h\|_{L_2(\Omega)}}$$

$$\leq \|\mathcal{A}^{-1/2}\phi\|_{L_2(\Omega)} \frac{\|\mathcal{A}^{1/2}\hat{\psi}_h\|_{L_2(\Omega)}}{\|\mathcal{A}^{1/2}\hat{\psi}_h\|_{L_2(\Omega)}} = \|\mathcal{A}^{-1/2}\phi\|_{L_2(\Omega)},$$

which proves (3.18ii). Next, we shall show that $\mathcal{A}_h$ complies with approximation property

(1.12i). Indeed,

$$(\mathcal{A}_h x_h, \xi_h)_{\hbar} = (\mathcal{A}_h^{-1} \mathcal{A}_h x_h, \xi_h)_{L_2(\Omega)} = (x_h, \xi_h)_{L_2(\Omega)} \tag{3.19}$$

$$= (\mathcal{A}^{-1} \mathcal{A} x_h, \xi_h)_{L_2(\Omega)} = (\mathcal{A} x_h, \xi_h)_{H^{-1}(\Omega)} = (\mathcal{A} x_h, \xi_h)_{\mathcal{H}}.$$

Since $\mathcal{A}$ is positive, self-adjoint on $\mathcal{H}$, the same properties are inherited by $\mathcal{A}_h$ with respect to the $\hbar$-inner product (not $\mathcal{H}$). Hence, $A_h^* = \begin{bmatrix} 0 & -\pi_h \\ \mathcal{A}_h & 0 \end{bmatrix} = -A_h$, where A_h^* is computed with respect to the H_h inner product, so

$$(A_h^* x_h, y_h)_{\mathcal{D}(\mathcal{A}^{1/2}) \times V_h} = (A_h y_h, x_h)_{\mathcal{D}(\mathcal{A}^{1/2}) \times V_h},$$

(notice that $(\xi_h, x_h)_{\mathcal{D}(\mathcal{A}^{1/2})} = (\mathcal{A}\xi_h, x_h)_{\mathcal{H}} = (by (1.12i)) = (A_h \xi_h, x_h)_{\hbar}$. Hence, $\| \xi_h \|^2_{\mathcal{D}(\mathcal{A}^{1/2})} = \| \mathcal{A}_h^{1/2} \xi_h \|^2_{\hbar} = \| \xi_h \|^2_{L_2(\Omega)}$. Note here that the discrete structure of $\mathcal{D}(\mathcal{A}^{1/2})$ on V_h is automatically obtained from the $\hbar$-norm.

Next, we compute $\mathcal{B}_h^*$, where

$$(\mathcal{B}_h^* x_h, u)_U = \int_\Gamma \mathcal{B}_h^* x_h u \, d\Gamma = (x_h, \mathcal{B}_h u)_{\hbar}$$

$$= \int_\Omega \mathcal{A}_h^{-1} x_h \mathcal{B}_h u \, d\Omega = \int_\Omega \mathcal{A}_h^{-1} x_h Q_h \mathcal{A} D u \, d\Omega = \int_\Omega \mathcal{A}_h^{-1} x_h \mathcal{A} D u \, d\Omega$$

$$= (\mathcal{A}_h^{-1} x_h, \mathcal{A} D u)_{L_2(\Omega)} = (D^* \mathcal{A}^* \mathcal{A}_h^{-1} x_h, u)_U = -\int_\Gamma \frac{\partial}{\partial \nu} \mathcal{A}_h^{-1} x_h u \, d\Gamma.$$

Hence,

$$\mathcal{B}_h^* x_h \equiv -\frac{\partial}{\partial \nu} \mathcal{A}_h^{-1} x_h \tag{3.20}$$

$$B_h^* \begin{bmatrix} x_{1h} \\ x_{2h} \end{bmatrix} = -\frac{\partial}{\partial \nu} \mathcal{A}_h^{-1} x_{2h}$$

We shall prove that approximation hypotheses (1.14) are verified for this problem.

Assumption (1.14i):

$$\| (\mathcal{B}_h^* \pi_h - \mathcal{B}^*) x \|_{L_2(\Gamma)} = \| \frac{\partial}{\partial \nu} (\mathcal{A}^{-1} x - \mathcal{A}_h^{-1} \pi_h x) \|_{L_2(\Gamma)}$$

From (3.13) and (3.16)

$$\le \| \frac{\partial}{\partial \nu} (\mathcal{A}^{-1} x - Q_h \mathcal{A}^{-1} x) \|_{L_2(\Gamma)} + \| \frac{\partial}{\partial \nu} (Q_h \mathcal{A}^{-1} x - \mathcal{A}_h^{-1} \pi_h x) \|_{L_2(\Gamma)}$$

$$\le C h^{1/2} \| \mathcal{A}^{-1} x \|_{H^2(\Omega)} + C h^{-1/2} \| Q_h \mathcal{A}^{-1} x - \mathcal{A}_h^{-1} \pi_h x \|_{H_0^1(\Omega)} \le C h^{1/2} \| x \|_{L_2(\Omega)} + C h^{-1/2} h \| x \|_{L_2(\Omega)}$$

$$\le C h^{1/2} \| x \|_{L_2(\Omega)} = C h^{1/2} \| x \|_{\mathcal{D}(\mathcal{A}^{1/2})}$$

Assumption (1.14ii):

Since convergence in (1.14i) is uniform, (1.14ii) follows by duality.

Assumption (1.14iii):

$$\| (\mathcal{A}_h^{-1} - \mathcal{A}^{-1})\mathcal{B}_h \|_{\mathcal{L}(U \to \mathcal{D}(\mathcal{A}^{1/2}))} = \| \frac{\partial}{\partial\nu} Q_h(\mathcal{A}_h^{-1} Q_h - \mathcal{A}^{-1}) \|_{\mathcal{L}(L_2(\Omega) \to L_2(\Gamma))} .$$

Indeed,

$$((\mathcal{A}_h^{-1} - \mathcal{A}^{-1})\mathcal{B}_h u, \phi)_{\mathcal{D}(\mathcal{A}^{1/2})} = ((\mathcal{A}_h^{-1} - \mathcal{A}^{-1})\mathcal{B}_h u, \phi)_{L_2(\Omega)}$$

$$= (\mathcal{B}_h u, (\mathcal{A}_h^{-1} Q_h - \mathcal{A}^{-1})\phi)_{L_2(\Omega)} = (Q_h \mathcal{A} D u, (\mathcal{A}_h^{-1} Q_h - \mathcal{A}^{-1})\phi)_{L_2(\Omega)}$$

$$(\mathcal{A} D u, (\mathcal{A}_h^{-1} Q_h - Q_h \mathcal{A}^{-1})\phi)_{L_2(\Omega)} = -\langle u, \frac{\partial}{\partial\nu}(\mathcal{A}_h^{-1} Q_h - Q_h \mathcal{A}^{-1})\phi \rangle_{L_2(\Gamma)} .$$

Hence, by the inverse approximation property (3.13),

$$\| \frac{\partial}{\partial\nu}(\mathcal{A}_h^{-1} Q_h - Q_h \mathcal{A}^{-1}) x \|_{L_2(\Gamma)} \le C h^{-1/2} \| (\mathcal{A}_h^{-1} Q_h - Q_h \mathcal{A}^{-1}) x \|_{H_0^1(\Omega)}$$

and by (3.16) with $s = 1$,

$$\le C h^{-1/2} h \| x \|_{L_2(\Omega)} \le C h^{1/2} \| x \|_{L_2(\Omega)}$$

which verifies the assumptions.

Thus, having verified all the hypotheses of Theorem 1.2, we are in a position to apply its conclusions to our problem. In this direction, we introduce the discrete Riccati equation,

$$(P_{h\varepsilon} x_h, A_h y_h)_{H_h} + (A_h x_h, P_{h\varepsilon} y_h)_{H_h} + \int_\Omega x_{h1} y_{h1} \, d\Omega = \int_\Gamma B_h^* P_{h\varepsilon} x_h B_h^* P_{h\varepsilon} y_h \, d\Gamma$$

$$+ \varepsilon_1 (\bar{P}_{h,\varepsilon,2} x_{h2}, y_{h2})_{V_h} + \varepsilon_1 (x_{h2}, \bar{P}_{h,\varepsilon,2} y_{h2})_{V_h} + \varepsilon_2 \int_\Gamma B_h^* P_{h\varepsilon} x_h B_h^* y_h \, d\Gamma + \varepsilon_2 \int_\Gamma B_h^* x_h B_h^* P_{h\varepsilon} y_h \, d\Gamma .$$

where

$$P_{h\varepsilon} \begin{bmatrix} x_{h1} \\ x_{h2} \end{bmatrix} = \begin{bmatrix} P_{h,\varepsilon,1} x_{h1} + P_{h,\varepsilon,2} x_{h2} \\ P_{h,\varepsilon,3} x_{h1} + P_{h,\varepsilon,4} x_{h2} \end{bmatrix} = \begin{bmatrix} \bar{P}_{h,\varepsilon,1}(x_h) \\ \bar{P}_{h,\varepsilon,2}(x_h) \end{bmatrix}$$

Using

$$(x_h, y_h)_{H_h} = \int_\Omega (x_h y_h + \mathcal{A}_h^{-1} x_h y_h) \, d\Omega,$$

$$(x_h, y_h)_{V_h} = \int_\Omega \mathcal{A}_h^{-1} x_h y_h \, d\Omega .$$

and the structure of A_h, B_h^*, we obtain the following explicit form of the $(ARE-\varepsilon-h)$,

$$\int_\Omega (P_{h,\varepsilon,1} x_h y_{h2} - P_{h,\varepsilon,2} x_h y_{h1} + x_{h2} P_{h,\varepsilon,1} y_h - x_{h1} P_{h,\varepsilon,2} + x_{h1} y_{h1} \, d\Omega =$$

$$\int_\Gamma \left(\frac{\partial}{\partial \nu}\mathcal{A}_h^{-1} P_{h,\varepsilon,2}x_h \frac{\partial}{\partial \nu}\mathcal{A}_h^{-1} P_{h,\varepsilon,2}y_h \right) d\Gamma + \varepsilon_1 \int_\Omega \left(P_{h,\varepsilon,2}x_h \mathcal{A}_h^{-1} y_{h2} + P_{h,\varepsilon,2}y_h \mathcal{A}_h^{-1} x_{h2} \right) d\Omega$$

$$+ \varepsilon_2 \int_\Gamma \left(\frac{\partial}{\partial \nu}\mathcal{A}_h^{-1} y_{h2} \frac{\partial}{\partial \nu}\mathcal{A}_h^{-1} P_{h,\varepsilon,2}x_h + \frac{\partial}{\partial \nu}\mathcal{A}_h^{-1} x_{h2} \frac{\partial}{\partial \nu}\mathcal{A}_h^{-1} P_{h,\varepsilon,2}y_h \right) d\Gamma, \ \forall \ x_h, y_h \in V_h .$$

Our main result is

Theorem 3.1:

Assume that the finite dimensional subspace V_h satisfies (3.13). Then there exists a unique solution $P_{h\varepsilon} \in \mathcal{L}(\hat{V}_h \times V_h)$ to the discrete Algebraic Riccati Equation such that

$$\lim_{\varepsilon_1 \to 0} \lim_{\varepsilon_2 \to 0} \lim_{h \to 0} \| P_{h\varepsilon} \Pi_h x - Px \|_{L_2(\Omega) \times H^{-1}(\Omega)} \to 0, \quad \forall \ x \in L_2(\Omega) \times H^{-1}(\Omega) ,$$

where

$$\Pi_h = \begin{bmatrix} Q_h & 0 \\ 0 & \pi_h \end{bmatrix}.$$

REFERENCES

[B-1] Balakrishnan, A.V.(1976). Applied Functional Analysis. Springer-Verlag. New York.

[B-I-W] H.T. Banks, K. Ito, C. Wang (1991), Exponentially Stable Approximations of Weakly Damped Wave Equations, ICASE, Report No. 91-47.

[B-K] H.T. Banks, K. Kunisch (1984), The linear regulator problem for parablic systems, *SIAM J. Control Vol. 22, No. 5*, 684-699.

[F-L-T] F. Flandoli, I. Lasiecka, R. Triggiani (1988), Algebraic Riccati Equations with nonsmoothing observation arising in hyperbolic and Euler-Bernoulli boundary control problems, *Ann. Mat. Pura e Appl.*, CLIII, 307-382.

[G-1] J. S. Gibson (1981), An analysis of optimal model regulation: Convergence and stability, *SIAM J. Control Optimiz. 19:686-707.*

[G-2] J. S. Gibson (1991), Approximation theory for linear quadratic Gaussian control of flexible structures, *SIAM J. Control Optimiz. 29:1-38.*

[G-3] P. Grisvard (1985), Elliptic Problems in Nonsmooth Domains, Pitman, Boston.

[H-1] E. Hendrickson (1992), Approximation and Regularization Methods for the Riccati Operator of the Undamped Kirchoff Plate, Master's Thesis, Univ. of Virginia.

[H-L] M. Ann Horn, I. Lasiecka, Asymptotic behavior with respect to thicknes of boundary stabilizing feedback for the Kirchoff plate, (to appear in *J. Diff. Eqns.*).

[K-1] S. Kesevan (1989). Topics in Functional Analysis and Applications. John Wiley and Sons.

[K-K] M. Kroller, K. Kunisch (1991), Convergence rates for the feedback operators arising in the linear quadratic problem governed by parabolic equations, *SIAM J. Numer. Anal. 28:1350-1385.*

[L-1] J. Lagnese, Infinite Horizon Linear-Quadratic Regulator Problem for Beams and Plates, *Lecture Notes LNCIS*, Springer-Verlag.

[L-2] I. Lasiecka (1990), Approximations of solutions to infinite-dimensional Algebraic Riccati equations with unbounded input operators, *Numer. Func. Anal. and Optimiz. 11:(304).*

[L-3] I. Lasiecka (1989), Stabilization of wave and plate like equations with nonlinear dissipation on the boundary, *J. Diff. Eqns. 78*, No. 2, 340-381.

[L-4] I. Lasiecka (1988), Stabilization of hyperbolic and parabolic systems with nonlinearly perturbed boundary conditions, *J. Diff. Eqns. 75:53-87.*

[L-5] I. Lasiecka (1992), Convergence Rates for the Approximations of the Solutions to Algebraic Riccati Equations with Unbounded Coefficients: Case of Analytic Semigroups, *Numer. Math. 63:357-390.*

[L-6] J.L. Lions (1988). Controlabilite Exacte. Vol. 2, Perturbations. Masson, Paris.

[L.L] J.E. Lagnese, J.-L. Lions (1989). Modelling Analysis and Control of Thin Plates. Vol. 6, Masson, Paris.

[L-L-T] I. Lasiecka, J.L. Lions, R. Triggiani (1986), Nonhomogeneous Boundary Value Problems for Second Order Hyperbolic Operators, *J. Math. Pure et Appl. 65:149-192.*

[L-T] I. Lasiecka and R. Triggiani (1991). Differential and Algebraic Riccati Equations with Applications to Boundary/Point Control Problems: Continuous Theory and Approximation Theory. Springer-Verlag Lecture Notes LNCIS, Vol. 164.

[L-T-1] I. Lasiecka and R. Triggiani (1991), Numerical approximations of Algebraic Riccati equations for abstract systems modelled by analytic semigroups, and applications, Mathematics of Computations, Vol. 57, No. 196, 639-662.

[L-T-2] I. Lasiecka and R. Triggiani (March 19987), Uniform Exponential Energy Decay of Wave Equations in a Bounded Region with $L_2(0,\infty;L_2(\Gamma))$-Feedback Control in the Dirichlet Boundary Conditions, *J. Diff. Eqns. Vol. 66, No. 3.*

[L-T-3] I. Lasiecka and R. Triggiani (1987), The Regulator Problem for Parabolic Equations with Dirichlet Boundary Control, *Applied Mathematics and Optimization Vol 16 No. 2* 147-168.

[L-T-4] I. Lasiecka and R. Triggiani (1983), Regularity of Hyperbolic equations under Boundary Terms, *Appl. Math. Optim. 10:275-286.*

[L-T-5] I. Lasiecka and R. Triggiani (1992), Uniform Stabilization of the Wave Equation with Dirichlet or Newmann Feedback Control Without Geometrical Conditions, *Appl. Math. Optim. 25:189-224.*

[P-1] A. Pazy (1986). Semigroups of operators and applications to Partial Differential Equations. Springer-Verlag.

[T-1] V. Thomee (1984). Galerkin finite element methods for parablic problems, Springer, Berlin Heidelberg New York.

10 Wellposedness and Uniform Decay Rates of Weak Solutions to a von Kármán System with Nonlinear Dissipative Boundary Conditions

Mary Ann Horn* University of Minnesota, Minneapolis, Minnesota

Irena Lasiecka† University of Virginia, Charlottesville, Virginia

Daniel Tataru Northwestern University, Evanston, Illinois

Abstract

A von Kármán system with nonlinear boundary feedback is considered. Existence and *uniqueness* of *weak* solutions is proven. Moreover, it is shown that these solutions *(i)* are limits (with respect to the thickness of the plate which is small) of solutions corresponding to a more general von Kármán model accounting for rotational forces, and *(ii)* decay to zero at a uniform rate.

*This material is based upon work partially supported under a National Science Foundation Mathematical Sciences Postdoctoral Research Fellowship.

†Partially supported by National Science Foundation Grant NSF DMS-9204338.

1 Introduction

1.1 Statement of the Problem

Let Ω be a bounded smooth domain in R^2 with boundary Γ. In $\Omega \times (0,T)$, where $T > 0$ is given, we consider the following von Kármán system with nonlinear boundary conditions:

$$w_{tt} + \Delta^2 w + b(x)w_t = [\mathcal{F}(w), w] \quad \text{in } Q_T = (0,T)\times\Omega \tag{1.1.a}$$

$$\left.\begin{aligned} w(0,\cdot) &= w_0 \\ w_t(0,\cdot) &= w_1 \end{aligned}\right\} \quad \text{in } \Omega \tag{1.1.b}$$

$$\Delta w + (1-\mu)B_1 w = -h_1(\frac{\partial}{\partial\nu}w_t) \quad \text{on } \Sigma_T = (0,T)\times\Gamma \tag{1.1.c}$$

$$\frac{\partial}{\partial\nu}\Delta w + (1-\mu)B_2 w - w = g(w_t) - \tfrac{\partial}{\partial\tau}h_2(\tfrac{\partial}{\partial\tau}w_t) \quad \text{on } \Sigma_T = (0,T)\times\Gamma, \tag{1.1.d}$$

and

$$\begin{aligned} \Delta^2\mathcal{F}(w) &= -[w,w] \quad \text{in } (0,T)\times\Omega \\ \mathcal{F}|_\Gamma &= \tfrac{\partial}{\partial\nu}\mathcal{F}|_\Gamma = 0 \quad \text{on } (0,T)\times\Gamma, \end{aligned} \tag{1.2}$$

where

$$[u,v] \equiv u_{xx}v_{yy} + u_{yy}v_{xx} - 2u_{xy}v_{xy}.$$

In (1.1), the boundary operators B_1 and B_2 are given by

$$\begin{aligned} B_1 w &= 2n_1n_2w_{xy} - n_1^2w_{yy} - n_2^2w_{xx}, \\ B_2 w &= \tfrac{\partial}{\partial\tau}[(n_1^2 - n_2^2)w_{xy} + n_1n_2(w_{yy} - w_{xx})], \end{aligned} \tag{1.3}$$

$0 < \mu < \frac{1}{2}$ is Poisson's ratio, $b(x) \in L^\infty(\Omega)$, and the functions, h_i and g are differentiable, monotone increasing, real-valued functions, zero at the origin, and are subject to the following constraints:

$$\left.\begin{array}{r} m \leq h_i'(s) \leq M \\ g'(s) \leq M(|s|^r + 1), \end{array}\right\} (H)$$

where r is any positive constant. This system describes the transversal displacement w and the Airy-stress function $\mathcal{F}(w)$ of a vibrating plate, whose boundary is subject to nonlinear damping in the form of moments and forces/shears applied to the edge of the plate.

Questions related to the existence of solutions for the von Kármán system (1.1.a), (1.1.b), (1.2), with the *homogeneous* boundary conditions:

$$w|_\Gamma = \frac{\partial}{\partial \nu} w|_\Gamma = 0 \qquad \text{on } (0,T) \times \Gamma, \tag{1.4}$$

have received considerable attention in the literature. Indeed, the existence of global *weak* (i.e., $(w(t), w_t(t)) \in H^2(\Omega) \times L_2(\Omega)$) solutions for (1.1.a), (1.1.b), (1.2), (1.4) has been proven by Faedo-Galerkin methods in Lions [16] and Vorovič [21]. In [19], [20], von Wahl gives a proof of existence *and* uniqueness of a local solution with higher regularity. Existence of *local classical* solutions has been established by Stahel in [17]. Arguments of [17] also prove the uniqueness property which is valid on a small time interval. More recently, Koch and Stahel [11] were able to establish appropriate a priori bounds for their classical solutions, hence proving the *global* existence of classical solutions. *Global* existence and uniqueness results for *strong* (i.e., $H^4(\Omega) \times H^2(\Omega)$) solutions to (1.1.a), (1.1.b), (1.2), (1.4) was proven by Chueskov in [3].

As already stated, the results quoted above refer to a situation when the boundary conditions are *homogeneous*, or, more precisely, of the form as in (1.4). The techniques used in these references can be adapted to treat the case of boundary conditions as in (1.1.c) and (1.1.d), but with *linear* boundary dissipation, i.e., g and h are linear (this fact was already noted by Lagnese in [12]). Instead, the presence of *nonlinear* functions in boundary conditions (1.1.c) and (1.1.d) raises a number of technical difficulties since the methods developed previously are not applicable any longer. Indeed, a standard Faedo-Galerkin method which has been used for proving existence of solutions of von Kármán systems with homogeneous boundary conditions (as in [3], [7], [21]) runs into difficulties (at the limit process) because of the appearance of nonlinearities on the boundary.

To our knowledge, the only results available in the literature and dealing with the *nonlinear boundary* damping are due to Lagnese and Leugering [13], where the one-dimensional model has been treated, and in [7], [14], where the rotational inertia of the plate, which induces a regularizing effect on the velocity, was accounted for in the model. In this last case, the mathematical nature of the problem is, obviously, very different.

Thus, the main distinctive feature of our paper is that we treat a two dimensional von Kármán system (1.1), (1.2) with fully nonlinear boundary damping. We shall prove that there exists a *unique weak* solution to (1.1), (1.2). We note that the uniqueness result for weak solutions is *new* even in the case of linear homogeneous boundary conditions. Precise statements of these results are given below.

Theorem 1.1 *(Uniqueness) Let* (w, w_t) *be any solution corresponding to (1.1) which satisfies* $(w, w_t) \in C(0, T; H^2(\Omega) \times L_2(\Omega))$. *Then, such a solution is unique.*

Theorem 1.2 *(Existence of weak solutions) For all initial data* $(w_0, w_1) \in H^2(\Omega) \times L_2(\Omega)$, *there exists a unique global solution,* $(w, w_t) \in C(0,T; H^2(\Omega)) \times C(0,T; L_2(\Omega))$.

Remark 1.1: The results of Theorems 1.1 and 1.2 hold true under somewhat weaker hypotheses imposed on h_i and g. Indeed, it would suffice to assume the growth condition (H) valid for the functions (rather than the derivatives) and holding for large values of s only. In order to keep the exposition as simple as possible and to focus attention on the key ideas, we shall dispense with this level of generality.

Our interest in studying wellposedness of this von Kármán system is motivated (among other things) by certain stabilization problems arising in control theory. Let us consider the following more general von Kármán model accounting for rotational forces (represented by the term with γ^2).

$$\begin{cases} w_{tt} - \gamma^2 \Delta w_{tt} + \Delta^2 w + b w_t = [w, \mathcal{F}(w)] \\ \quad w(0) = w_0 \in H^2(\Omega); \quad w_t(0) = w_1 \in H^1(\Omega) \\ \Delta w + (1-\mu) B_1 w = -h_1(\frac{\partial}{\partial \nu} w_t) \\ \frac{\partial}{\partial \nu} \Delta w + (1-\mu) B_2 w - \gamma^2 \frac{\partial}{\partial \nu} w_{tt} - w = g(w_t) - \frac{\partial}{\partial \tau} h_2(\frac{\partial}{\partial \tau} w_t), \end{cases} \tag{1.5}$$

Here, $\mathcal{F}(w)$ satisfies (1.2), the functions h_i and g are under the same hypothesis (H), and $b(x) > 0$ a.e. in Ω.

With (1.5), we associate the energy

$$E_{\gamma,w}(t) \equiv \frac{1}{2} \int_\Omega \{|w_t^2 + \gamma^2 |\nabla w_t|^2 + |\Delta \mathcal{F}(w)|^2\} d\Omega + \frac{1}{2} a(w,w), \tag{1.6}$$

where

$$a(w,w) \equiv \int_{\Omega} \{|\Delta w|^2 + (1-\mu)(2w_{xy}^2 - w_{xx}^2 - w_{yy}^2)\} d\Omega + \frac{1}{2}\int_{\Gamma} w^2 d\Gamma. \tag{1.7}$$

The parameter γ in (1.1.a) is proportional to the thickness of the plate and it is usually assumed to be small. It can be shown that problem (1.5) admits a unique solution in the energy norm (notice the regularizing effect of the term $\gamma^2 \Delta w_{tt}$). Moreover a special case of the uniform decay result presented in [9] implies that for initial data in a bounded set of $H^2(\Omega) \times H^1(\Omega)$, the energy of weak solutions decays exponentially to zero. This is to say that: $\forall (w_0, w_1) \in \mathcal{B}(0,R) \subset H^2(\Omega) \times H^1(\Omega)$,

$$E_{\gamma,w}(t) \leq Ce^{\omega t}, \qquad t \geq 0, \tag{1.8}$$

where the constants C, $\omega > 0$ *do not* depend on γ. Thus the decay estimate in (1.8) is "robust" with respect to the small parameter γ.

An interesting question that arises is: What happens to solutions of (1.5) when $\gamma \to 0$? A *formal* limit of system (1.5) is, of course, equation (1.1). Moreover, the uniformity (with respect to γ) of the estimate in (1.8) indicates that a similar estimate should hold for solutions to (1.1) However, this limit process is not obvious at all, due to nonlinearities appearing on the boundary. Indeed, (1.8) implies weak convergence (on a subsequence) of solutions $w_\gamma(t)$. This, however, is not sufficient to pass with the limit on nonlinear boundary conditions. On the other hand, techniques used in this paper will allow us to justify, in a rigorous manner, the limit process taken on (1.5) and leading to system (1.1). Precise formulation of this result is given below.

Theorem 1.3 *Let $w_\gamma(t)$ be any solution to (1.5) such that (without loss of generality) $E_{\gamma,w}(t) \leq 1$. Then, as $\gamma \to 0$,*

$$
\begin{aligned}
w_\gamma &\to w \quad in\ C(0,\infty; H^2(\Omega)) \\
w_{\gamma,t} &\to w_t \quad in\ C(0,\infty; L_2(\Omega)),
\end{aligned}
\tag{1.9}
$$

where (w, w_t) satisfy equation (1.1). Moreover, the following uniform stability estimate holds:

$$
\|w(t)\|_{H^2(\Omega)} + \|w_t(t)\|_{L_2(\Omega)} \leq Ce^{-\omega t}, \tag{1.10}
$$

for some constants $C, \omega > 0$.

Remark 1.2: A direct proof of uniform decay rates for *smooth* solutions of the von Kármán system with $\gamma = 0$ subject to *linear* dissipation acting on the boundary of *Ω-star shaped* domain was proven in [12]. Extension of this result to *(i) nonlinear* (sublinear) boundary feedbacks, *(ii)* general domains and *(iii)* all *weak* solutions of the von Kármán equation corresponding to $H^2(\Omega) \times L_2(\Omega)$ initial data is given in [10]. The advantage of the decay estimate in (1.10) in the context of equation (1.1) is that it also holds for a *superlinear* boundary feedback, g. The price to pay is a stronger smoothness requirement imposed on the initial velocity ($H^1(\Omega)$ rather than $L_2(\Omega)$).

Remark 1.3: Results related to an existence of compact (and, in some cases, finite dimensional) attractors associated with a von Kármán system subject to nonlinear boundary dissipation are given in [15].

2 Uniqueness: Proof of Theorem 1.1

In what follows, in (1.1), we shall assume $b \equiv 0$ to simplify exposition. (Addition of this lower order term does not alter arguments.) We consider the map $G : H^2(\Omega) \times H^2(\Omega) \to L_2(\Omega)$ given by

$$\begin{aligned} \Delta^2 G &= -[u,v] \quad in\ \Omega \\ G &= \tfrac{\partial}{\partial \nu} G = 0 \quad on\ \Gamma \end{aligned} \tag{2.1}$$

The following result is critical to the proof.

Lemma 2.1 *The map $(u,v) \to G[u,v]$ is bounded from $H^2(\Omega) \times H^2(\Omega) \to W^2_\infty(\Omega)$.*

As the proof of Lemma 2.1, based on the compensated compactness method, is technical and is relegated to section 5.

Remark 2.1: Notice that the result of Lemma 2.1 improves by "ϵ" a known regularity result stating that this map is bounded from $H^{2+\epsilon}(\Omega) \times H^2(\Omega) \to W^2_\infty(\Omega)$. As we shall see later, this improvement by "ϵ" is critical.

Let w_1 and w_2 be two solutions of (1.1). Set $\tilde{w} \equiv w_1 - w_2$. Then $\tilde{w}$ satisfies

$$\begin{cases} \tilde{w}_{tt} + \Delta^2 \tilde{w} = [\mathcal{F}(w_1) - \mathcal{F}(w_2), w_1] + [\mathcal{F}(w_2), \tilde{w}] & \text{in } Q_T \\ \Delta \tilde{w} + (1-\mu) B_1 \tilde{w} = -[h_1(\tfrac{\partial}{\partial \nu} w_{1,t}) - h_1(\tfrac{\partial}{\partial \nu} w_{2,t})] & \text{on } \Sigma_T \\ \tfrac{\partial}{\partial \nu} \Delta \tilde{w} + (1-\mu) B_2 \tilde{w} = g(w_{1,t}) - g(w_{2,t}) - \tfrac{\partial}{\partial \tau}[h_2(\tfrac{\partial}{\partial \tau} w_{1,t}) - h_2(\tfrac{\partial}{\partial \tau} w_{2,t})] & \text{on } \Sigma_T. \end{cases} \tag{2.2}$$

Multiplying (2.2) by $\tilde{w}_t$ and integrating (first formally) over Q_T yields

$$\begin{aligned}\|\tilde{w}(t)\|^2_{H^2(\Omega)} + \|\tilde{w}_t(t)\|^2_{L_2(\Omega)} &+ \int_0^t \int_\Gamma [h_1(\tfrac{\partial}{\partial\nu} w_{1,t}) - h_1(\tfrac{\partial}{\partial\nu} w_{2,t})](\tfrac{\partial}{\partial\nu} w_{1,t} - \tfrac{\partial}{\partial\nu} w_{2,t}) d\Gamma dt \\ &+ \int_0^t \int_\Gamma [h_2(\tfrac{\partial}{\partial\tau} w_{1,t}) - h_2(\tfrac{\partial}{\partial\tau} w_{2,t})](\tfrac{\partial}{\partial\tau} w_{1,t} - \tfrac{\partial}{\partial\tau} w_{2,t}) d\Gamma dt \\ &+ \int_0^t \int_\Gamma [g(w_{1,t}) - g(w_{2,t})](w_{1,t} - w_{2,t}) d\Gamma dt \\ &\le \int_0^t \int_\Omega |[\mathcal{F}(w_1) - \mathcal{F}(w_2), w_1]\tilde{w}_t + [\mathcal{F}(w_2), \tilde{w}]\tilde{w}_t| d\Omega dt.\end{aligned} \tag{2.3}$$

By Lemma 2.1, $\mathcal{F}(w_i) \in W^2_\infty(\Omega)$, hence the right-hand side of (2.2) belongs to $L_2(\Omega)$. Therefore, the formal computation leading to (2.3) can be made rigorous (see [6]) by treating the nonlinear bracket, $[w, \chi(w)]$ as a perturbation of a nonlinear semigroup of contractions generated by (1.1) with $\mathcal{F} = 0$ and applying the usual (see [2]) approximation arguments in the context of nonlinear semigroup theory.

From monotonicity of h and g and using the Cauchy-Schwartz inequality, we obtain

$$\|\tilde{w}(t)\|^2_{H^2(\Omega)} + \|\tilde{w}_t(t)\|^2_{L_2(\Omega)} \le C \int_0^t \{\|[\mathcal{F}(w_1) - \mathcal{F}(w_2), w_1]\|^2_{L_2(\Omega)} + \|[\mathcal{F}(w_2), \tilde{w}]\|^2_{L_2(\Omega)}\} dt. \tag{2.4}$$

From Lemma 2.1,

$$\|\mathcal{F}(w_2)\|_{W^2_\infty(\Omega)} \le C\|w_2\|_{H^2(\Omega)}\|w_2\|_{H^2(\Omega)}. \tag{2.5}$$

Moreover,

$$\begin{aligned}\|[\mathcal{F}(w_2), \tilde{w}]\|_{L_2(\Omega)} &\le C\|\tilde{w}\|_{H^2(\Omega)}\|\mathcal{F}(w_2)\|_{W^2_\infty(\Omega)} \\ &\le C\|\tilde{w}\|_{H^2(\Omega)}\|w_2\|_{H^2(\Omega)}\|w_2\|_{H^2(\Omega)}.\end{aligned} \tag{2.6}$$

We can rewrite $\mathcal{F}(w_1) - \mathcal{F}(w_2)$ in the following way:

$$\mathcal{F}(w_1) - \mathcal{F}(w_2) = G([w_1, w_1] - [w_2, w_2]) = G[\tilde{w}, w_1 + w_2]. \tag{2.7}$$

Hence, again from Lemma 2.1,

$$\|G[\tilde{w}, w_1 + w_2]\|_{W^2_\infty(\Omega)} \leq C\|\tilde{w}\|_{H^2(\Omega)}[\|w_1\|_{H^2(\Omega)} + \|w_2\|_{H^2(\Omega)}], \tag{2.8}$$

$$\begin{aligned} \Longrightarrow \|[\mathcal{F}(w_1) - \mathcal{F}(w_2), w_1]\|_{L_2(\Omega)} &\leq C\|w_1\|_{H^2(\Omega)}\|\mathcal{F}(w_1) - \mathcal{F}(w_2)\|_{W^2_\infty(\Omega)} \\ &\leq C\|w_1\|_{H^2(\Omega)}\|\tilde{w}\|_{H^2(\Omega)}[\|w_1\|_{H^2(\Omega)} + \|w_2\|_{H^2(\Omega)}]. \end{aligned} \tag{2.9}$$

Combining (2.3), (2.6), and (2.9) proves that the solution must be unique. □

3 Existence of Weak Solutions: Proof of Theorem 1.2

Lemma 3.1 *(Local existence) Assume hypothesis (H). Then there exists $T_0 > 0$ such that for any initial data $w_0 \in H^2(\Omega)$, $w_1 \in L_2(\Omega)$, there exists a unique solution to (1.1), $w \in C(0, T_0; H^2(\Omega))$, $w_t \in C(0, T_0; L_2(\Omega))$.*

Proof: We consider the following nonlinear equation:

$$\begin{aligned} w_{tt} + \Delta^2 w = [f, w] &\quad \text{in } Q_T = (0,T) \times \Omega \\ \left.\begin{aligned} w(0,\cdot) &= w_0 \\ w_t(0,\cdot) &= w_1 \end{aligned}\right\} &\quad \text{in } \Omega \\ \Delta w + (1-\mu)B_1 w = -h_1(\tfrac{\partial}{\partial\nu} w_t) &\quad \text{on } \Sigma_T = (0,T) \times \Gamma \\ \tfrac{\partial}{\partial\nu}\Delta w + (1-\mu)B_2 w - w = g(w_t) - \tfrac{\partial}{\partial\tau} h_2(\tfrac{\partial}{\partial\tau} w_t) &\quad \text{on } \Sigma_T = (0,T) \times \Gamma, \end{aligned} \tag{3.1}$$

where the function f is a given element of $L_1(0, T; W^2_\infty(\Omega))$.

Applying a standard perturbation theorem for nonlinear semigroups (see Barbu [2]), we obtain the existence and uniqueness of the solutions to (3.1) on the space $H^2(\Omega) \times L_2(\Omega)$.

Multiplying (3.1) by w_t, integrating over Q_T and using nonlinear semigroup arguments to justify formal calculations (see [14]) gives

$$\begin{aligned}\|w_t(t)\|^2_{L_2(\Omega)} & +a(w(t),w(t)) + 2\int_0^t\int_\Gamma h_1(\tfrac{\partial}{\partial\nu}w_t)\tfrac{\partial}{\partial\nu}w_t d\Gamma dt \\ & +2\int_0^t\int_\Gamma g(w_t)w_t d\Gamma dt + 2\int_0^t\int_\Gamma h_2(\tfrac{\partial}{\partial\tau}w_t)\tfrac{\partial}{\partial\tau}w_t d\Gamma dt \\ & = \|w_1\|^2_{L_2(\Omega)} + a(w_0,w_0) + 2\int_0^t\int_\Omega [f,w]w_t d\Omega dt,\end{aligned} \tag{3.2}$$

where

$$m_1\|w\|^2_{H^2(\Omega)} \leq a(w,w) \leq M_1\|w\|^2_{H^2(\Omega)}. \tag{3.3}$$

We construct a map

$$v \longrightarrow \mathcal{T}v, \tag{3.4}$$

defined on a Banach space, X, where

$$X \equiv C(0,T;H^2(\Omega)) \cap C^1(0,T;L_2(\Omega)), \tag{3.5}$$

and $\mathcal{T}v$ is defined as the solution to (3.1) with $f(t)$ given by

$$f(t) \equiv \mathcal{F}(v(t)). \tag{3.6}$$

To prove Lemma 3.1, it suffices to construct a unique fixed point for the map $\mathcal{T}$ introduced in (3.4) and defined on $C(0,T_0;B_R)$, where $B_R \equiv \{u \in H^2(\Omega)\times L_2(\Omega) : \|u\|_{H^2(\Omega)\times L_2(\Omega)} \leq R\}$.

We shall prove that for sufficiently small values of T_0 and sufficiently large values of R, the map $\mathcal{T}$ is a contraction on $C(0,T_0;B_R)$. To accomplish this, we first note that by virtue of

Lemma 2.1, the Airy's stress function $\mathcal{F}(w)$ is locally Lipschitz: $H^2(\Omega) \to W^2_\infty(\Omega)$. Indeed, from Lemma 2.1, it follows that

$$\begin{aligned} \|\mathcal{F}(w_1) - \mathcal{F}(w_2)\|_{W^2_\infty(\Omega)} &= \|G[w_1 - w_2, w_1 + w_2]\|_{W^2_\infty(\Omega)} \\ &\le C\|w_1 - w_2\|_{H^2(\Omega)}[\|w_1\|_{H^2(\Omega)} + \|w_2\|_{H^2(\Omega)}]. \end{aligned} \tag{3.7}$$

Applying (3.2) with $w \equiv \mathcal{T}v$, where $v \in C(0,T;B_R)$, and $f \equiv \mathcal{F}(v)$, we obtain for $t \le T_0$,

$$\begin{aligned} \|\mathcal{T}v(t)\|^2_{H^2(\Omega)} + \|\tfrac{d}{dt}\mathcal{T}v(t)\|^2_{L_2(\Omega)} &\le C\{\|w_0\|^2_{H^2(\Omega)} + \|w_1\|^2_{L_2(\Omega)} + \textstyle\int_0^t \|[\mathcal{F}(v(\tau)), w(\tau)]\|^2_{L_2(\Omega)} d\tau\} \\ &\le C\{\|w_0\|^2_{H^2(\Omega)} + \|w_1\|^2_{L_2(\Omega)} + \textstyle\int_0^t \|v(\tau)\|^4_{H^2(\Omega)}\|\mathcal{T}v(\tau)\|^2_{H^2(\Omega)} d\tau\} \\ &\le C\{\|w_0\|^2_{H^2(\Omega)} + \|w_1\|^2_{L_2(\Omega)} + R^4 T_0 \|\mathcal{T}v\|^2_{C(0,T_0;B_R)}\}, \end{aligned} \tag{3.8}$$

where the first inequality follows from Lemma 2.1. Selecting R so that

$$\|w_0\|^2_{H^2(\Omega)} + \|w_1\|^2_{L_2(\Omega)} \le \frac{R^2}{4C},$$

and taking T_0 sufficiently small yields

$$\mathcal{T}(C(0,T_0;B_R)) \subset C(0,T_0;B_R). \tag{3.9}$$

Let $v_1, v_2 \in C(0,T;B_R)$. Similar arguments as above yield

$$\begin{aligned}
\|(\mathcal{T}v_1 - \mathcal{T}v_2)(t)\|^2_{H^2(\Omega)} &+ \|\tfrac{d}{dt}(\mathcal{T}v_1 - \mathcal{T}v_2)(t)\|^2_{L_2(\Omega)} \\
&\le C\int_0^t \{\|\mathcal{F}(v_1) - \mathcal{F}(v_2)\|^2_{W^2_\infty(\Omega)} \|\mathcal{T}v_1\|^2_{H^2(\Omega)} \\
&\qquad + \|\mathcal{F}(v_2)\|^2_{W^2_\infty(\Omega)} \|\mathcal{T}v_1 - \mathcal{T}v_2\|^2_{H^2(\Omega)}\} dt \\
&\le CR^4 T_0 \{\|v_1 - v_2\|^2_{C(0,T_0;H^2(\Omega))} + \|\mathcal{T}v_1 - \mathcal{T}v_2\|^2_{C(0,T_0;H^2(\Omega))}\},
\end{aligned} \tag{3.10}$$

where we have used (2.1), (2.2) and Lemma 2.1. Hence,

$$\|\mathcal{T}v_1 - \mathcal{T}v_2\|^2_{C(0,T_0;B_R)} \le \frac{CR^4T_0}{1 - CR^4T_0} \|v_1 - v_2\|^2_{C(0,T_0;H^2(\Omega))}. \tag{3.11}$$

Taking T_0 small enough yields contraction property for the map $\mathcal{T}$. The result of Lemma 3.1 now follows from the Contraction Mapping Principle. □

To complete the proof of Theorem 1.2, it suffices to establish the following a priori bound.

Lemma 3.2 *(A priori bounds) Assume hypothesis (H). Let $(w, \dot{w}_t)$ be any local solution to (1.1) such that $w \in C(0,T_0;H^2(\Omega))$, $w_t \in C(0,T_0;L_2(\Omega))$. Then the following a priori bound holds.*

$$\|w(t)\|^2_{H^2(\Omega)} + \|w_t(t)\|^2_{L_2(\Omega)} \le C(\|w_0\|^2_{H^2(\Omega)}, \|w_1\|^2_{L_2(\Omega)}) \qquad t \ge 0, \tag{3.12}$$

where $C(u,v)$ denotes a continuous function of u, v.

Proof: By using energy methods, the a priori bound in (3.12) can be proven in a standard way for *"smooth"* solutions. On the other hand, the existence of such smooth solutions for "smooth" initial data has been shown in [6]. (Here is the point where we use the differentiability of f and g.) We need to extend this bound to hold for all *weak* solutions. To this end, we select a suitable approximation of the initial data $w_0 \in H^2(\Omega)$, $w_1 \in L_2(\Omega)$ such that

$$\begin{cases} H^4(\Omega) \ni w_{0m} \to w_0 & in\ H^2(\Omega) \\ H^2(\Omega) \ni w_{1m} \to w_1 & in\ L_2(\Omega), \end{cases} \tag{3.13}$$

and w_{0m}, w_{1m} satisfy appropriate compatibility conditions. Let $w_m(t)$ denote a solution to (1.1) corresponding to initial data (w_{0m}, w_{1m}). Then by the result of Theorem 1.3 in [6], $w_m(t)$ is "smooth," hence the following inequality holds.

$$\begin{aligned} \|w_m(t)\|^2_{H^2(\Omega)} + \|w_{m,t}(t)\|^2_{L_2(\Omega)} & + \int_0^t \int_\Gamma h_1(\tfrac{\partial}{\partial \nu} w_{m,t}) \tfrac{\partial}{\partial \nu} w_{m,t} d\Gamma dt \\ & + \int_0^t \int_\Gamma g(w_{m,t}) w_{m,t} d\Gamma dt + \int_0^t \int_\Gamma h_2(\tfrac{\partial}{\partial \tau} w_{m,t}) \tfrac{\partial}{\partial \tau} w_{m,t} d\Gamma dt \\ & \leq C(\|w_0\|_{H^2(\Omega)}, \|w_1\|_{L_2(\Omega)}). \end{aligned} \tag{3.14}$$

We shall show that

$$\begin{cases} w_m \to w & in\ C(0, T_0; H^2(\Omega)) \\ w_{m,t} \to w_t & in\ C(0, T_0; L_2(\Omega)), \end{cases} \tag{3.15}$$

where, we recall, w is a weak solution to (1.1). Indeed,

$$\begin{aligned}\|(w_n - w_m)(t)\|^2_{H^2(\Omega)} &+ \|(w_{n,t} - w_{m,t}(t)\|^2_{L_2(\Omega)} + \int_0^t \int_\Gamma [h_1(\tfrac{\partial}{\partial\nu} w_{n,t}) \\ &- h_1(\tfrac{\partial}{\partial\nu} w_{m,t})] \tfrac{\partial}{\partial\nu}(w_{n,t} - w_{m,t}) d\Gamma dt \\ &+ \int_0^t \int_\Gamma [g(w_{n,t}) - g(w_{m,t})](w_{n,t} - w_{m,t}) d\Gamma dt \\ &+ \int_0^t \int_\Gamma [h_2(\tfrac{\partial}{\partial\tau} w_{n,t}) - h_2(\tfrac{\partial}{\partial\tau} w_{m,t})] \tfrac{\partial}{\partial\tau}(w_{n,t} - w_{m,t}) d\Gamma dt \\ &\le C(\|w_{0n} - w_{0m}\|^2_{H^2(\Omega)}, \|w_{1n} - w_{1m}\|^2_{L_2(\Omega)}) \\ &+ C \int_0^t \|\mathcal{F}(w_n) - \mathcal{F}(w_m)\|^2_{W^2_\infty(\Omega)} \|w_n\|^2_{H^2(\Omega)} d\tau \\ &+ C \int_0^t \|\mathcal{F}(w_m)\|^2_{W^2_\infty(\Omega)} \|w_n - w_m\|^2_{H^2(\Omega)} d\tau.\end{aligned} \tag{3.16}$$

From (3.7) and (3.14),

$$\begin{aligned}\int_0^t \|\mathcal{F}(w_n) - \mathcal{F}(w_m)\|^2_{W^2_\infty(\Omega)} & \|w_n\|^2_{H^2(\Omega)} d\tau \\ &\le C(\|w_0\|_{H^2(\Omega)}, \|w_1\|_{L_2(\Omega)}) \int_0^t \|w_n(\tau) - w_m(\tau)\|^2_{H^2(\Omega)} d\tau.\end{aligned} \tag{3.17}$$

From Lemma 2.1 and (3.14),

$$\begin{aligned}\int_0^t \|\mathcal{F}(w_m(\tau))\|^2_{W^2_\infty(\Omega)} \|w_n(\tau) & - w_m(\tau)\|^2_{H^2(\Omega)} d\tau \\ &\le C(\|w_0\|_{H^2(\Omega)}, \|w_1\|_{L_2(\Omega)}) \int_0^t \|w_n(\tau) - w_m(\tau)\|^2_{H^2(\Omega)} d\tau.\end{aligned} \tag{3.18}$$

Inserting (3.17) and (3.18) into (3.16) and applying Gronwall's inequality with (3.13) gives

$$\begin{cases} w_m \to w^* & in\ C(0,T; H^2(\Omega)) \\ w_{m,t} \to w_t^* & in\ C(0,T; L_2(\Omega)). \end{cases} \tag{3.19}$$

Moreover, from (H) and (3.16), (3.19), we also obtain that

$$\lim_{n,m\to\infty}\{\int_0^t\int_\Gamma[h(\tfrac{\partial}{\partial\nu}w_{n,t})-h(\tfrac{\partial}{\partial\nu}w_{m,t})]\tfrac{\partial}{\partial\nu}(w_{n,t}-w_{m,t})d\Gamma dt$$
$$+\int_0^t\int_\Gamma[h_2(\tfrac{\partial}{\partial\tau}w_{n,t})-h_2(\tfrac{\partial}{\partial\tau}w_{m,t})]\tfrac{\partial}{\partial\tau}(w_{n,t}-w_{m,t})d\Gamma dt\}\to 0. \tag{3.20}$$

Hence, by (H),

$$\|\nabla(w_{n,t}-w_{m,t})\|_{L_2(\Sigma)}\to 0. \tag{3.21}$$

From (3.7), (3.19), and properties of the bracket, [,], it follows that

$$[\mathcal{F}(w_m(t)),w_m(t)]\to[\mathcal{F}(w^*(t)),w^*(t)]\ in\ L_2(\Omega). \tag{3.22}$$

The result in (3.21), (3.22) allows us to deduce (by standard arguments) that w^* satisfies the original equation (1.1). Thus, by Theorem 1.1, $w^*\equiv w$ is a unique solution to (1.1).

Convergence in (3.15) together with passage to the limit on inequality (3.14) yields the desired a priori bound in Lemma 3.2. □

4 Proof of Lemma 2.1

Let $f=G[u,v]$, i.e.,

$$\Delta^2 f=[u,v];\quad f=0,\ \tfrac{\partial}{\partial\nu}f=0\ on\ \Gamma. \tag{4.1}$$

By the Closed Graph Theorem, it suffices to prove that for u, $v\in H^2(\Omega)$, $f\in W^2_\infty(\Omega)$. By using partition of unity, it is enough to consider the case when both u and v are supported in the neighborhood of the point $x_0\in\bar{\Omega}$. We shall consider two cases: *(i)* $x_0\in Int(\Omega)$, *(ii)*

$x_0 \in \Gamma$.

Case *(i)*: We introduce a bilinear continuous operator $\mathcal{B}$ from $C_0^\infty(R^2) \times C_0^\infty(R^2)$ into $\mathcal{D}'(R^2)$ defined as

$$\mathcal{B}(u,v) \equiv [\Delta^{-1}u, \Delta^{-1}v],$$

where Δ^{-1} denotes the inverse of the Laplacian on R^2. Let $b(\xi,\eta)$ be a symbol associated with $\mathcal{B}$ (see [4], page 29) defined by

$$\mathcal{B}(e^{i\xi x}e^{i\eta x}) = b(\xi,\eta)e^{i(\xi+\eta)x}, \quad \xi \in R^2 \eta \in R^2, x \in R^2.$$

Then

$$b(\xi,\eta) = \frac{\xi_1^2\eta_2^2 + \xi_2^2\eta_1^2 - 2\xi_1\xi_2\eta_1\eta_2}{(\xi_1^2+\xi_2^2)(\eta_1^2+\eta_2^2)}. \tag{4.2}$$

Since $b(\xi,-\xi) = 0$ for all $\xi \neq 0$, the result of Theorem VI in [4] applies and tells us that $\mathcal{B}$ is bounded from $L_2(R^2) \times L_2(R^2) \to H_1(R_2)$, where H_1 is a real Hardy space (see [5]). This means that for $u \in H^2(\Omega)$, $v \in H^2(\Omega)$ (where, we recall, u and v are supported away from Γ),

$$[u,v] \in H_1. \tag{4.3}$$

Let $f_0 \equiv G_0[u,v]$, where $G_0 \in OPS^{-4}$ with a symbol $g_0 = \frac{1}{(\xi_1^2+\xi_2^2)^2}$. By OPS^n, we denote, as usual (see [18]), a class of pseudodifferential operators of order n.

Since $[u,v]$ is supported away from Γ, f_0 differs from f by a C^∞ function. From

$$D^\alpha G_0 \in OPS^0 \; for \; |\alpha| \leq 4, \tag{4.4}$$

where D^α stands for the differential operator of order $|\alpha|$, we obtain, by Theorem 26 of [5] (see page 121), (4.3) and (4.4), that

$$D^\alpha G_0[u,v] \subset L_1(R^2),\ for\ |\alpha| \leq 4. \tag{4.5}$$

By Sobolev's Imbedding (see [1], page 97, (7)),

$$f_0 \in W_1^4(\Omega) \subset W_\infty^2(\Omega)\ \ and\ \ f \in W_\infty^2(\Omega). \tag{4.6}$$

Case *(ii)*: Let $x_0 \in \Gamma$. By a local change of coordinates, we flatten the boundary and consider the extended functions $u, v \in H^2(R^2)$ with compact support (i.e., supported in the neighborhood of x_0). We denote by $\mathcal{L}$ the differential operator (with variable coefficients) defined on R^2 which corresponds to Δ^2 in new coordinates. Writing $[u,v]$ in new coordinates produces a symbol whose principal part satisfies the conditions of Theorem VI in [4] and the lower-order terms are the products of $D^\alpha u D^\beta v$, $|\alpha| + |\beta| \leq 3$. Thus, the contribution of lower-order terms is in $L_{1+\epsilon}(R^2)$, where $\epsilon > 0$. Therefore, a straightforward modification (to account for variable coefficients) of the compensated compactness result of [4] applied to the principal part of the symbol of $[u,v]$ gives

$$[u,v] = w + z,\ w \in H_1,\ z \in L_{1+\epsilon}(R^2). \tag{4.7}$$

It follows from the definition of H_1 space and boundedness of the Riesz transform on $L_p(R^2)$ $(p > 1)$ spaces, that the element $z \in L_{1+\epsilon}(R^2)$ with compact support can be written as

$z = z_1 + z_2$, where $z_1 \in H_1$, $z_2 \in C^\infty$. Thus, without loss of generality, we may write that

$$[u, v] = w + z, \; w \in H_1, \; z \in C^\infty. \tag{4.8}$$

We denote by W a pseudodifferential operator in OPS^{-4} whose symbol is compactly supported in R^2 and is an approximation of $\mathcal{L}^{-1}$ in a small neighborhood of x_0. This is to say that $(W\mathcal{L} - I)u \in C^\infty$ whenever u is supported in a neighborhood of x_0. (Recall $\mathcal{L}$ represents the biharmonic operator written in the new coordinates and extended to R^2.) We define

$$g \equiv W[u, v]. \tag{4.9}$$

By (4.8),

$$g = W[u, v] = Ww + Wz \in Ww + C^\infty, \; w \in H_1. \tag{4.10}$$

From the construction of W and $\mathcal{L}$, it follows that

$$\mathcal{L}f - \mathcal{L}g \in C^\infty(R^2). \tag{4.11}$$

We shall show that

$$g \in W^2_\infty(R^2). \tag{4.12}$$

Indeed, since $W \in OPS^{-4}$, $D^\alpha W \in OPS^0$ for $|\alpha| \leq 4$. Theorem 4 in [8] then gives

$$D^\alpha Ww \in h_1, \quad |\alpha| \leq 4, \tag{4.13}$$

where h_1 is a local Hardy space (see [8]). By using Lemma 4 in [8] together with (4.13), we obtain

$$D^\alpha Ww \in H_1 + C^\infty, \quad |\alpha| \leq 4. \tag{4.14}$$

Combining (4.10) with (4.14) gives

$$D^\alpha g \in H_1 + C^\infty, \quad |\alpha| \leq 4. \tag{4.15}$$

This, in particular, implies

$$g \in W_1^4(R^2) \subset W_\infty^2(R^2), \tag{4.16}$$

which proves (4.12).

We next consider the following elliptic problem:

$$\begin{cases} \mathcal{L}h = 0 & in\ \Omega_{x0} \\ h = g_1 & on\ \Gamma_{x0} \\ \frac{\partial}{\partial\nu} = g_2 & on\ \Gamma_{x0}, \end{cases} \tag{4.17}$$

where $g_1 \equiv -g|_\Gamma$, $g_2 \equiv -\frac{\partial}{\partial\nu}g|_\Gamma$, and Ω_{x_0} denotes $\Omega \cap U(x_0)$, $\Gamma_{x_0} = \Gamma \cap U(x_0)$. From (4.15) we infer $D^\alpha g \in L_1(R^2)$, $|\alpha| \leq 4$, and from (4.10), $D^\alpha R_\tau g \in L_1(R^2)$, $|\alpha| \leq 4$, where $R_\tau \in OPS^0$ denotes the Riesz transform in the tangential direction (see [5]). Hence, both g, $R_\tau g \in W_1^4(R^2)$. Applying trace theory on L_1 spaces yields $g|_\Gamma \in W_1^3(R)$, $R_\tau g|_\Gamma \in W_1^3(R)$,

$\frac{\partial}{\partial\nu}g|_\Gamma \in W_1^2(R)$, $\frac{\partial}{\partial\nu}R_\tau g|_\Gamma \in W_1^2(R)$. From the definition of Hardy's spaces [5], we infer

$$\begin{cases} D^\alpha g_1 \in H_1(R^1) + C^\infty(R^1), & |\alpha| \le 3 \\ D^\alpha g_2 \in H_1(R^1) + C^\infty(R^1), & |\alpha| \le 2. \end{cases} \tag{4.18}$$

Proposition 4.1 *Let h satisfy (4.17) with g_1, g_2 subject to (4.18). Then $h \in C^2(\Omega)$.*

Proof: We shall use a decoupling procedure as in [18], Chapter 5. Let x represent the normal outward direction to the boundary Γ and y represent the tangential direction. By the collar neighborhood theorem, it suffices to consider $x \in [0,1]$, $y \in R^1$.

By Proposition 2.1 in [18], the solution h to (4.17) can be written as

$$h(x,y) = A_1(x,y,D_y)P^{cx}g_1 + A_2(x,y,D_y)P^{cx}g_2 + h_0, \tag{4.19}$$

where $h_0 \in C^\infty(\Omega_{x0})$ and

$$(P^{cx}g_i)(y) \equiv \int_{R^1} e^{-x|\xi|}e^{iy\xi}\hat{g}_i(\xi)d\xi,$$

$\hat{g}_i$ are Fourier transforms of g_i, and the constant $c \ge 0$ is determined from the ellipticity constant of $\mathcal{L}$.

The pseudodifferential operators, A_1, A_2 satisfy (see Proposition 2.1 in [19])

$$\begin{cases} D_x^j A_1(x,y,D_y) \in OPS^j \\ D_x^j A_2(x,y,D_y) \in OPS^{j-1} \end{cases} \quad j = 0,1,2,... \tag{4.20}$$

Taking derivatives up to the order 3 in (4.19) yields

$$D^\alpha h = B_1(x,y,D_y)P^{cx}D_y^3 g_1 + B_2(x,y,D_y)P^{cx}D_y^2 g_2, \quad |\alpha| \le 3, \tag{4.21}$$

and $B_i(x,y,D_y) \in OPS^0$.

From (4.18) and the structure of the operator P^{cx}, it follows that

$$\begin{cases} P^{cx}D_y^3 g_1 \in C([0,1]; H_1(R^1)) \\ P^{cx}D_y^2 g_2 \in C([0,1]; H_1(R^1)). \end{cases} \tag{4.22}$$

By Theorem 26 in [5], and (4.21), (4.22), we infer

$$D^\alpha h \in C([0,1]; L_1(R^1)), \quad |\alpha| \le 3. \tag{4.23}$$

Hence, by standard Sobolev's Imbeddings, $D^2 h \in C([0,1]; C(\Gamma_{x_0}))$, which implies $h \in C^2(\Omega_{x_0})$, as desired. □

From (4.11), it follows that

$$\begin{cases} \mathcal{L}(f-g) \in C^\infty(\Omega_{x_0}) \\ (f-g)|_\Gamma = g_1 & on\ \Gamma_{x_0} \\ \frac{\partial}{\partial\nu}(f-g)|_\Gamma = g_2 & on\ \Gamma_{x_0}. \end{cases} \tag{4.24}$$

Hence, by the result of Proposition 8.1 and standard elliptic regularity,

$$f - g \in C^2(\Omega_{x_0}). \tag{4.25}$$

Combining (4.25) with (4.12) yields $f \in W^2_\infty(\Omega_{x_0})$. □

5 Uniform Decay Rates: Proof of Theorem 1.3

Let $w_\gamma(t)$ be a solution to (1.5) corresponding to the initial data $(w_0, w_1) \in H^2(\Omega) \times H^1(\Omega)$. Without loss of generality, we assume $E_{\gamma,w}(0) \leq 1$.

By arguments identical to those used in Lemma 3.2 (see (3.16)-(3.19)), we obtain

$$\begin{aligned}
&\|w_{\gamma_1}(t) - w_{\gamma_2}(t)\|^2_{H^2(\Omega)} + \gamma^2\|\nabla(w_{\gamma_1,t}(t) - w_{\gamma_2,t}(t))\|^2_{L_2(\Omega)} + \|w_{\gamma_1,t}(t) - w_{\gamma_2,t}(t)\|^2_{L_2(\Omega)} \\
&\quad + \textstyle\int_0^t \int_\Gamma [h_1(\frac{\partial}{\partial\nu} w_{\gamma_1,t}) - h_1(\frac{\partial}{\partial\nu} w_{\gamma_2,t})]\frac{\partial}{\partial\nu}(w_{\gamma_1,t} - w_{\gamma_2,t})d\Gamma dt \\
&\quad + \textstyle\int_0^t \int_\Gamma [h_2(\frac{\partial}{\partial\tau} w_{\gamma_1,t}) - h_2(\frac{\partial}{\partial\tau} w_{\gamma_2,t})]\frac{\partial}{\partial\tau}(w_{\gamma_1,t} - w_{\gamma_2,t})d\Gamma dt \qquad (5.1) \\
&\quad + \textstyle\int_0^t \int_\Gamma [g(w_{\gamma_1,t}) - g(w_{\gamma_2,t})](w_{\gamma_1,t} - w_{\gamma_2,t})d\Gamma dt \\
&\longrightarrow 0 \quad as\ \gamma_1, \gamma_2 \longrightarrow 0.
\end{aligned}$$

Hence, in particular,

$$\begin{aligned}
w_\gamma &\rightarrow w^* \quad in\ C(0,T;H^2(\Omega)) \\
w_{\gamma,t} &\rightarrow w^*_t \quad in\ C(0,T;L_2(\Omega)),
\end{aligned} \qquad (5.2)$$

$$\lim_{\gamma_1,\gamma_2\rightarrow 0}\int_0^T\int_\Gamma [h_1(\tfrac{\partial}{\partial\nu} w_{\gamma_1,t}) - h_1(\tfrac{\partial}{\partial\nu} w_{\gamma_2,t})]\tfrac{\partial}{\partial\nu}(w_{\gamma_1,t} - w_{\gamma_2,t})d\Gamma dt \longrightarrow 0, \qquad (5.3)$$

$$\lim_{\gamma_1,\gamma_2\rightarrow 0}\int_0^T\int_\Gamma [h_2(\tfrac{\partial}{\partial\tau} w_{\gamma_1,t}) - h_2(\tfrac{\partial}{\partial\tau} w_{\gamma_2,t})]\tfrac{\partial}{\partial\tau}(w_{\gamma_1,t} - w_{\gamma_2,t})d\Gamma dt \longrightarrow 0, \qquad (5.4)$$

$$\lim_{\gamma_1,\gamma_2\rightarrow 0}\int_0^T\int_\Gamma [g(w_{\gamma_1,t}) - g(w_{\gamma_2,t})](w_{\gamma_1,t} - w_{\gamma_2,t})d\Gamma dt \longrightarrow 0. \qquad (5.5)$$

Moreover, from (H), we also obtain

$$\|\nabla(w_{\gamma_1,t} - w_{\gamma_2,t})\|_{L_2(\Sigma_T)} \rightarrow 0. \qquad (5.6)$$

Hence,

$$h_1(\frac{\partial}{\partial\nu}w_{\gamma,t}) \longrightarrow h_1(\frac{\partial}{\partial\nu}w_t^*) \quad in\ L_2(0,T;\Gamma) \tag{5.7}$$

$$h_2(\frac{\partial}{\partial\tau}w_{\gamma,t}) \longrightarrow h_2(\frac{\partial}{\partial\tau}w_t^*) \quad in\ L_2(0,T;\Gamma) \tag{5.8}$$

From $E_{w_\gamma}(t) \leq 1$ and from Sobolev's imbeddings, we obtain

$$\sup_{t\geq 0} \|w_{\gamma,t}(t)\|_{L_p(\Gamma)} \leq C \qquad for\ all\ p \geq 1. \tag{5.9}$$

Moreover, (5.6) and Sobolev's imbeddings give

$$\int_0^T \|w_{\gamma_1,t} - w_{\gamma_2,t}\|^2_{L_p(\Gamma)} dt \to 0 \qquad for\ all\ p \geq 1. \tag{5.10}$$

This together with (5.9) and hypothesis (H) gives

$$g(w_{\gamma,t}) \to g(w_t^*) \qquad in\ L_2(\Sigma_T). \tag{5.11}$$

Convergence in (5.2), (5.7), (5.8), and (5.11), together with

$$[\mathcal{F}(w_\gamma(t)), w_\gamma(t)] \longrightarrow [\mathcal{F}(w^*(t)), w^*(t)], \tag{5.12}$$

(which in turn follows from (5.2) and (3.7), (3.19)), allow us to pass with a limit on equation (1.5). Thus, we conclude that $w^* = w$ satisfies the original equation (1.1).

Convergence in (5.2) together with (1.8) then implies (1.10). □

References

[1] R. A. Adams. *Sobolev spaces.* Academic Press, New York, 1975.

[2] V. Barbu. *Nonlinear semigroups and differential equations in Banach spaces.* Editura Academiei, Bucuresti, 1976.

[3] I. D. Chueshov. Strong solutions and the attractor of the von Kármán equations. *Math. USSR Sbornik*, 69(1):25–36, 1991.

[4] R. Coiffman, P. L. Lions, Y. Meyer, and S. Semmes. Compensated compactness and Hardy spaces. *J. Math. Pure Appl.*, 72:247–286, 1993.

[5] R. Coiffman and Y. Meyer. Au dela des operateurs pseudodifferentiels. *Asterisque*, 57, 1978.

[6] A. Favini, M. A. Horn, I. Lasiecka, and D. Tataru. Global existence, uniqueness and regularity of solutions to the dynamic von Kármán system with nonlinear boundary dissipation. *Matematisch Annalen.* Submitted.

[7] A. Favini and I. Lasiecka. Second-order abstract equations with nonlinear boundary conditions: Applications to a von Kármán system with nonlinear boundary damping. In G. Dore, A. Favini, E. Obrecht, and A. Venni, editors, *Differential Equations in Banach Spaces*, volume 148 of *Lecture Notes in Pure and Applied Mathematics.* Marcel Dekker, 1993.

[8] D. Goldberg. A local version of real Hardy spaces. *Duke Mathematics Journal*, 46:25–42, 1979.

[9] M. A. Horn and I. Lasiecka. Global stabilization of a dynamic von Kármán plate with nonlinear boundary feedback. *Applied Mathematics and Optimization.* To appear.

[10] M. A. Horn and I. Lasiecka. Uniform decay of weak solutions to a von Kármán plate with nonlinear boundary dissipation. To appear.

[11] H. Koch and A. Stahel. Global existence of classical solutions to the dynamical von Kármán equations. November 1991. Preprint.

[12] J. E. Lagnese. *Boundary Stabilization of Thin Plates.* Society for Industrial and Applied Mathematics, Philadelphia, 1989.

[13] J. E. Lagnese and G. Leugering. Uniform stabilization of a nonlinear beam by nonlinear boundary feedback. *Journal of Differential Equations*, 91(2):355–388, 1991.

[14] I. Lasiecka. Existence and uniqueness of the solutions to second order abstract equations with nonlinear and nonmonotone boundary conditions. *Journal of Nonlinear Analysis, Methods and Applications.* To appear.

[15] I. Lasiecka. Finite dimensionality of attractors associated with von Kármán plate equations and boundary damping. *Journal of Differential Equations.* To appear.

[16] J. L. Lions. *Quelques methodes de resolution des problemes aux limites nonlineaires.* Dunod, Paris, 1969.

[17] A. Stahel. A remark on the equation of a vibrating plate. *Proceedings of the Royal Society of Edinburgh*, 106A:307–314, 1987.

[18] M. E. Taylor. *Pseudodifferential Operators.* Princeton University Press, Princeton, New Jersey, 1981.

[19] W. von Wahl. On nonlinear evolution equations in a Banach space and on nonlinear vibrations of the clamped plate. *Bayreuther Mathematische Schriften*, 7:1–93, 1981.

[20] W. von Wahl. Corrections to my paper: On nonlinear evolution equations in a Banach space and on nonlinear vibrations of the clamped plate. *Bayreuther Mathematische Schriften*, 20:205–209, 1985.

[21] I. I. Vorovič. On some direct methods in the nonlinear theory of vibrations of curved shells. *Izv. Akad. Nauk. SSSR Ser. Mat.*, 21:747–784, 1957.

11 Optimal Control Hyperbolic Systems with Bounded Variation of Controls

Dariusz Idczak Łódź University, Łódź, Poland

Stanisław Walczak Łódź University, Łódź, Poland

ABSTRACT. In this paper we consider a hyperbolic control system with a cost functional depending on the total variation of controls. We prove Helly's principle of choice for functions of several variables with bounded variation and sufficient condition for the existence of an optimal solution for some hyperbolic systems with distributed parameters. Moreover, an application is given for some chemical process.

1. INTRODUCTION

Let us consider a control system of the form

$$\dot{x} = f(t, x, u), \tag{1.1}$$

$$u(\cdot) \in U, \qquad (x(0), x(T)) \in V, \tag{1.2}$$

where $f : [0,T] \times R^n \times R^m \to R^n$, U is a given set of controls and V is a subset of R^{2n}. In a standard optimal control problem, the cost functional has the form

$$I(x,u) = \int_0^T f^0(t,x,u)\,dt \tag{1.3}$$

where $f^0 : [0,T] \times R^n \times R^m \to R$. The problem consists in the determination of the minimum of the functional $I(x,u)$ on the set of all admissible processes (x,u) satisfying conditions (1.1)–(1.2).

It is well known that apparently well-posed problem (1.1)–(1.3) may not have solutions in the classical sense, even if f and f^0 are smooth (for details, see [3, 7, 12]). To eliminate this awkwardness of the theory, many authors considered special classes of controls ([2, 11, 15]) or a special form of the cost functional which depends not only on the state $(x(t), u(t))$ but also on the variation of controls ([8, 10, 11,

Supported by KBN grant 211029101 .

12, 13]). Moreover, in many technical questions related to automatic control, it is necessary to take account of the costs connected with the number of switchings of a control and the speed of its changes (in other words it is necessary to take account of the total variation of control).

In recent papers [10, 12, 13] the authors consider system (1.1–1.2) with a performance index of the form

$$I(x,u) = \int_0^T f^0(t,x,u)dt + \bigvee_0^T (Du) \tag{1.4}$$

where D is a nonsingular $m \times m$ matrix and $\bigvee_0^T(Du)$ denotes the total variation of the function Du on the interval $[0,T]$.

In papers [12, 13], effective methods of the numerical solving of such problems are given.

In this paper we consider an optimization problem, analogous to problem (1.1), (1.2), (1.4), for systems with distributed parameters of the form

$$\frac{\partial^2 z}{\partial x \partial y} = f(x,y,z,\frac{\partial z}{\partial x},\frac{\partial z}{\partial y},u). \tag{1.5}$$

with a cost functional

$$I(z,u) = \int_0^1 \int_0^1 f^0(x,y,z,u)dxdy + \bigvee(Du). \tag{1.6}$$

The basic results of this paper is a theorem on the existence of an optimal solution for system (1.5), (1.6) (cf. th. 4.3). The proof of this theorem is based on Helly's principle choice for functions of two variables (cf. th. 4.2). In section 3 we consider some chemical interpretation of the optimal control problem described by system (1.5), (1.6).

2. Preliminaries

Let z be a function defined on P^2 with values in R where $P^2 = \{(x,y) \in R^2;\ 0 \leq x \leq 1,\ 0 \leq y \leq 1\}$. By F_z we shall denote a function of an interval defined by the formula

$$F_z(Q) = z(x^2,y^2) - z(x^1,y^2) + z(x^1,y^1) - z(x^2,y^1) \tag{2.1}$$

where $Q = [x^1, y^1] \times [x^2, y^2] \subset P^2$.

The function F_z is called a function of an interval, associated with the function z (cf. [9]).

We shall say that a function F of an interval $Q \subset P^2$ is nonnegative if $F(Q) \geq 0$ for each interval Q, and has a bounded (or finite) variation if

$$\text{(2.2)} \qquad \operatorname{var} F = \sup\{\sum_{i=1}^{n} |F(Q_i)| : \bigcup_{i=1}^{n} Q_i = P^2, \quad n \geq 1\} < \infty$$

where Q_i are intervals such that $\operatorname{Int} Q_i \cap \operatorname{Int} Q_j = \emptyset$ for $i \neq j$, $i, j = 1, 2, \dots, n$ (cf. [9]).

A function $z : P^2 \to R$ will be called a nondecreasing function if $z(\cdot, 0), z(0, \cdot)$ are nondecreasing functions of one variable and the function F_z associated with z is nonnegative.

A function $z : P^2 \to R$ will be called a function of finite variation if the functions $z(\cdot, 0), z(0, \cdot)$ are functions of finite variation on $[0, 1]$ and F_z has a finite variation (cf. (2.1) and (2.2)).

By the variation of z on the interval P^2 we shall mean the number (cf. [1, 4])

$$\text{(2.3)} \qquad \operatorname{var} z = \operatorname{var} z(\cdot, 0) + \operatorname{var}(0, \cdot) + \operatorname{var} F_z.$$

The space of all functions of finite variation will be denoted by $BV(P^2, R)$.

One can prove the following theorems which characterize functions of finite variation.

Theorem 2.1 (cf. [4]). *If $z \in BV(P^2, R)$, then, for each $(x_0, y_0) \in \operatorname{Int} P^2$, there exist*

$$\lim z(x, y) \;\; \textit{if} \;\; (x, y) \to (x_0, y_0), \quad x < x_0 \;\; \text{and} \;\; y < y_0,$$
$$\lim z(x, y) \;\; \textit{if} \;\; (x, y) \to (x_0, y_0), \quad x_0 < x \;\; \text{and} \;\; y_0 < y.$$

Theorem 2.2 (cf. [1]). *A function of finite variation is continuous on P^2 a.e. Its discontinuity points lie on at most countable number of intervals of the form $x = x_k$, $y = y_l$, $0 \leq x_k \leq 1$, $0 \leq y_l \leq 1$, where k, l are positive integers.*

Theorem 2.3 (The Jordan distribution theorem, cf. [4, 5]). *If $z \in BV(P^2, R)$, then z may be represented in the form $z = \varphi - \psi$ where φ and ψ are nondecreasing functions of two variables, and*

$$\varphi(x,y) = g^1(x) + g^2(y) - \frac{1}{2}f(0,0) + F_f^+([0,x] \times [0,y]),$$

$$(2.4) \qquad \psi(x,y) = h^1(x) + h^2(y) + \frac{1}{2}f(0,0) + F_f^-([0,x] \times [0,y]),$$

$$z(x,0) = g^1(x) - h^1(x), \qquad z(0,y) = g^2(y) - h^2(y),$$

g^1, g^2, h^1, h^2 are nondecreasing functions on $[0,1]$, F_z^+ and F_z^- denote the upper and the lower variations of the function F_z, respectively (cf. [9]).

Now, we shall recall the definition of absolutely continuous functions of two variables (cf [16]).

A function $z : P^2 \to R$ is called absolutely continuous on P^2 if the associated function $F_z(Q)$ given by formula (2.1) is an absolutely continuous function of the interval $Q \subset P^2$ (cf. [9]) and each of the functions $z(x,0)$ and $z(0,y)$ is an absolutely continuous function of one variable on the interval $[0,1]$. The space of absolutely continuous functions on P^2 will be denoted by $AC(P^2, R)$.

One can prove that a necessary and sufficient condition for z to belong to $AC(P^2, R)$ is the following integral representation:

$$(2.5) \qquad z(x,y) = \int_0^x \int_0^y l^{1,2}(x,y) + \int_0^x l^1(x) + \int_0^y l^2(y) + c$$

where $l^{1,2} \in L^1(P^2)$, $l^1, l^2 \in L^1([0,1])$, $c \in R$ (cf. [16]).

Moreover, the function $z \in AC(P^2, R)$ possesses the classical partial derivatives $\dfrac{\partial z}{\partial x} = \int_0^y l^{1,2} + l^1$, $\dfrac{\partial z}{\partial y} = \int_0^x l^{1,2} + l^2$ and $\dfrac{\partial^2 z}{\partial x \partial y} = l^{1,2}$ defined for $(x,y) \in P^2$ a.e. (cf. [17]).

The space $AC(P^2, R)$ with the norm

$$(2.6) \qquad \| z \|_{AC} = \| l^{1,2} \|_{L^1(P^2)} + \| l^1 \|_{L^1([0,1])} + \| l^2 \|_{L^1([0,1])} + |c|.$$

is a Banach space.

3. Formulation of the Problem and Its Chemical Interpretation

Denote by $AC(P^2, R^m)$ the space of vector functions $z = (z^1, z^2, \dots, z^m)$ where $z^i \in AC(P^2, R)$ for $i = 1, 2, \dots, m$. The space $AC(P^2, R^m)$ will be referred to as a space of trajectories.

Let $M \subset R^r$ be any compact set. Denote by U the set of all functions $u = (u^1, u^2, \dots, u^r) : P^2 \to R^r$ such that $u(x,y) \in M$ for $(x,y) \in P^2$ a.e. In the class of admissible controls $u \in U$ and admissible trajectories $z \in AC(P^2, R^m)$, let us consider a hyperbolic control system of the form

$$(3.1) \qquad \frac{\partial^2 z}{\partial x \partial y} = z(x, y, z, \frac{\partial z}{\partial x}, \frac{\partial z}{\partial y}, u), \qquad z(x,0) = 0, \quad z(0,y) = 0,$$

with a cost functional

$$(3.2) \qquad I(z,u) = \int_{P^2} f^0(x, y, z(x,y), u(x,y)) dx\, dy + \sum_{i=1}^{r} \lambda_i \operatorname{var} u^i$$

where $f = (f^1, f^2 \dots, f^m)$, $\lambda_i > 0$, and $\operatorname{var} u^i$ denotes the variation of the function u^i on the interval P^2, $i = 1, 2, \dots, r$ (cf. formula (2.3)).

We shall assume that:

(3.3) there exists a constant $L > 0$ such that

$$|f(x, y, w_0, w_1, w_2, u) - f(x, y, z_0, z_1, z_2, u)| \le L(|w_0 - z_0| + |w_1 - z_1| + |w_2 - z_2|)$$

for any $(x,y) \in P^2$, $w_i, z_i \in R^m$, $i = 0, 1, 2$;

(3.4) for any $z_i \in R^m$ and $u \in M$, the function $f(\cdot, \cdot, z_0, z_1, z_2, u)$ is measurable on P^2; for any $(x,y) \in P^2$ and $z_i \in R^m$, the function $f(x, y, z_0, z_1, z_2, \cdot)$ is continuous on M,

(3.5) $f^0(\cdot, \cdot, z, u)$ is measurable with any $z \in R^m$ and $u \in M$; $f(x, y, \cdot, \cdot)$ is continuous for $(x,y) \in P^2$; for any $R > 0$, there exists an integrable function $\varphi_R = \varphi_R(x,y)$ such that $|f^0(x, y, z, u)| \le \varphi_R(x,y)$ where $(x,y) \in P^2$, $|z| \le R$, $u \in M$.

Basing oneself on the Banach contradiction principle, one can prove the following lemma (cf. [6]).

Lemma 3.1. *If the function f satisfies conditions (3.3) and (3.4), then, for each measurable function $u : P^2 \to M$, there exists a unique solution $z \in AC(P^2, R^m)$ of system (3.1).*

From Lemma 3.1 it follows that, for any admissible control $u \in U$, there exists exactly one solution of system (3.1) in the space of absolutely continuous functions

$AC(P^2, R^m)$. This solution will be denoted by z_u and called the trajectory of system (3.1), corresponding to the control u.

The following lemma holds (cf. [6]).

Lemma 3.2. *If assumptions (3.3)–(3.4) are satisfied and the sequence of controls* $u_n \in U$ *tends to* $u_0 \in U$ *at each point* $(x, y) \in P^2$*, then the sequence of trajectories* z_{u_n} *tends to* z_{u_0} *in the space* $AC(P^2, R^m)$ *(cf. (2.3)).*

From the above lemmas it follows that control system (3.1) is well-posed. In Section 5 we shall prove that optimal control problem (3.1)–(3.2) possesses a solution in the set of admissible controls U. Now, we shall present some chemical interpretation of system (3.1)–(3.2).

Consider a gas filter made in the form of a pipe filled up with a substance S which absorbs a poison gas. Through the filter a mixture of air and gas is pressed at a speed $v = v(x,t) > a > 0$ with the aid of an aggregation A. Denote by $\bar{u} = \bar{u}(x,t)$ the quantity of the poison gas being present in the capacity unit of the substance S at a distance x from the inlet of the filter and at a moment t. Assume the speed $v = v(x,t)$ to be so great that the diffusion process plays no essential role in the motion of the gas. In this case, the process of the absorption of the poison gas by the filter filled up with the substance S is described by a differential equation of the form

$$\bar{u}_{xt}(x,t) + \frac{\beta}{v(x,t)}\bar{u}_t(x,t) + \beta\gamma\bar{u}_x(x,t) = 0 \tag{3.6}$$

with the boundary conditions

$$\begin{aligned} \bar{u}(x,0) &= \bar{u}_0 \exp\left(\frac{-\beta}{v_0}x\right), \\ \bar{u}(0,t) &= \bar{u}_0, \end{aligned} \tag{3.7}$$

where $\bar{u}_0$ is the gas concentration at the inlet to the filter ($\bar{u}_0$ = const.), $v(x,t)$ denotes the speed of the flow of the mixture of air and gas through the filter at the moment t and the distance x from the inlet of the filter, $v_0 = v(0,0)$, β and γ are physical quantities characterizing the given gas (for details, see [14], chapter II). With the aid of the aggregation A we are able to control the flow speed $v = v(x,t)$ in the interval $[a,b]$, i.e. $v(x,t) \in [a,b]$ where $0 < a < b < \infty$.

Without loss of generality we may assume that $x \in [0,1]$ and $t \in [0,1]$. Put

$$\bar{u}(x,t) = z(x,t) + \bar{u}_0 \exp(-\frac{\beta}{v_0}x).$$

It is easy to demonstrate that system (3.6)–(3.7) is equivalent to a system of the form

$$z_{xt}(x,t) + \beta\gamma z_x(x,t) + \frac{\beta}{v(x,t)} z_t(x,t) - \frac{\gamma\beta^2\bar{u}_0}{v_0}\exp(\frac{-\beta}{v_0}x) = 0, \tag{3.8}$$

$$z(x,0) = 0, \qquad z(0,t) = 0. \tag{3.9}$$

The function $v = v(x,t)$ will be treated as a control in system (3.8)–(3.9). Suppose that each rapid or sudden change in the speed of the flow of the gas through the filter is expensive and should be taken into account in the general costs of air filtration. Consequently, the cost functional ought to have the form

$$I(z,v) = \int_0^1 \int_0^1 f^0(x,t,z(x,t),v(x,t))dxdt + \lambda \bigvee v \tag{3.10}$$

where $\lambda > 0$, and $\bigvee v$ denotes the variation of v.

Assume that the function f^0 satisfies conditions (3.5). In the next section we shall prove the general existence theorem for system (3.1)–(3.2) (cf. th. 4.3). This theorem will imply that control system (3.8)–(3.10), describing the chemical process of filtering the air, possesses an optimal solution.

4. Helly's Principle of Choice and an Existence Theorem

To begin with, we shall prove two theorems which are some generalizations of the well-known Helly theorems for functions of one variable. First of them deals with nondecreasing functions of two variables (cf. preliminaries).

Theorem 4.1. *Let $\{f_n\}$ be a sequence of nondecreasing functions on P^2, commonly bounded. Then there exists some subsequence $\{f_{n_k}\} \subset \{f_n\}$ such that $f_{n_k}(x,y) \to \tilde{f}(x,y)$ for each $(x,y) \in P^2$, and the function $\tilde{f}$ is nondecreasing on P^2.*

Proof. Let $\{f_n\}$ satisfy the conditions of the above theorem. By Z let us denote the set of all points $(x,y) \in P^2$ where x and y are rational numbers. Applying the diagonal method of choice (cf. [9]), one can find a subsequence $\{f_{\alpha_n}\} \subset \{f_n\}$ such that $f_{\alpha_n}(x,y)$ tends to some limit $g(x,y)$ for any $(x,y) \in Z$.

It is easy to check that the function $g : Z \to \mathbb{R}$ thus defined possesses the following properties:

- if $(\tilde{x},\tilde{y}),(\bar{x},\bar{y}) \in Z$ and $(\tilde{x},\tilde{y}) \leq (\bar{x},\bar{y})$ (i.e. $\tilde{x} \leq \bar{x}$ and $\tilde{y} \leq \bar{y}$), then $g(\tilde{x},\tilde{y}) \leq g(\bar{x},\bar{y})$,
- the function F_g of an interval, associated with g, defined in the closed subintervals of P^2 whose vertices belong to the set Z, is nonnegative.

Let us now define the following function:

$$f : P^2 \ni (x,y) \mapsto \sup\{g(s,t),\ (s,t) \in ([0,x] \times [0,y]) \cap Z\} \in \mathbb{R}.$$

The function z is, of course, an extension of g. Without loss of generality, in view of Helly's theorem for a nondecreasing function of one variable (cf. [9]), we may assume that $f(0,\cdot), f(\cdot,0)$ are nondecreasing functions of one variable on $[0,1]$. For any interval $[x_1,x_2] \times [y_1,y_2] \subset P^2$ and for $\varepsilon > 0$, we can choose rational numbers $s_1 \leq x_1$, $s_2 \leq x_2$, $t_1 \leq y_1$, $t_2 \leq y_2$, $s_1 < s_2$, $t_1 < t_2$, in such a way that

$$\begin{aligned} F_f([x_1,x_2] \times [y_1,y_2]) + \varepsilon &= f(x_2,y_2) - f(x_1,y_2) - f(x_2,y_1) + f(x_1,y_1) + \varepsilon \\ &\geq g(s_2,t_2) - g(s_1,t_2) - g(s_2,t_1) + g(s_1,t_1) = F_g([s_1,s_2] \times [t_1,t_2]) \geq 0 \end{aligned}$$

From the arbitrariness of $\varepsilon > 0$ we have

$$F_f([x_1,x_2] \times [y_1,y_2]) \geq 0.$$

Consequently, f is a nondecreasing function of two variables on P^2.

Let now $(x,y) \in \operatorname{Int} P^2$. There exist sequences $((s_\nu,t_\nu)), ((u_\nu,v_\nu))$ of points of the set Z, such that

$$\begin{aligned} s_\nu &\to x, & s_\nu &< x, \\ t_\nu &\to y, & t_\nu &< y, \\ u_\nu &\to x, & x &< u_\nu, \\ v_\nu &\to y, & y &< v_\nu. \end{aligned}$$

On account of the fact that the functions f_{α_n} are nondecreasing on P^2, we have

$$f_{\alpha_n}(s_\nu, t_\nu) \leq f_{\alpha_n}(x, y) \leq f_{\alpha_n}(u_\nu, v_\nu)$$

for any $n \in \mathbb{N}$. Consequently,

$$f(s_\nu, t_\nu) = g(s_\nu, t_\nu) \leq \liminf_{n\to\infty} f_{\alpha_n}(x, y)$$
$$\leq \limsup_{n\to\infty} f_{\alpha_n}(x, y) \leq g(u_\nu, v_\nu) = f(u_\nu, v_\nu).$$

Now, passing with ν to ∞, we get

$$\lim_{\substack{\bar{x}\to x,\ \bar{x}<x \\ \bar{y}\to y,\ \bar{y}<y}} f(\bar{x}, \bar{y}) \leq \liminf_{n\to\infty} f_{\alpha_n}(x, y) \leq \limsup_{n\to\infty} f_{\alpha_n}(x, y) \leq \lim_{\substack{\bar{x}\to x,\ x<\bar{x} \\ \bar{y}\to y,\ y<\bar{y}}} f(\bar{x}, \bar{y}).$$

From the above inequality it follows that, at each point $(x, y) \in \operatorname{Int} P^2$ being a point of continuity of the function f, we have

$$f(x, y) = \lim_{n\to\infty} f_{\alpha_n}(x, y).$$

The set of all such points has the full measure in P^2 (cf. th. 2.2).

Let now $x = x_k$, $y = y_l$, $0 \leq x_k \leq 1$, $0 \leq y_l \leq 1$, $k, l = 1, 2, \ldots$, be segments on which points of discontinuity of the function f lie. From the sequence (f_{α_n}) let us now choose a subsequence (f_{β_n}) so that it converge everywhere (in $[0, 1]$) on $x = x_1$ to a nondecreasing function $\bar{f}(x_1, \cdot)$ of one variable. Further, from the sequence (f_{β_n}) choose a subsequence (f_{γ_n}) so that it converge everywhere on $x = x_2$ to a nondecreasing function $\bar{f}(x_2, \cdot)$ of one variable. Proceeding in the same way with the remaining hyperplanes $x = x_k$ and $y = y_l$, we obtain, on the basis of the diagonal method of choice, a subsequence (f_{δ_n}) of the sequence (f_{α_n}), converging everywhere on P^2 to a function $\tilde{f}$ which, for $x = x_k$ or $y = y_l$, is identical with the functions $\bar{f}(x_k, \cdot)$, $\bar{f}(\cdot, y_l)$ and, apart from these segments, with the function f. Moreover,

$$F_{\tilde{f}}([x_1, x_2] \times [y_1, y_2]) = \tilde{f}(x_2, y_2) - \tilde{f}(x_1, y_2) - \tilde{f}(x_2, y_1) + \tilde{f}(x_1, y_1)$$
$$= \lim_{n\to\infty} (f_{\delta_n}(x_2, y_2) - f_{\delta_n}(x_1, y_2) - f_{\delta_n}(x_2, y_1) + f_{\delta_n}(x_1, y_1))$$
$$= \lim_{n\to\infty} F_{f_{\delta_n}}([x_1, x_2] \times [y_1, y_2]) \geq 0.$$

In view of the fact that $f(0, \cdot), f(\cdot, 0)$ are nondecreasing functions of one variable, we deduce that $\tilde{f}$ is a nondecreasing function of two variables.

Basing ourselves on the above theorem, we shall now prove Helly's theorem for functions of two variables with finite variation on P^2.

Theorem 4.2. *Let (f_n) be a sequence of functions commonly bounded on P^2. If the sequence of variations $\operatorname{var} f_n$ is bounded by some number $c \geq 0$, then from the sequence (f_n) one can choose a subsequence (f_{n_k}) converging to some function f at each point $(x,y) \in P^2$. Moreover, $\operatorname{var} f \leq c$.*

Proof. Let, for any $n \in \mathbb{N}$, $f_n = \varphi_n - \psi_n$ be a Jordan distribution of the function f_n into the difference of two nondecreasing functions (cf. th. 2.3). Then, on the basis of th. 1 [9, §4, ch. I] and formulae (2.4), we obtain, for any point $(x,y) \in P^2$,

$$\begin{aligned}
|\varphi_n(x,y)| &\leq |g_n^1(x)| + |g_n^2(y)| + \frac{1}{2}|f_n(0,0)| + |F_n^+([0,x]\times[0,y])| \\
&= \frac{1}{2}|\operatorname{var}_{[0,x]} f_n(\cdot,0) + f_n(x,0)| + \frac{1}{2}|\operatorname{var}_{[0,y]} f_n(0,\cdot) + f_n(0,y)| \\
&+ \frac{1}{2}|f_n(0,0)| + |F_n^+([0,x]\times[0,y])| \\
&\leq \frac{1}{2}\operatorname{var}_{[0,1]} f_n(\cdot,0) + \frac{1}{2}|f_n(x,0)| + \frac{1}{2}\operatorname{var}_{[0,1]} f_n(0,\cdot) \\
&+ \frac{1}{2}|f_n(0,y)| + \frac{1}{2}|f_n(0,0)| + \operatorname{var}_{P^2} F_{f_n} \\
&\leq \frac{1}{2}\operatorname{var}_{[0,1]} f_n(\cdot,0) + \frac{1}{2}|f_n(0,0)| + \frac{1}{2}\operatorname{var}_{[0,1]} f_n(\cdot,0) + \frac{1}{2}\operatorname{var}_{[0,1]} f_n(0,\cdot) \\
&+ \frac{1}{2}|f_n(0,0)| + \frac{1}{2}\operatorname{var}_{[0,1]} f_n(0,\cdot) + \frac{1}{2}|f_n(0,0)| + \operatorname{var}_{P^2} F_{f_n} \leq d
\end{aligned}$$

where d is some positive constant common for all $n \in \mathbb{N}$, and

$$\begin{aligned}
f_n(\cdot,0) &= g_n^1 - h_n^1, \\
f_n(0,\cdot) &= g_n^2 - h_n^2, \\
F_{f_n} &= F_n^+ - F_n^-
\end{aligned}$$

are suitable Jordan distributions.

Similarly,

$$|\psi_n(x,y)| \leq d$$

for $(x,y) \in P^2$ and $n \in \mathbb{N}$.

Applying th. 4.1 twice, first - to the sequence (φ_n), and next - to the sequence (ψ_n), we shall obtain subsequences $(\varphi_{\beta_n}), (\psi_{\beta_n})$ converging in P^2 to nondecreasing functions φ, ψ. Consequently, for any point $(x,y) \in P^2$, we have

$$f_{\beta_n}(x,y) = \varphi_{\beta_n}(x,y) - \psi_{\beta_n}(x,y) \xrightarrow[n\to\infty]{} \varphi(x,y) - \psi(x,y).$$

Put $f = \varphi - \psi$. It can easily be noticed that f is a function with finite variation on P^2.

Thus the first part of the theorem has been proved.

Let us now fix an arbitrary number $\varepsilon > 0$. For β_n sufficiently large, we have

$$\begin{aligned}
\operatorname{var}_{P^2} f &= \operatorname{var}_{[0,1]} f(0,\cdot) + \operatorname{var}_{[0,1]} f(\cdot,0) + \operatorname{var}_{P^2} F_f \\
&\le \sum_j |f(0,y_j) - f(0,y_{j-1})| + \frac{\varepsilon}{6} + \sum_i |f(x_i,0) - f(x_{i-1},0)| + \frac{\varepsilon}{6} \\
&+ \sum_l |f(x_2^l,y_2^l) - f(x_1^l,y_2^l) - f(x_2^l,y_1^l) + f(x_1^l,y_1^l)| + \frac{\varepsilon}{6} \\
&\le \sum_j |f_{\beta_n}(0,y_j) - f_{\beta_n}(0,y_{j-1})| + \frac{\varepsilon}{3} + \sum_i |f_{\beta_n}(x_i,0) - f_{\beta_n}(x_{i-1},0)| + \frac{\varepsilon}{3} \\
&+ \sum_l |f_{\beta_n}(x_2^l,y_2^l) - f_{\beta_n}(x_1^l,y_2^l) - f_{\beta_n}(x_2^l,y_1^l) + f_{\beta_n}(x_1^l,y_1^l)| + \frac{\varepsilon}{3} \\
&\le \operatorname{var}_{[0,1]} f_{\beta_n}(0,\cdot) + \operatorname{var}_{[0,1]} f_{\beta_n}(\cdot,0) + \operatorname{var}_{P^2} F_{\beta_n} + \varepsilon = \operatorname{var}_{P^2} f_{\beta_n} + \varepsilon.
\end{aligned}$$

So, we have shown that, for any $\varepsilon > 0$, there exists an n_0 such that, for any $\beta_n \ge n_0$,

$$\operatorname{var} f \le \operatorname{var} f_{\beta_n} + \varepsilon.$$

Hence, in particular, it appears that

$$\operatorname{var}_{P^2} f \le c + \varepsilon,$$

which, in view of the arbitrariness of $\varepsilon > 0$, completes the proof of the theorem. □

Basing ourselves on theorem 4.2, we shall prove the following existence theorem.

Theorem 4.3. *If the functions f and f^0 satisfy conditions (3.3)–(3.5), then optimal control problem (3.1)–(3.2) possesses an optimal solution (z^*, u^*) where $z^* \in AC(P^2, R^m)$ and $u^* \in U$.*

Proof. Let $\{z_n, u_n\}$ be a minimizing sequence for problem (3.1)–(3.2), i.e. $\lim I(z_n, u_n) \to m$, where m is the lower bound of the functional $I(z,u)$. By assumption (3.5), there exists a number $a > -\infty$ such that

$$a \le \int_{P^2} f^0(x,y,z_n(x,y),u_n(x,y))\,dxdy, \qquad n = 1,2,\dots .$$

It is easy to notice that, for n sufficiently large, we have

$$a \leq \int_{P^2} f^0(x, y, z_n(x,y), u_n(x,y))\, dx dy + \sum_{i=1}^{r} \lambda_i \operatorname{var} u_n^i \leq m + 1.$$

Since $\lambda_i > 0$, therefore

$$\operatorname{var} u_n^i \leq \frac{m+1-a}{\lambda_i} \leq c_1$$

for some c_1, $i = 1, 2, \ldots, r$, and $n > N$.

We have proved that the sequence of functions u_n^i is commonly bounded and the sequence of variations is bounded, too. By theorem 4.2, there exists u_0 such that $u_n(x,y) \to u_0(x,y)$ for $(x,y) \in P^2$. (Subsequences will be denoted by some symbols as the original sequences.) It is easy to notice that $u_0 \in U$. By Lemma 3.2, the sequence of trajectories $z_n = z_{u_n}$ tends to a trajectory $z_0 = z_{u_0}$. Assumption (3.5) and theorem 4.2 imply that the functional $I(z, u)$ is lower semicontinuous in the topology of pointwise convergence. Hence, for any $\varepsilon > 0$, there exists N such that

$$m \leq I(z_0, u_0) \leq I(z_n, u_n) + \varepsilon \leq m + 2\varepsilon$$

for $n > N$. Therefore $I(z_0, u_0) = m$. □

REFERENCES

1. P. Antosik, *The investigation of the continuity of functions of several variables*, Annales Soc. Math. Pol., Comment. Math. Ser. I (1966).
2. J.M. Blatt, *Optimal control with a cost of switching control*, J. Austral. Math. Soc. Ser. B **19** (197), 316–332.
3. L. Cesari, *Optimization Theory and Applications*, Springer-Verlag, New York, 1983.
4. T.H. Hildebrandt, *Introduction to the Theory of Integration*, Academic Press, New York-London, 1963.
5. D. Idczak, *Functions of several variables of finite variation and their differentiability*, to appear in Annales Pol. Math..
6. D. Idczak, K. Kibalczyc, S. Walczak, *On some optimization problem with cost of rapid variation of control*, to appear in J. Austral. Math. Soc. Ser. B.
7. A.D. Ioffe, V.M. Tikhomirov, *Theory of Extremal Problems*, New York, 1979.
8. K. Kibalczyc, S. Walczak, *Necessary optimality conditions for a problem with costs of rapid variation*, J. Austral. Math. Soc. Ser. B **26** (1984), 45–55.
9. S. Lojasiewicz, *An Introduction to the Theory of Real Functions*, John Wiley and Sons, Chichester, 1988.
10. J. Matula, *On an extremum problem*, J. Austral. Math. Soc. Ser. B **28** (1987), 376–392.
11. E.S. Noussair, *On the existence of piecewise continuous optimal controls*, ibid. **20** (1977), 31–37.
12. D.E. Stewart A numerical algorithm for optimal control problems with switching costs, ibid. **34** (1992), 212–228.
13. K.L. Teo, L.S. Jennings, *Optimal control with a cost on changing control*, J. Optim. Theory Applic. **68** no. 2 (1991), 335–357.

14. A.N. Tikhonov, A.A. Samarski, *Equations of Mathematical Physics*, Moscow, 1978.
15. S. Walczak, *Euler-Lagrange's conditions for controls with bounded variations*, Bull. Polish Acad. Sci. Math. **26** no. 2 (1978), 125–128.
15. ———, *Absolutely continuous functions of several variables and their applications to differential equations*, ibid. **35** no. 11–12 (1987), 733–744.
17. ———On the differentiability of absolutely continuous functions of several variables, ibid. **36** no. 9–10 (1988), 513–520.

12 Further Regularity Properties in Quadratic Cost Problems for Parabolic Equations with Boundary Control and Non-Smoothing Final State Penalization

Irena Lasiecka University of Virginia, Charlottesville, Virginia

Roberto Triggiani University of Virginia, Charlottesville, Virginia

Abstract

We consider the optimal quadratic cost problem for abstract parabolic equations over a finite horizon and with non-smoothing finite state penalization. Complementing [L-T.2], we prove regularity results for the first time derivatives of the optimal control and of the optimal trajectory, in terms of functions spaces, introduced in [D-I.1], which measure the degree of singularity at the endpoints. The setting includes all boundary control problems for parabolic, or parabolic-like, partial differential equations [L-T.3], [L-T.4].

1 Mathematical setting and formulation of the problem

Dynamical model. In this paper we consider the following abstract differential equation,

$$\dot{y} = Ay + Bu \text{ on, say,} [\mathcal{D}(A^*)]', \quad y(0) = y_0 \in Y, \tag{1.1}$$

subject to the following assumptions, to be maintained throughout:

(H.1): A is the infinitesimal generator of a strongly continuous analytic semigroup, denoted by e^{At}, on the Hilbert space Y. Without loss of generality for the problem here considered, where the dynamics (1.1) is studied over a finite interval $[0, T]$, $T < \infty$, we

Research partially supported by National Science Foundation under Grant NSF-DMS-8902811.

may assume that the semigroup is of negative type so that A is boundedly invertible, i.e., $A^{-1} \in \mathcal{L}(Y)$. Then, the fractional powers $(-A)^{\theta}$, $0 < \theta < 1$, are well defined.

(H.2): B is a linear continuous operator: $U = \mathcal{D}(B) \to [\mathcal{D}(A^*)]'$ (generally unbounded as an operator from U to Y), where U is another Hilbert space, such that

$$A^{-\gamma} B \in \mathcal{L}(U;Y), \text{ or } \|A^{-\gamma}B\|_{\mathcal{L}(U;Y)} = \|B^* A^{*-\gamma}\|_{\mathcal{L}(Y;U)} \leq c_{\gamma} \quad \text{for some fixed } \gamma, 0 \leq \gamma < 1. \tag{1.2}$$

Generally, dependence on γ will not necessarily be explicitly noted in the sequel. In (1.1), A^* is the Y-adjoint of A, and $[\mathcal{D}(A^*)]'$ denotes the dual space of $\mathcal{D}(A^*)$ with respect to the Y-inner product, so that $\|y\|_{[\mathcal{D}(A^*)]'} = \|A^{-1}y\|_Y$, and $B^* \in \mathcal{L}(\mathcal{D}(A^*); U)$.

Remark 1.1. The above abstract setting comprises parabolic, or parabolic-like, mixed problems for partial differential equations, with point/boundary control, see [L-T.3]. □

Optimal control problem on $[0,T]$. With the dynamics (1.1), we associate the following quadratic functional cost over a preassigned fixed time interval $[0,T]$, $0 < T < \infty$:

$$J(u,y) \equiv \int_0^T \left[\|Ry(t)\|_W^2 + \|u(t)\|_U^2\right] dt + \|Gy(T)\|_Z^2, \tag{1.3}$$

where, in (1.3), $y(t) = y(t; y_0)$; W, Z are other Hilbert spaces, and

(H.3):

$$R \in \mathcal{L}(Y;W) \text{ and } G \in \mathcal{L}(Y;Z). \tag{1.4}$$

The corresponding optimal control problem is

$$\text{minimize } J(u,y) \text{ over all } u \in L_2(0,T;U), \text{ where } y \text{ is the solution of (1.1) due to } u. \tag{1.5}$$

Optimal control problem on $[s,T]$. If the initial time for the dynamics (1.1) is $t = s \geq 0$, with corresponding initial condition $y(s) = y_0$, the resulting solution is now denoted by $y(t) \equiv y(t; s; y_0)$. The corresponding optimal control problem over the interval $[s,T]$, $s < T < \infty$, is then:

Minimize over all $u \in L_2(s,T;U)$ the functional cost

$$\int_s^T \left[\|Ry(t)\|_W^2 + \|u(t)\|_U^2\right] dt + \|Gy(T)\|_Z^2. \tag{1.6}$$

The above problem was studied in [B-1] for $G = 0$, and in [L-T.1] for $G = \lambda^2 I$, $\lambda \in \mathbf{R}$, by variational methods; in [F.1] with $(-A^*)^{\gamma} G^* G \in \mathcal{L}(Y)$ by direct method; in [D-I.1] for $(-A^*)^{\beta} G^* G \in \mathcal{L}(Y)$, $\beta > 2\gamma - 1$ also by direct method. These results were recently further generalized in [F.2] by direct method, and in [L-T.2] by variational methods. The most general results are in [L-T.2] and are sharp. Some will be recalled (and used) below. The presence of a non-smoothing G adds technical difficulties to the problem, and the final results (e.g., the issue of uniqueness of the solution of the Differential Riccati equation) are depending on it. Generally, a non-smoothing G is responsible for the development of singularities of the relevant quantities of the optimal control problem at $t = T$. In this paper, we provide additional regularity results, pointwise in t, on the time derivatives of the optimal pair. Besides of being of interest in itself, they have additional implications. For instance, the mere existence of the time derivatives a.e. of the optimal solution (trajectory) is needed in the derivation of the Riccati equation [L-T.2]. The present paper complements and completes [L-T.2], by providing additional, relevant, technical regularity results of the optimal pair.

Preliminaries. The solution to Eqn. (1.1) with initial datum $y_0 \in Y$ at the initial time s, $0 \leq s \leq t \leq T$, is given by

$$y(t, s; y_0) = e^{A(t-s)} y_0 + (L_s u)(t), \tag{1.7}$$

$$(L_s u)(t) = \int_s^t e^{A(t-\tau)} Bu(\tau) d\tau \tag{1.8a}$$

$$: \text{ continuous } L_2(s, T; U) \to L_2(s, T; \mathcal{D}((-A)^{1-\gamma})). \tag{1.8b}$$

The adjoint operator L_s^* defined by $(L_s u, f)_{L_2(s,T;Y)} = (u, L_s^* f)_{L_2(s,T;U)}$ is given explicitly by

$$(L_s^* f)(t) = \int_t^T B^* e^{A^*(\tau-t)} f(\tau) d\tau, \quad s \leq t \leq T \tag{1.9a}$$

$$: \text{ continuous } L_2(s, T; [\mathcal{D}((-A)^{1-\gamma})]') \to L_2(s, T; U). \tag{1.9b}$$

We shall also need the (unbounded) operator L_{sT}

$$L_{sT} u = (L_s u)(T) = \int_s^T e^{A(T-t)} Bu(t) dt, \tag{1.10}$$

with domain $\mathcal{D}(L_{sT}) = \{u \in L_2(s, T; U) : L_{sT} u \in Y\}$, and its adjoint $L_{sT}^* : (L_{sT} u, y)_Y = (u, L_{sT}^* y)_{L_2(s,T;U)}$ given by

$$\{L_{sT}^* y\}(t) = B^* e^{A^*(T-t)} y, \quad s \leq t \leq T, \; y \in Y, \tag{1.11}$$

with domain $\mathcal{D}(L_{sT}^*) = \{y \in Y : L_{sT}^* y \in L_2(s, T; U)\}$, which is unbounded from $Y \supset \mathcal{D}(L_{sT}^*)$ into $L_2(s, T; U)$. We note that $H^1(s, T; U) \subset \mathcal{D}(L_{sT})$, so that $\mathcal{D}(L_{sT})$ is dense in $L_2(s, T; U)$, and that L_{sT} is a closed operator.

1.1 Main results

We begin by reviewing results of the literature.

Results from [L-T.2]. The following theorem from [L-T.2] refers to the general case where the operator G of the terminal condition is non-smoothing, i.e., satisfies only (1.4). Accordingly, the optimal control, the optimal trajectory, the gain operator, and other relevant quantities display all a singularity at the terminal time $t = T$, which can be quantitatively measured in terms of the Banach spaces introduced first in [D-I], defined below in (1.22)-(1.24).

Theorem 1.2. [L-T.2] Assume (H.1)-(H.3). Moreover, let the (densely defined) operator GL_{sT} be closed (or closable), as an operator $L_2(s,T;U) \supset \mathcal{D}(GL_{sT}) \to Z$.

(a) Then, there exists a unique optimal pair $\{u^0(t,s;y_0), y^0(t,s;y_0)\}$ of the optimal control problem (1.1)-(1.5), (1.6), with $T < \infty$. The optimal pair satisfies the optimality condition

$$u^0(\,\cdot\,,s;x) + L_s^*R^*Ru^0(\,\cdot\,,s;x) = -L_{sT}^*G^*Gy^0(T,s;x) - L_s^*R^*R(e^{A(\,\cdot\,-s)}x). \tag{1.12}$$

(b) Moreover, there exists a non-negative, self-adjoint operator $P(t) = P^*(t) \geq 0$, defined explicitly in terms of the data in (vii) = (1.19) below such that

(i)

$$P(\,\cdot\,) \in \mathcal{L}(Y;C([0,T];Y)); \tag{1.13}$$

(ii) for $0 \leq \theta < 1$,

$$\|(-A^*)^\theta P(t)\|_{\mathcal{L}(Y)} \leq \frac{C_{T\gamma\theta}}{(T-t)^\theta}; \tag{1.14}$$

(iii) for any $0 < \epsilon \leq T$,

$$(-A^*)^\theta P(t) \in \mathcal{L}(Y;C([0,T-\epsilon];Y)), \quad 0 \leq \theta < 1; \tag{1.15}$$

(iv)

$$\|B^*P(t)\|_{\mathcal{L}(Y;U)} \leq \frac{C_{T\gamma}}{(T-t)^\gamma}, \quad 0 \leq t < T; \tag{1.16}$$

(v) for any $0 < \epsilon \leq T$,

$$B^*P(\,\cdot\,) \in \mathcal{L}(Y;C([0,T-\epsilon];Y)); \tag{1.17}$$

(vi) for each $y_0 \in Y$, the optimal control $u^0(t,0;y_0)$ is given in pointwise feedback form by

$$u^0(t,0;y_0) = -B^*P(t)y^0(t,0;y_0), \quad 0 \leq t < T; \tag{1.18}$$

(vii) the operator $P(t)$ is given (explicitly) by

$$P(t)x = \int_t^T e^{A^*(\tau-t)}R^*Ry^0(\tau,t;x)d\tau + e^{A^*(T-t)}G^*Gy^0(T,t;x); \tag{1.19}$$

the expression in (1.19) defines $P(t)$ constructively solely in terms of the data of the problem, see [L-T.2];

(viii) the optimal cost of the corresponding optimal control problem on $[t,T]$ initiating at the time t at the point $x \in Y$ is

$$J(u^0(\,\cdot\,,t;x), y^0(\,\cdot\,,t;x)) = (P(t)x,x)_Y; \tag{1.20}$$

(ix) for all $0 < t < T$, and for all $x, y \in \mathcal{D}((-A)^\epsilon)$, $\forall\, \epsilon > 0$, the operator $P(t)$ satisfies the following Differential Riccati Equation

$$\begin{aligned}(\dot{P}(t)x,y)_Y &= -(R^*Rx,y)_Y - (P(t)x,Ay)_Y \\ &\quad -(P(t)Ax,y)_Y + (B^*P(t)x, B^*P(t)y)_U.\end{aligned} \tag{1.21}$$

□

Next, we recall regularity properties of the optimal pair $\{u^0(\,\cdot\,,s;x), y^0(\,\cdot\,,s;x)\}$. To state them, we need to introduce appropriate Banach spaces to measure the degree of singularity at the point $t = T$ (see [D-I.1]).

If X is a Hilbert space and r, r_1, r_2, are any real numbers, then we define the following Banach spaces

(a)

$$C_r([s,T];X) = \{f(t) \in C([s,T);X) : \|f\|_{C_r([s,T];X)} = \sup_{s \le t < T} (T-t)^r \|f(t)\|_X < \infty\}; \tag{1.22}$$

(b)

$${}_rC([s,T];X) = \{f(t) \in C((s,T];X) : \|f\|_{{}_rC([s,T];X)} = \sup_{s \le t \le T} (t-s)^r \|f(t)\|_X < \infty\}; \tag{1.23}$$

(c)

$$\begin{aligned}{}_{r_1}C_{r_2}([s,T];X) &= \{f(t) \in C((s,T);X) : \|f\|_{{}_{r_1}C_{r_2}([s,T];X)} \\ &= \sup_{s \le t \le T} (t-s)^{r_1}(T-t)^{r_2}\|f(t)\|_X < \infty\}.\end{aligned} \tag{1.24}$$

The regularity of the optimal pair $\{u^0(\cdot\,; s; x), y^0(\cdot\,; s; x)\}$ is then given by the following result:

Theorem 1.2 [L-T.2]. Under the assumptions of Theorem 1.1, the following regularity properties hold true for the optimal pair with $0 \leq s < T$:

$$\|u^0(\,\cdot\,, s; x)\|_{L_2(s,T;U)} + \|y^0(\,\cdot\,, s; x)\|_{L_2(s,T;U)} \leq c_T \|x\|_Y; \tag{1.25}$$

$$\|Gy^0(T, s; x)\|_Z \leq c_T \|x\|_Y; \tag{1.26}$$

$$\|u^0(\,\cdot\,, s; x)\|_{C_\gamma([s,T];U)} \leq c_{T\gamma} \|x\|_Y; \tag{1.27}$$

$$\begin{cases} \|y^0(\,\cdot\,, s; x)\|_{C([s,T];Y)} \leq c_{T\gamma} \|x\|_Y & \text{if } 0 \leq \gamma < \frac{1}{2}; \quad (1.28a) \\[2ex] \|y^0(\,\cdot\,, s; x)\|_{C_{2\gamma-1+\epsilon}([s,T];Y)} \leq c_{T\gamma} \|x\|_Y & \text{if } \frac{1}{2} \leq \gamma < 1. \quad (1.28b) \end{cases}$$

New results. The goal of the present paper is to complement the above results by providing pointwise regularity results for the time derivatives

$$\frac{du^0(t, s; x)}{dt} \text{ and } \frac{dy^0(t, s, x)}{dt},$$

$0 < s < T$, $x \in Y$ of the optimal pair. The pointwise existence of $\frac{dy^0(t,s;x)}{dt}$ is needed in the derivation of the Riccati equation. Indeed, the next theorem shows more: not only that these two quantities are, respectively, U-valued and Y-valued functions continuous in t for $s < t < T$, but it also measures their order of singularity at $t = s$. In the special case $\gamma < \frac{1}{2}$ (see Remark 1.2 below), we shall also obtain a quantitative description of their order of singularity at $t = T$.

Theorem 1.3. For $x \in Y$, the time derivatives $\frac{du^0(t,s;x)}{dt}$ and $\frac{dy^0(t,s;x)}{dt}$ exist as, respectively, U-valued and Y-valued functions, continuous in $s < t < T$. Moreover,

(i)

$$\left\| \frac{du^0(t, s; x)}{dt} \right\|_U \leq \frac{C_{T\epsilon_1\gamma}}{(t-s)^{\gamma+\epsilon_1}} \|x\|_Y, \; x < t < T, \; \forall\, \epsilon_1 > 0, \tag{1.29a}$$

so that recalling (1.23),

$$\frac{du^0(t, s; x)}{dt} \in {}_{(\gamma+\epsilon_1)}C([s, T-\epsilon]; U), \qquad \forall\, \epsilon, \epsilon_1 > 0. \tag{1.29b}$$

Furthermore,

(ii)

$$\left\| \frac{dy^0(t,s;x)}{dt} - Ae^{A(t-s)}x \right\|_Y \leq \frac{C_{T\epsilon_1\gamma}}{(t-s)^\gamma}\|x\|_Y, \epsilon_1 \text{ as in (1.29)}, \ s < t < T, \quad (1.30a)$$

so that recalling (1.23),

$$\frac{dy^0(t,s;x)}{dt} - Ae^{A(t-s)}x \in \ {}_\gamma C([s, T-\epsilon]; Y), \quad \forall \epsilon > 0. \quad (1.30b)$$

Remark 1.2. *In the special case* $\gamma < \frac{1}{2}$, we can also obtain a measure of the order of singularity at $t = T$, more precisely:

$$\left\| \frac{du^0(t,s;x)}{dt} \right\|_U \leq \frac{C_T}{(T-t)^{1+\gamma}}\|x\|_Y, \quad (1.31a)$$

hence

$$\frac{du^0(t,s;x)}{dt} \in \ {}_{(\gamma+\epsilon_1)}C_{1+\gamma}([s,T];U), \quad \forall \epsilon_1 > 0. \quad (1.31b)$$

Moreover,

$$\left\| \frac{dy^0(t,s;x)}{dt} \right\|_Y \leq \frac{C_T}{(T-t)^{2\gamma+\epsilon}} \frac{1}{(t-s)^\gamma}\|x\|_Y, \ \forall \epsilon > 0; \quad (1.32a)$$

hence

$$\frac{dy^0(t,s;x)}{dt} \in \ {}_\gamma C_{2\gamma+\epsilon}([s,T];Y), \quad \forall \epsilon > 0. \quad (1.32b)$$

□

In order to obtain results analogous to (1.31)-(1.32) in the case when $\gamma \geq \frac{1}{2}$, we need to impose an additional "smoothing" hypothesis on the observation G,

$$(-A^*)^\beta G^* G \in \mathcal{L}(Y), \quad (1.33)$$

for $\beta > 2\gamma - 1$ and $\gamma \geq \frac{1}{2}$.

The following regularity results under the additional hypothesis (1.33) hold true.

Theorem 1.4. [L-T.2] In addition to the hypotheses of Theorem 1.1, assume (1.33). Then, for $x \in Y$, and $0 \leq \theta < 1$:

$$\|y^0(\cdot,s;x)\|_{C([s,T];Y]} \leq C_{T_\gamma}\|x\|_Y; \quad (1.34)$$

$$y^0(T,s;x) = \Phi(T,\cdot)x \in C([s,T];Y); \quad (1.35)$$

$$\|u^0(\cdot,s;x)\|_{C_{\theta-\beta}([s,T];U)} \leq C_{T_\gamma}\|x\|_Y; \quad (1.36)$$

$$\|(-A^*)^\theta P(t)\|_{\mathcal{L}(Y)} \leq \frac{C_{T_\gamma}}{1-\theta} \frac{1}{(T-t)^{\theta-\beta}}; \quad (1.37)$$

$$\lim_{t\to T} P(t)x = G^* G x. \quad (1.38)$$

New results. We shall show that:

(i) the time derivatives $\frac{du^0}{dt}(t,s;x)$ and $\frac{dy^0}{dt}(t,s;x)$ are continuous functions on (s,T) with *controlled* singularity at $t=s$ and $t=T$;

(ii) the solution to the Riccati Equation is classical, in the sense that, for $t < T$, $\frac{dP(t)}{dt} \in \mathcal{L}(Y)$, and the operators $A^*P(t)+P(t)A$ are in $\mathcal{L}(Y)$ [see [B-D-M-P]).

Theorem 1.5. In addition to the hypotheses of Theorem 1.1, assume (1.33). Then, for any $\epsilon > 0$:

$$\left\| \frac{du^0(t,s;x)}{dt} \right\|_U \leq \frac{C_T}{(T-t)^{1+\gamma-\beta}(t-s)^{\gamma+\epsilon}} \|x\|_Y, \tag{1.39a}$$

hence

$$\frac{du^0(t,s;x)}{dt} \in \ _{\gamma+\epsilon}C_{1+\gamma-\beta}([s,T];U). \tag{1.39b}$$

Moreover,

$$\left\| \frac{dy^0(t,s;x)}{dt} - Ae^{At}x \right\|_Y \leq \frac{C_T}{(T-t)^{2\gamma-\beta}(t-s)^{\gamma}} \|x\|_Y, \tag{1.40a}$$

hence

$$\frac{dy^0(t,s;x)}{dt} - Ae^{At}x \in \ _{\gamma}C_{2\gamma-\beta}([s,T];Y). \tag{1.40b}$$

Theorem 1.6. Under the assumptions of Theorem 1.1 and, moreover, under (1.33) for $\gamma \geq \frac{1}{2}$, we obtain

$$\left\| \frac{d}{dt}P(t) \right\|_{\mathcal{L}(Y)} \leq \frac{C_T}{(T-t)^{\alpha}}, \quad t < T; \tag{1.41}$$

$$\|A^*P(t)+P(t)A\|_{\mathcal{L}(Y)} \leq \frac{C_T}{(T-t)^{\alpha}}, \tag{1.42}$$

$$\|B^*P(t)\|_{\mathcal{L}(Y,U)} \leq \frac{C_T}{(T-t)^{\alpha}}, \tag{1.43}$$

where $\alpha = \max[1-\beta, 3\gamma-1-\beta]$ (recall that $\beta = 0$ for $\gamma < \frac{1}{2}$).

Theorem 1.6 implies the following corollary.

Corollary 1.7. The solution $P(t)$ to the Riccati equation (1.21) is classical; i.e.,

$$\begin{cases} \dot{P}(t) = -R^*R - \Lambda P(t) + (P(t)B^*)^*B^*P(t) & \text{for } t < T, \\ P(T) = G^*G. \end{cases}$$

where $\Lambda P \equiv A^*P + PA$ has a bounded extension as an operator in $\mathcal{L}(Y)$ and equality (1.44) holds on the whole space Y.

2 Proof of Theorem 1.3

The proof of Theorem 1.3 proceeds through several steps.

Step 1. We return to a characterization (1.12) of the optimal pair which we rewrite here as

$$\begin{aligned} u^0(t,s;x) + \{L_s^* R^* R L_s u^0(\,\cdot\,,s;x)\}(t) &= -\{L_{sT} G^* G y^0(T,s;x)\}(t) \\ &\quad -\{L_s^* R^* R e^{A(\,\cdot\,-s)}x\}(t). \end{aligned} \tag{2.1}$$

Our first task is to analyze the time derivative of the right-hand side of (2.1).

Step 2. Recalling (1.11), we compute for $x \in Y$,

$$\frac{d}{dt}\{L_{sT} G^* G y^0(T,s;x)\}(t) = -B^* A^* e^{A^*(T-t)} G^* G y^0(T,s;x) \tag{2.2a}$$

$$\in C_{1+\gamma}([s,T];U), \tag{2.2b}$$

i.e., the U-function in (2.2a) is continuous for $s \le t < T$ (recall from (1.26) that $Gy^0(T,s;x) \in Y$) and is norm-bounded by $C/(T-t)^{1+\gamma}$ for t near T, as seen from (2.2a) by use of (1.2) on B and analyticity of e^{A^*t}

Step 3. Lemma 2.1. For $x \in Y$, we have

$$\frac{d}{dt}\{L_s^* R^* R e^{A(\,\cdot\,-s)}x\}(t) = f_1(t;x) + f_2(t;x); \tag{2.3}$$

$$f_1(t;x) \equiv B^* e^{A^*(T-t)} R^* R e^{A(T-s)} x \in C_\gamma([s,T];U); \tag{2.4}$$

$$f_2(t;x) \equiv -\int_t^T B^* e^{A^*(\tau-t)} R^* R A e^{A(\tau-s)} x \, d\tau \tag{2.5a}$$

$$\equiv -\{L_s^* R^* R A e^{A(\,\cdot\,-s)} x\}(t) \in {}_{(\gamma+\epsilon)}C([s,T];U). \tag{2.5b}$$

Proof. By recalling (1.9) and changing here variable $\tau - t = \sigma$, we write

$$\{L_s^* R^* R e^{(A(\,\cdot\,-s)}x\}(t) = \int_0^{T-t} B^* e^{A^*\sigma} R^* R e^{A(\sigma+t-s)} x \, d\sigma, \tag{2.6}$$

whose time derivative then produces the expression given by (2.3)-(2.5a). The regularity of f_1 in (2.4) follows from (1.2) and analyticity of e^{A^*t} as usual. As to the regularity of

f_2 described in (2.5b), U-continuity for $s < t < T$ is obvious, while for the singularity at $t = s$ we compute from (2.5a) via (1.2) and analyticity of e^{At}:

$$\begin{aligned} \|f_2(t;x)\| &\leq C_T\|x\|_Y \int_t^T \frac{d\tau}{(\tau-t)^\gamma(\tau-s)} \\ &= C_T\|x\|_Y \int_t^T \frac{d\tau}{(\tau-t)^\gamma(\tau-s)^{1-\gamma-\epsilon}(\tau-s)^{\gamma+\epsilon}}, \quad s < t < T \end{aligned}$$

(using $(\tau-s)^{1-\gamma-\epsilon} \geq (\tau-t)^{1-\gamma-\epsilon}$ and $(\tau-s)^{\gamma+\epsilon} > (t-s)^{\gamma+\epsilon}$)

$$\leq \frac{C_T}{(t-s)^{\gamma+\epsilon}}\|x\|_Y \int_t^T \frac{d\tau}{(\tau-t)^{1-\epsilon}} = \frac{C_T(T-t)^\epsilon}{\epsilon(t-s)^{\gamma+\epsilon}}\|x\|_Y, \tag{2.7}$$

which is the required bound via (1.23). □

Using the regularity in (2.2b), (2.4), and (2.5b), we obtain

Corollary 2.2. For $x \in Y$, the following result follows for the right-hand side (R.H.S.) of (2.1),

$$\frac{d}{dt}\{\text{R.H.S. of (2.1)}\}(t) = B^*A^*e^{A^*(T-t)}G^*Gy^0(T,s;x) - f_1(t;x) - f_2(t;x) \equiv \rho(t;x); \tag{2.8a}$$

$$\rho(t;x) \in {}_{\gamma+\epsilon}C_{1+\gamma}([s,T];U). \tag{2.8b}$$

Step 4. We next take the time derivative of $\{L_s^*R^*RL_su^0(\,\cdot\,,s;x)\}(t)$ for $s < t < T-\epsilon$, $\epsilon > 0$ arbitrary. To express the final result, we note the following. The operators L_s and L_s^* in (1.8) and (1.9) are based on the time interval $s \leq t \leq T$. In the next result, we shall need to take these same operators and restrict them, however, as to act on the time interval $s < t \leq T-\epsilon$, $\epsilon > 0$. These restrictions are needed only in this step. They will be denoted by $L_{[s,T-\epsilon]}$ and $L_{[s,T-\epsilon]}^*$. Thus, by definition

$$(L_{[s,T-\epsilon]}u)(t) = \int_s^t e^{A(t-\tau)}Bu(\tau)d\tau, \; s \leq t \leq T-\epsilon, \tag{2.9}$$

whose adjoint is then

$$(L_{[s,T-\epsilon]}^*f)(t) = \int_t^{T-\epsilon} B^*e^{A^*(\tau-t)}f(\tau)d\tau, \; s \leq t \leq T-\epsilon. \tag{2.10}$$

We note from (1.8) and (2.10) that plainly

$$(L_{[s,T-\epsilon]}u)(t) \equiv (L_su)(t), \; s \leq t \leq T-\epsilon. \tag{2.11}$$

With the above notation we have

Proposition 2.3. Let $T > \epsilon > 0$ arbitrary, and let $\mu(t) \in C_\gamma([s,T];U)$. Then

$$\frac{d}{dt}\{L_s^* R^* R L_s \mu(\,\cdot\,)\}(t) = \{L_{[s,T-\epsilon]}^* R^* R L_{[s,T-\epsilon]} \frac{d}{d\sigma}\mu(\sigma)\}(t) + w_\epsilon(t) \in H^{-1}(s, T-\epsilon; U), \tag{2.12}$$

where we have set

$$\begin{aligned} w_\epsilon(t) \;\equiv\; & \{L_{[s,T-\epsilon]}^* R^* R e^{A(\,\cdot\,-s)} B\mu(s)\}(t) + B^* e^{A^*(T-\epsilon-t)} R^* R\{[L_s\mu(\,\cdot\,)](T-\epsilon)\} \\ & - B^* A^* e^{A^*(T-\epsilon-t)} \int_{T-\epsilon}^{t} e^{A^*(\tau-(T-\epsilon))} R^* R L_s \mu(\tau) d\tau, \end{aligned} \tag{2.13}$$

and where we note that $[L_s\mu(\,\cdot\,)](T-\epsilon) \in Y$. Also, $\forall \epsilon_1 > 0$:

$$w_\epsilon(t) \in {}_{(2\gamma-1+\epsilon_1)}C_{\gamma+\epsilon_1}([s, T-\epsilon]; U) \subset L_1(s, T-\epsilon; U). \tag{2.14}$$

Proof. We take initially $\mu \in C^1([s, T-\epsilon]; U) \cap C_\gamma([s,T];U)$, $\epsilon > 0$. We shall later extend the results to $\mu \in C_\gamma([s,T];U)$ after Lemma 2.5 below. If we set for convenience,

$$v(t) \equiv \{R^* R L_s \mu(\,\cdot\,)\}(t), \tag{2.15}$$

we rewrite (1.9) for $s < t \le T - \epsilon$ as

$$\begin{aligned} \{L_s^* R^* R L_s \mu(\,\cdot\,)\}(t) &= \{L_s^* v(\,\cdot\,)\}(t) \\ &= \int_t^{T-\epsilon} B^* e^{A^*(\tau-t)} v(\tau) d\tau + \int_{T-\epsilon}^{T} B^* e^{A^*(\tau-t)} v(\tau) d\tau, \end{aligned} \tag{2.16}$$

and after changing variable $\tau - t = \sigma$ in the first integral

$$\begin{aligned} \{L_s^* v(\,\cdot\,)\}(t) &= \int_0^{T-\epsilon-t} B^* e^{A^*\sigma} v(t+\sigma) d\sigma \\ & \quad + B^* e^{A^*(T-\epsilon-t)} \int_{T-\epsilon}^{T} e^{A^*(\tau-(T-\epsilon))} v(\tau) d\tau, \quad s < t < T-\epsilon. \end{aligned} \tag{2.17}$$

Taking the time derivative of (2.17) yields

$$\begin{aligned} \frac{d}{dt}\{L_s^* R^* R L_s \mu(\,\cdot\,)\}(t) &= \frac{d}{dt}\{L_s^* v(\,\cdot\,)\}(t) \\ &= B^* e^{A^*(T-\epsilon-t)} v(T-\epsilon) + \int_t^{T-\epsilon} B^* e^{A^*(\tau-t)} \frac{dv(\tau)}{d\tau} d\tau \\ & \quad - B^* A^* e^{A^*(T-\epsilon-t)} \int_{T-\epsilon}^{T} e^{A^*(\tau-(T-\epsilon))} v(\tau) d\tau. \end{aligned} \tag{2.18}$$

We next compute $\frac{dv(t)}{dt}$, as required by the second term in (2.18). By (2.15) and (1.8), we compute after a change of variable $\tau - t = \sigma$,

$$\begin{aligned}\frac{dv(t)}{dt} &= \frac{d}{dt}\{R^*RL_s\mu(\,\cdot\,)\}(t)\\ &= R^*R\frac{d}{dt}\int_0^{t-s} e^{A\sigma}B\mu(t-\sigma)d\sigma\\ &= R^*Re^{A(t-s)}B\mu(s) + \left\{R^*RL_s\left[\frac{d\mu(\sigma)}{d\sigma}\right]\right\}(t),\ s < t < T-\epsilon. \end{aligned} \quad (2.19)$$

Finally, inserting (2.19) into (2.18) yields (2.12) as desired, via (2.13), (2.10), (2.11), (2.15), at least for $\mu \in C^1([s,T-\epsilon];U) \cap C_\gamma([s,T];U)$. We shall later extend our result to μ only in $C_\gamma([s,T];U)$. But first we show the regularity of $w_\epsilon(t)$ noted in (2.14) as a consequence of the definition of $w_\epsilon(t)$ in (2.13) and $\mu \in C_\gamma([s,T];U)$. For the first term in the definition of $w_\epsilon(t)$ we have

$$L^*_{[s,T-\epsilon]}R^*Re^{A(\,\cdot\,-s)}B\mu(s) \in {}_{(2\gamma-1+\epsilon_1)}C([s;T-\epsilon];U),\ \forall\, \epsilon, \epsilon_1 > 0. \quad (2.20)$$

This is so since $\mu(s) \in Y$ and since

$$e^{A(\,\cdot\,-s)}B\mu(s) \in {}_\gamma C([s;T-\epsilon];Y) \quad (2.21)$$

by (1.2) on B and analyticity of e^{At}. Next, one applies the (independent) Proposition 2.8(ii), Eqn. (2.37), whereby the operator L^*_s, equivalently $L^*_{[s,T-\epsilon]}$, reduces the order of singularity of $t = s$ by $(1-\gamma-\epsilon_1)$ from γ to $\gamma+\gamma-1+\epsilon_1 = 2\gamma-1+\epsilon_1$, and (2.20) follows from (2.21).

The second term in the definition of $w_\epsilon(t)$ in (2.13) is plainly in $C_\gamma([s,T-\epsilon];U)$, $\forall\epsilon, \epsilon_1 > 0$. This one sees upon using (1.2) on B, analyticity of e^{A^*t} and the following

Claim. For any $r < 1$, the vector

$$z_r \equiv \int_{T-\epsilon}^T (-A^*)^r e^{A^*(\tau-(T-\epsilon))}R^*RL_s\mu(\tau)d\tau \in Y \quad (2.22)$$

is a well-defined vector in Y. Indeed, by assumption, $\mu \in C_\gamma([s,T];U)$, hence by Proposition 3.1(ii), Eqn. (3.3) with $r = \gamma$ of [L-T.2], we obtain that $L_s\mu \in C_{2\gamma-1+\epsilon_2}C([s,T];Y)$, $\forall\, \epsilon_2 > 0$, and hence, since $2\gamma - 1 + \epsilon_2 < 1$ and $r < 1$, we have

$$\|z_r\|_Y \le C_T \int_{T-\epsilon}^T \frac{d\tau}{(\tau-(T-\epsilon))^r(T-\tau)^{2\gamma-1+\epsilon_2}} < \infty, \quad (2.23)$$

as desired, and our Claim follows.

Thus the regularity of $w_\epsilon(t)$ in (2.14) is proved.

The same proof gives in particular

Proposition 2.4. Let $T > \epsilon > 0$ arbitrary, and let $\mu \in C^1([s,T-\epsilon];U)$. Then for $s < t < T-\epsilon$

$$\frac{d}{dt}\left\{L^*_{[s,T-\epsilon]}R^*RL_{[s,T-\epsilon]}\mu(\,\cdot\,)\right\}(t) = \left\{L^*_{[s,T-\epsilon]}R^*RL_{[s,T-\epsilon]}\left(\frac{d}{d\sigma}u(\sigma)\right)\right\}(t)+w_0(t), \quad (2.24)$$

where $w_0(t)$ is the function $w_\epsilon(t)$ in (2.13) after setting $\epsilon = 0$: $w_0(t) = w_\epsilon(t)|_{\epsilon=0}$ (whereby the third term on the right of (2.18) vanishes), so that

$$w_0 \in {}_{(2\gamma-1+\epsilon_1)}C_\gamma([s,T-\epsilon];U) \subset L_1(s,T-\epsilon;U), \quad (2.25)$$

and moreover,

$$\|w_0\|_{L_1(s,T-\epsilon;U)} \le C_{T-\epsilon}\|\mu\|_{C([s,T-\epsilon];U)} \quad (2.26)$$

as it follows from (2.13) for $\epsilon = 0$. □

To complete the proof of Proposition 2.3, we not extend identity (2.12) to all μ just in $C_\gamma([s,T];U)$. To this end we need the following consequence of Proposition 2.4.

Lemma 2.5. Let $\epsilon > 0$. With reference to (2.9), (2.10) we have

$$L^*_{[s,T-\epsilon]}R^*RL_{[s,T-\epsilon]}\frac{d}{dt} : \text{ continuous } C([s,T-\epsilon];U) \to H^{-1}(s,T;U). \quad (2.27)$$

Proof. We take, say, $C^1([s,T-\epsilon],U)$ to be the domain of the operator in (2.27), dense in $C([s,T-\epsilon];U)$. If $\mu \in C^1([s,T-\epsilon];U)$, then by (2.24), (2.25) we obtain

$$\begin{aligned}&\left\|L^*_{[s,T-\epsilon]}R^*RL_{[s,T-\epsilon]}\left(\frac{d}{dt}\mu(\,\cdot\,)\right)\right\|_{H^{-1}(s,T-\epsilon;U)}\\ &\qquad \le \|L^*_{[s,T-\epsilon]}R^*RL_{[s,T-\epsilon]}\mu(\,\cdot\,)\|_{L_2(s,T-\epsilon;U)} + \|w_0\|_{H^{-1}(s,T-\epsilon;U)}.\end{aligned} \quad (2.28)$$

But

$$L_1(s,T-\epsilon;U) \hookrightarrow H^{-1}(s,T-\epsilon;U), \quad (2.29a)$$

as it follows by duality,

$$\begin{aligned}\|f\|_{H^{-1}(s,T_1)} &= \sup_\phi \frac{|(f,\phi)|}{\|\phi\|_{H^1_0(s,T_1)}}\\ &\le \sup_\phi \frac{|\phi|_{L_\infty(s,T_1)}\|f\|_{L_1(s,T)}}{\|\phi\|_{H^1(s,T_1)}} \le C_T\|f\|_{L_1(s,T_1)},\end{aligned} \quad (2.29b)$$

after using the one-dimensional embedding $\|\phi\|_{L_\infty(s,T_1)} \le C_T\|\phi\|_{H^1_0(s,T_1)}$. Then, using (2.29) and (2.26) in (2.27) yields

$$\left\|L^*_{[s,T-\epsilon]}R^*RL_{[s,T-\epsilon]}\frac{d}{dt}\mu(\,\cdot\,)\right\|_{H^{-1}(s,T-\epsilon;U)} \le C_{T-\epsilon}\|\mu\|_{C([s,T-\epsilon];U)}, \quad (2.30)$$

$\forall \mu \in C^1([s, T-\epsilon]; U)$, since $L^*_{[s,T-\epsilon]} R^* R_{[s,T-\epsilon]}$ is *a-fortiori* continuous $C([s, T-\epsilon]; U) \to L_2(s, T-\epsilon; U)$. Inequality (2.30) says that the operator in (2.27) with domain $C^1([s, T-\epsilon]; U)$ is closable. Taking closure, inequality (2.30) can be extended to all of $C([s, T-\epsilon]; U)$ and conclusion (2.27) follows. □

We can now complete the proof of Proposition 2.3. Identity (2.12) was initially shown for $\mu \in C^1([s, T-\epsilon]; U) \cap C_\gamma([s, T]; U)$, where the term $w_\epsilon(t)$ requires, however, only $\mu \in C_\gamma([s, T]; U)$. By virtue of Lemma 2.5 we can now extend the validity of identity (2.12) to μ just in $C_\gamma([s, T]; U)$. Proposition 2.3 is proved. □

Next we specialize Proposition 2.3 to $\mu(t) = u^0(t, s; x) \in C_\gamma([s, T]; U)$ by (1.27). We obtain

Corollary 2.6. Identity (2.12) of Proposition 2.3 holds true in $H^{-1}(s, T-\epsilon; U)$ for $\mu(\cdot) = u^0(\cdot, s; x)$. Thus, we have the following identity for any $x \in Y$:

$$\frac{d}{dt}\{L_s^* R^* R L_s u^0(\cdot, s; x)\}(t) = \left\{ L^*_{[s,T-\epsilon]} R^* R L_{[s,T-\epsilon]} \left(\frac{d}{d\sigma} u^0(\sigma, s; x) \right) \right\}(t) + w_\epsilon(t; x) \tag{2.31a}$$

$$\in H^{-1}(s, T-\epsilon; U), \tag{2.31b}$$

where $w_\epsilon(t; x)$ is equal to the function $w_\epsilon(t)$ in (2.18) when $\mu(\cdot)$ there is replaced by $u^0(\cdot, s; x)$, so that

$$w_\epsilon(t; x) \in {}_{(2\gamma-1+\epsilon_1)}C_{\gamma+\epsilon_1}([s, T-\epsilon]; U) \tag{2.32}$$

$$= \subset L_1(s, T-\epsilon; U) \subset H^{-1}(s, T-\epsilon; U). \tag{2.33}$$

Step 5. Using Corollary 2.2 and Corollary 2.6, we obtain directly from identity (2.1) the following result.

Theorem 2.7. Let $T > \epsilon > 0$ arbitrary, and let $x \in Y$. Then for $s < t < T - \epsilon$,

$$\frac{du^0(t, s; x)}{dt} + \left\{ L^*_{[s,T-\epsilon]} R^* R L_{[s,T-\epsilon]} \frac{du^0}{d\sigma}(\sigma, s; x) \right\}(t) = \nu_\epsilon(t; x), \tag{2.34}$$

where $\frac{du^0}{dt}(t, s; x) \in H^{-1}(s, T-\epsilon; U)$, and where $\nu(t; x)$ satisfies (recall (2.32) for $w_\epsilon(t; x)$ and (2.8) which gives $\rho(t; x)$ continuous for $t < T$):

$$\nu_\epsilon(t;x) \equiv -w_\epsilon(t;x) + \rho(t;x) \in {}_{(\gamma+\epsilon_1)}C_{\gamma+\epsilon_1}([s,T-\epsilon];U), \ \forall \epsilon_1 > 0, \tag{2.35}$$

since $2\gamma - 1 + \epsilon_1 < \gamma + \epsilon_1$. □

Step 6. Eqn. (2.34) poses the (usual) problem of the inversion of the operator $I + L^*_{[s,T-\epsilon]}R^*RL_{[s,T-\epsilon]}$. This operator is clearly boundedly invertible on $L_2(s,T-\epsilon;U)$. The crux now is to assert that it is likewise boundedly invertible on the space identified in (2.35) of functions which are continuous in $s < t < T-\epsilon$ and which have the indicated order of singularity at $t = s$ and $t = T-\epsilon$. After this inversion has been justified, (2.34) then yields $\frac{du^0(t,s;x)}{dt} \in {}_{(\gamma+\epsilon_1)}C_{\gamma+\epsilon_1}([s,T-\epsilon];U)$, hence a U-valued function continuous in t, for $s < t < T-\epsilon$, $\epsilon > 0$ arbitrary, as desired. To show such an inversion, we employ the same strategy used to prove Theorem 3.4 of [L-T.2] when only singularity at the right-hand point occurred. Accordingly, it suffices to prove that the operators $L_{[s,T-\epsilon]}$ and $L^*_{[s,T-\epsilon]}$, or equivalently, the operators L_s and L^*_s, having a smoothing effect which reduces the singularity at the left-hand point $t = s$ as well. The next result is the counterpart of Proposition 3.1 in [L-T.2] from singularity at the right-hand point to singularity at the left-hand point.

Proposition 2.8. With reference to the operators L_s and L^*_s defined in (1.8) and (1.9) and the spaces defined in (1.23), we have

(i) Let $0 < r < 1$.
Then L_s: continuous ${}_rC([s,T;U) \rightarrow_{(r+\gamma-1)} C([s,T];Y)$

$$\|L_s u\|_{(r+\gamma-1)C([s,T];Y)} \le C_{T\gamma r}\|u\|_{rC([s,T];U)}, \tag{2.36}$$

so that the bound in (2.36) may be made independent of s, $0 \le s \le T$.

(ii) Let $r > 0$.
Then L^*_s: continuous ${}_rC([s,T];Y) \rightarrow_{(r+\gamma-1+\epsilon)} C([s,T];U)$,

$$\|L^*_s f\|_{(r+\gamma-1+\epsilon)C([s,T];U)} \le C_{T\epsilon\gamma r}\|f\|_{rC([s,T];Y)}. \tag{2.37}$$

(iii) Thus, *a-fortiori*, if $0 < r < 1$, there exists a positive integer $m = m(r)$ such that

$$(L^*_s R^* R L_s)^m : \text{ continuous } {}_rC([s,T];U) \to C([s,T];Y) \tag{2.38}$$

with bound on the uniform norm which may be taken independent of s, $0 \le s \le T$. We can take the smallest integer m such that $2(1-\gamma)m > r$. □

Proof. (Similar to the proof of Proposition 3.1 of [L-T.2].) Continuity of $(L_s u)(t)$ and $(L_s^* f)(t)$ for $s < t \leq T$ is seen directly. We need to show the appropriate order of singularity at $t = s$.

(i) Let $u \in {}_rC([s,T];U)$. By (1.2) on B and analyticity of e^{At}, we compute for from (1.8)

$$
\begin{aligned}
\|(L_s u)(t)\|_Y &\leq C_{T\gamma} \int_s^t \frac{\|u(\tau)\|_U (\tau - s)^r d\tau}{(t-\tau)^\gamma (\tau - s)^r} \\
&\leq C_{T\gamma r} \|u\|_{{}_rC([s,T];U)} \left\{ \int_s^{s+(t-s)/2} \frac{d\tau}{(t-\tau)^\gamma (\tau - s)^r} \right. \\
&\quad \left. + \int_{s+(t-s)/2}^t \frac{d\tau}{(t-\tau)^\gamma (\tau-s)^r} \right\} \qquad (2.39) \\
&\leq C_{T\gamma r} \|u\|_{{}_rC([s,T];U)} \left\{ \left(\frac{2}{(t-s)} \right)^{\gamma + r - 1} + \left(\frac{2}{t-s} \right)^{\gamma + r - 1} \right\}. \qquad (2.40)
\end{aligned}
$$

In the first integral in (2.39) we have used $(t - \tau) \geq (t - s)/2$, and $r < 1$; in the second integral we have used $(\tau - s) \geq (t - s)/2$. Eqn. (2.40) shows (2.36) via the definition in (1.23).

(ii) Similarly, from (1.9) we compute with $f \in {}_rC([s,T];Y)$:

$$
\begin{aligned}
\|(L_s^* f)(t)\|_U &\leq C_{T\gamma} \int_t^T \frac{\|f(\tau)\|_Y (\tau - s)^r d\tau}{(\tau - t)^\gamma (\tau - s)^r}, \quad s \leq t \leq T \\
&\leq C_{T\gamma r} \|f\|_{{}_rC([s,T];Y)} I(t), \qquad (2.41)
\end{aligned}
$$

where with $r_1 + r_2 = r$, $\gamma + r_1 = 1 - \epsilon$, $\epsilon > 0$, we have

$$
I(t) = \int_t^T \frac{d\tau}{(\tau - t)^\gamma (\tau - s)^{r_1} (\tau - s)^{r_2}} \leq \frac{1}{(t-s)^{r_2}} \int_t^T \frac{d\tau}{(\tau - t)^{1-\epsilon}} = \frac{(T-t)^\epsilon}{\epsilon (t-s)^{r_2}}, \qquad (2.42)
$$

upon using $(\tau - s)^{r_1} \geq (\tau - t)^{r_1}$ and $(\tau - s)^{r_2} \geq (t - s)^{r_2}$. Estimate (2.42) used in (2.39) yields (2.37) as desired, with $r_2 = r + \gamma - 1 + \epsilon$, via the definition (1.23).

(iii) Iteration of part (i) and (ii) with R bounded as assumed in (1.4) yields part (iii), Eqn. (2.38). As integer m, we can take the smallest integer such that $2(1-\gamma)m > r$, $(1-\gamma)$ or $(1 - \gamma - \epsilon)$ being the reduction of singularity at each application of L_s or L_s^*. □

Step 7. The crucial result is then the following extension of Theorem 3.4 of [L-T.2].

Theorem 2.9. Let $\epsilon, \epsilon_1 > 0$. The operator (see (2.9) and (2.10)) $[I + L_{[s,T-\epsilon]}^* R^* R L_{[s,T-\epsilon]}]$ is boundedly invertible on the space ${}_{(\gamma+\epsilon_1)}C_{\gamma+\epsilon_1}([s, T - \epsilon]; U)$:

$$
\left[I + L_{[s,T-\epsilon]}^* R^* R L_{[s,T-\epsilon]} \right]^{-1} \in \mathcal{L} \left({}_{(\gamma+\epsilon_1)}C_{\gamma+\epsilon_1}([s, T-\epsilon]; U) \right) \qquad (2.43)
$$

with uniform bound which depends on ϵ, ϵ_1, T, but which may be taken independent of s, $0 \le s \le T - \epsilon$.

Proof. (Similar to the proof of Theorem 3.4 of [L-T.2].) Let $h \in_{(\gamma+\epsilon_1)} C_{\gamma+\epsilon_1}([s, T - \epsilon]; U)$. We seek a unique $g \in_{(\gamma+\epsilon_1)} C_{\gamma+\epsilon_1}([s, T - \epsilon]; U)$ such that

$$g + L^*_{[s,T-\epsilon]} R^* R L_{[s,T-\epsilon]} g = h. \tag{2.44}$$

To simplify notation we may take $R = I$ below. Since $\gamma + \epsilon_1 < 1$, we may apply both Proposition 2.8 for the left-hand singularity at $t = s$ and Corollary 3.2 of [L-T.2] for the right-hand singularity at $t = T - \epsilon$. Let N be a positive integer greater than both m and n_0, where m is the positive integer provided by Proposition 2.8(iii), Eqn. (2.38), while n_0 is the positive integer provided by Corollary 3.2 of [L-T.2], Eqn. (3.9). Then there exists a unique $v \in L_2(s, T - \epsilon; U)$ such that

$$v + L^*_{[s,T-\epsilon]} L_{[s,T-\epsilon]} v = (L^*_{[s,T-\epsilon]} L_{[s,T-\epsilon]})^N h \in C([s,T]; U) \subset L_2(s, T; U). \tag{2.45}$$

As in the proof of Step 2 of Theorem 3.4 of [L-T.2], we know that, in fact, $v \in C([s, T - \epsilon]; U)$, by relying on Theorem 3.3 of [L-T.2]. Finally, we proceed as in the last steps of the proof of Theorem 3.4 of [L-T.2] by defining recursively the vectors $g_{n_0 - 1}, \ldots, g$ which now belong all to ${}_{(\gamma+\epsilon_1)}C_{(\gamma+\epsilon_1)}([s, T - \epsilon]; U)$, precisely as h in the present case. The last vector g is the unique sought-after solution of (2.44). □

Step 8. We return to identity (2.34) with right-hand side as in (2.35), and we apply Theorem 2.9. We then obtain with $x \in Y$, and $\forall\, \epsilon, \epsilon_1 > 0$:

$$\frac{du^0(t, s; x)}{dt} \in_{(\gamma+\epsilon_1)} C_{\gamma+\epsilon_1}([s, T - \epsilon]; U); \tag{2.46}$$

explicitly

$$\frac{du^0(t, s; x)}{dt} \in C(s, T - \epsilon); U, \ \forall\, \epsilon > 0, \tag{2.47a}$$

and

$$\left\| \frac{du^0(t, s; x)}{dt} \right\|_U \le \frac{C_{T\gamma\epsilon_1\epsilon}}{[(t-s)(t-(T-\epsilon))]^{\gamma+\epsilon_1}}, \quad s < t < T - \epsilon. \tag{2.47b}$$

Since in (2.47a), $\epsilon > 0$ is arbitrary, we then have $\frac{du^0}{dt}(t, s; x) \in C((s, T); U)$ as desired. The statement (1.29) of Theorem 1.3 concerning $\frac{du^0(t,s;x)}{dt}$ is thus proved.

Step 9. With the regularity (1.29) of $\frac{du^0(t,s;x)}{dt}$ at hand, we return to

$$\frac{dy^0(t, s; x)}{dt} - Ae^{A(t-s)}x = e^{A(t-s)} B u^0(s, s; x) + \{L_s \dot{u}^0(\cdot, s; x)\}(t). \tag{2.48}$$

We then apply Proposition 2.8(i) with $r = \gamma + \epsilon < 1$ and obtain

$$L_s u^0(\cdot, s; x) \ \in \ {}_{2\gamma-1+\epsilon}\, C([s, T - \epsilon]; Y), \tag{2.49}$$

while, since $u^0(s,s;x) \in U$ by (1.27), where

$$\|(-A)^\gamma e^{A(t-s)} A^{-\gamma} B u^0(s,s;x)\| \le \frac{C_T}{(t-s)^\gamma} \|u^0(s,s;x)\|_U, \quad t > s. \tag{2.50}$$

Since $2\gamma - 1 + \epsilon < \gamma$, we obtain the statement (1.30) for $\frac{dy^0(t,s;x)}{dt}$ as desired by combining (2.49) and (2.50) in (2.48). Theorem 1.3 is proved. □

Remark 2.1. We now prove the content of Remark 1.2 in the special case $\gamma < \frac{1}{2}$ noted there. In this case, $L_{sT}u^0(\,\cdot\,,s;x) = y^0(T,s;x) - e^{A(T-s)}x$ is well defined as a vector in Y, by (1.28a). Setting $\epsilon = 0$ in the preceding computations we obtian first

$$\frac{d}{dt}\{L_s^* R^* R L_s u^0(\,\cdot\,,s;x)\} = \{L_s^* R^* R L_s \frac{d}{d\sigma} u^0(\sigma,s;x)\}(t) + w_0(t;x) \tag{2.51}$$

from (2.31), with (see (2.25))

$$w_0(t;x) \in_{(2\gamma-1+\epsilon_1)} C_\gamma([s,T];U), \tag{2.52}$$

where $w_0(t;x)$ is the function $w_0(t)$ in (2.24) when $\mu(\cdot)$ there is replaced by $u^0(\,\cdot\,,s;x)$. Next, from (2.34) with $\epsilon = 0$ we obtain

$$\frac{du^0(t,s;x)}{dt} + L_s^* R^* R L_s \frac{du^0}{d\sigma}(\sigma,s;x) = \nu_0(t;x), \tag{2.53}$$

where, since $\gamma + \epsilon > 2\gamma - 1 + \epsilon$, we have

$$\nu_0(t,x) = -w_0(t,x) + \rho(t,x) \in_{(\gamma+\epsilon)} C_{1+\gamma}([s,T];U), \ \forall\, \epsilon > 0. \tag{2.54}$$

So far, the assumption that $\gamma < \frac{1}{2}$ to obtain (2.53), (2.54) is not really crucial. Several other cases with $\frac{1}{2} \le \gamma < 1$ may still lead to $L_{sT}u^0(\,\cdot\,,s;x) \in Y$, hence to (2.53), (2.54) (e.g., in the smoothing case of Subsection 5.1 of [L-T.4]. However, it is at the level of asserting, from (2.53), that

$$[I_s + L_s^* R^* R L_s]^{-1} \in \mathcal{L}\left({}_{(\gamma+\epsilon)}C_{1+\gamma}([s,T];U)\right) \tag{2.55}$$

that the assumption $\gamma < \frac{1}{2}$ is critical. Indeed, (2.55) can be shown the by the same boot-strap argument employed in the proof of Theorem 2.9: this requires applying $(L_s^* R^* R L_s)$ repeatedly to ν_0. But with ν_0 as in (2.54) we obtain that the order of singularity of $L_s\mu$ is $(1+\gamma) + (\gamma - 1 + \epsilon) = 2\gamma + \epsilon$ by Proposition 3.1(ii), Eqn. (3.3) in [L-T.2]. However, further application of L_s^* to $R^* R L_s \mu$ requires that $2\gamma + \epsilon < 1$, i.e., $\gamma < \frac{1}{2}$, in accordance with the requirement of Proposition 3.1(iii) of [L-T.2]. Once (2.55) is established, i.e., for $\gamma < \frac{1}{2}$, then we obtain from (2.55) that

$$\frac{du^0}{dt}(t,s;x) \in_{(\gamma+\epsilon)} C_{(1+\gamma)}([s,T];U), \tag{2.56}$$

an estimate (1.31) is proved. Then, if follows from Proposition 3.1(ii), Eqn. (3.3) of [L-T.2], since $1+\gamma > 1$, that $\frac{dy^0}{dt}(t,s;x)$ has a singularity at $t = T$ of order $(1+\gamma) + (\gamma - 1 + \epsilon) = 2\gamma + \epsilon$, and estimate (1.32) follows. □

3 Proof of Theorem 1.5

It is conceptually identical to that of Theorem 1.3 after taking into account the "improved" regularity results listed in Theorem 1.4. We shall only point out the technical differences. In the result of Step 2, using (1.34) and assumption (1.33), we obtain

$$\frac{d}{dt}\{L_{sT}G^*Gy^0(T,s;x)\}(t) \in C_{1+\gamma-\beta}([s,T];U). \tag{3.1}$$

Consequently, the regularity in Corollary 2.2 should be replaced by

$$\rho(t,x) \equiv \frac{d}{dt}\{\text{R.H.S. of (2.1)}\}(t) \ \in \ {}_{\gamma+\epsilon}C_{1+\gamma-\beta}([s,T];U). \tag{3.2}$$

Finally (2.54) should be replaced by

$$\nu_0(t,x) = -\omega_0(t,x) + \rho(t,x) \ \in \ {}_{\gamma+\epsilon}C_{1+\gamma-\beta}([s,T];U) \tag{3.3}$$

and (2.55) replaced by

$$[I_s + L_s^* R^* R L_s]^{-1} \in \mathcal{L}({}_{\gamma+\epsilon}C_{1+\gamma-\beta}([s,T];Y)). \tag{3.4}$$

Noticing that $1+\gamma-\beta+\gamma-1+\epsilon = 2\gamma-\beta+\epsilon < 1$, after taking $\beta = 2\gamma-1+2\epsilon$, we are in a position to apply the same argument as in the previous case, which then yields

$$\frac{du^0}{dt}(t,s;x) \ \in \ {}_{\gamma+\epsilon}C_{1+\gamma-\beta}([s,T];U), \tag{3.5}$$

and this proves (1.39).

The remaining estimate for $\frac{dy^0}{dt}$ follows from (2.48), (2.50). □

Remark 3.1. With reference to Theorem 1.6, in the special case where $AR = RA$ and $A = A^*$ we also obtain that

$$\|A^*P(t)\|_{\mathcal{L}(Y)} \le \frac{C_T}{(T-t)^\alpha}, \quad \|P(t)A\|_{\mathcal{L}(Y)} \le \frac{C_T}{T-t}, \quad t < T. \tag{3.6}$$

In this case, a solution $P(t)$ to the Riccati Equation (1.21) can be written as

$$\begin{cases} \dot{P}(t) = -R^*R - A^*P(t) - P(t)A + (P(t)B^*)^*B^*P(t) \\ P(T) = G^*G. \end{cases} \tag{3.7}$$

4 Proof of Theorem 1.6

We shall use the representation formula in (1.19) to obtain

$$\begin{aligned} A^*P(t)x \ &= \ \int_t^T A^*e^{A^*(\tau-t)}R^*Ry^0(\tau,t;x)d\tau \\ &+ A^*e^{A^*(T-t)}G^*Gy^0(T,t,x) \equiv I + II. \end{aligned} \tag{4.1}$$

By (1.34) and (1.33) for $\gamma \geq \frac{1}{2}$,

$$\|I\|_Y = \|A^* e^{A^*(T-t)} G^* G y^0(T,t;x)\|_Y \leq \frac{C_T}{(T-t)^{1-\beta}} \|x\|_Y, \tag{4.2}$$

where $\beta = 0$ if $\gamma < \frac{1}{2}$.

As for the second term in (4.1), we write

$$\begin{aligned} II &= -\int_t^T e^{A^*(\tau-t)} R^* R \frac{d}{d\tau} y^0(\tau,t;x) d\tau \\ &\quad + e^{A^*(T-t)} R^* R y^0(T,t;x) - R^* R x \\ &= -\int_t^T e^{A^*(\tau-t)} R^* R \frac{d}{d\tau} \left[y^0(\tau,t;x) - e^{A(\tau-t)} x \right] d\tau \\ &\quad - \int_t^T e^{A^*(\tau-t)} R^* R e^{A(\tau-t)} x \, d\tau \\ &\quad + e^{A^*(T-t)} R^* R y^0(T,t;x) - R^* R x. \end{aligned} \tag{4.3}$$

By (1.34),

$$\left\| e^{A^*(T-t)} R^* R y^0(T,t;x) - R^* R x \right\|_Y \leq C_T \|x\|_Y. \tag{4.4}$$

From (1.32b),

$$\begin{aligned} &\left\| \int_t^T e^{A^*(\tau-t)} R^* R \frac{d}{d\tau} \left[y^0(\tau,t;x) - e^{A(\tau-t)} x \right] d\tau \right\|_Y \\ &\leq C \|x\|_Y \int_t^T \frac{1 \, d\tau}{(T-\tau)^{2\gamma+\epsilon} (\tau-t)^\gamma} \leq C_T \frac{1}{(T-t)^{3\gamma-1+\epsilon}} \|x\|_Y, \text{ if } \gamma < \frac{1}{2}. \end{aligned} \tag{4.5}$$

Finally, from (1.40a),

$$\begin{aligned} &\left\| \int_t^T e^{A^*(\tau-t)} R^* R \frac{d}{d\tau} \left[y^0(\tau,t;x) - e^{A(\tau-t)} x \right] d\tau \right\|_Y \\ &\leq C \|x\|_Y \int_t^T \frac{d\tau}{(\tau-t)^\gamma (T-\tau)^{2\gamma-\beta}} \\ &\leq \frac{C\|x\|_Y}{(T-t)^{3\gamma-\beta-1}} \int_t^{t+\frac{T-t}{2}} \frac{d\tau}{(\tau-t)^\gamma} + \frac{C\|x\|_Y}{(T-t)^\gamma} \int_{t+\frac{T-t}{2}}^T \frac{d\tau}{(T-\tau)^{2\gamma-\beta}} \\ &\text{(since } 2\gamma - \beta < 1, \ \gamma < 1) \\ &\leq \frac{C_T \|x\|_Y}{(T-t)^{3\gamma-1-\beta}}, \qquad \gamma \geq \frac{1}{2}. \end{aligned} \tag{4.6}$$

Combining (4.3)-(4.6) yields

$$\left\| II + \int_t^T e^{A^*(\tau-t)} R^* R A e^{A(\tau-t)} x \, d\tau \right\|_Y \leq \frac{C_T \|x\|_Y}{|T-t|^{3\gamma-1-\beta}}. \tag{4.7}$$

Moreover, from (4.2), (4.1),

$$\left\| A^*P(t)x + \int_t^T e^{A^*(\tau-t)}R^*RAe^{A(\tau-t)}x\,d\tau \right\|_Y \leq \frac{C_T\|x\|_Y}{(T-t)^\alpha}$$
$$\alpha = \max(1-\beta, 3\gamma - 1 - \beta). \tag{4.8}$$

Hence, by duality

$$\left\| P(t)Ax + \int_t^T A^*e^{A^*(\tau-t)}R^*Re^{A(\tau-t)}x\,d\tau \right\|_Y \leq \frac{C_T\|x\|_Y}{(T-t)^\alpha}, \quad t < T. \tag{4.9}$$

Then, (4.8) and (4.9) now give

$$\begin{aligned}
\|A^*P(t)x + P(t)Ax\| &\leq \frac{C_T\|x\|_Y}{(T-t)^\alpha} + \left\| \int_t^T e^{A^*(\tau-t)}R^*RAe^{A(\tau-t)}x\,d\tau \right. \\
&\quad \left. + \int_t^T A^*e^{A^*(\tau-t)}R^*Re^{A(\tau-t)}x\,d\tau \right\|_Y \\
&\leq \frac{C_T\|x\|_Y}{(T-t)^\alpha} + \left\| \int_t^T \frac{d}{d\tau}\left[e^{A^*(\tau-t)}R^*Re^{A(\tau-t)}x\right]d\tau \right\|_Y \\
&\leq \frac{C_T\|x\|_Y}{(T-t)^\alpha} + \left\| e^{A^*(T-t)}R^*Re^{A(T-t)}x \right\|_Y \\
&\quad + \|R^*Rx\|_Y \leq C_T\|x\|_Y\left[\frac{1}{(T-t)^\alpha} + 1\right],
\end{aligned} \tag{4.10}$$

which proves assertion (1.42) in Theorem 1.6.

The inequality in (1.43) then follows from Theorem 1.1. Finally, the regularity in (1.41) is the result of (1.42), (1.43) and the structure of the Differential Riccati Equation in (1.21). □

Proof of Remark 3.1. If, in addition, $RA = AR$ and $A^* = A$, then

$$\begin{aligned}
&\int_t^T Ae^{A(\tau-t)}R^*Re^{A(\tau-t)}x\,d\tau \\
&\quad = \frac{1}{2}\int_t^T \frac{d}{d\tau}\left[e^{A(\tau-t)}R^*Re^{A(\tau-t)}x\right]d\tau \\
&\quad = \frac{1}{2}e^{A(T-t)}R^*Re^{A(T-t)}x - \frac{1}{2}R^*Rx.
\end{aligned} \tag{4.11}$$

Thus, combining (4.8) and (4.11), yields

$$\|AP(t)\|_{\mathcal{L}(Y)} \leq \frac{C_T}{(T-t)^\alpha}, \quad t < T,$$

as claimed by Remark 3.1. □

References

[B.1] A. V. Balakrishnan, Boundary control of parabolic equations: L-Q-R theory, in *Theory of Nonlinear Operators,* Prof. Fifth Intern. Summer School, Berlin (1977).

[B-D-D-M] A. Bensoussan, G. DaPrato, M. Delfour, S. Mitter, *Representations of Infinite Dimensional Systems*, Birkhäuser, 1992.

[D-I.1] G. DaPrato and A. Ichikawa, Riccati equations with unbounded coefficients, *Ann. di Matem. Pura e Appl.* 140(1985), 209-221.

[F.1] F. Flandoli, Algebraic Riccati equations arising in boundary control problems with distributed parameters, *SIAM J. Control and Optimiz.* 22(1984), 76-78.

[F.1] F. Flandoli, On the direct solution of Riccati equations arising in boundary control theory, *Ann. di Matem. Pura e Appl.* (iv), vol. CLXII (1993), 93-131.

[L-T.1] I. Lasiecka and R. Triggiani, Dirichlet boundary control problem for parabolic equation with quadratic cost: Analyticity and Riccati's feedback synthesis, *SIAM J. Control and Optimiz.* 21(1983), 41-67.

[L-T.2] Riccati differential equations with unbounded coefficients and non-smoothing terminal condition—The case of analytic semigroup, *SIAM J. Mathematical Analysis* 25(1992), 449-81.

[L-T.3] Differential and Algebraic Riccati equations with application to boundary/point control problems: Continuous theory and approximation theory, Volume 164 in the *Springer-Verlag Lecture Notes Series in Control and Information Sciences* (1991), p. 160.

[L-T.4] I. Lasiecka and R. Triggiani, Monograph for the series, *Encyclopedia of Mathematics*, Cambridge University Press, in preparation.

13 Necessary and Sufficient Conditions for Optimality for Nonlinear Control Problems in Banach Spaces

Urszula Ledzewicz Southern Illinois University at Edwardsville, Edwardsville, Illinois

Andrzej Nowakowski Łódź University, Łódź, Poland

Abstract. We consider the problem to minimize an integral functional defined on the space of absolutely continuous functions and measurable control functions with values in infinite dimensional real Banach spaces. The states are governed by abstract first order semilinear differential equations and are subject to periodic or anti-periodic type boundary conditions. We derive necessary conditions for optimality and introduce the notion of a dual field of extremals to obtain sufficient conditions for optimality. Such a dual field of extremals is constructed and a dual optimal synthesis is proposed.

1. Introduction

Let X and U be real separable reflexive Banach spaces. Let X^* be the dual to X and let $\langle x^*, x\rangle$ denote the duality pairing between X^* and X . Let $A(t) : X \to X^*$ be a family of densely defined linear operators with domains $D(A(t))$ and let $U(t)$ be a family of subsets of U. We consider the problem to minimize

$$J(x,u) = \int_0^T L(t, x(t), u(t))dt + l(x(0), x(T)) \tag{1.1}$$

over all absolutely continuous functions $x : [0,T] \to X$ and Lebesgue measurable functions $u : [0,T] \to U$, the controls, subject to

$$\dot{x}(t) + A(t)x(t) = f(t, x(t), u(t)) \qquad \text{a.e. in } [0,T], \tag{1.2}$$

Supported in part by the National Science Foundation under Grant No. DMS-9109324 and by SIUE Research Scholar Award

and

$$u(t) \in U(t) \qquad \text{a.e. in } [0,T]. \tag{1.3}$$

The maps are defined between the following spaces $L : [0,T] \times X \times U \to R$, $f : [0,T] \times X \times \to X$, and

$$l(a,b) = \begin{cases} +\infty & \text{if } a \neq b, \\ 0 & \text{if } a = b \end{cases} \quad \text{or} \quad l(a,b) = \begin{cases} +\infty & \text{if } a \neq -b, \\ 0 & \text{if } a = -b, \end{cases} \tag{1.4}$$

depending on whether periodic or anti-periodic boundary conditions are imposed.

Necessary conditions for optimal control problems in abstract spaces have been considered in several papers starting with references [4,5]. In [7] Lions considered a general distributed parameter control problem in a Banach space without a priori given initial condition, so-called systems with insufficient data. Papageorgiou extended this result in [12] to the case of a general convex integral functional by using the Dubovitskii-Milyutin method (see [3]). His results found continuations in the analysis of problems with terminal data [8] as well as for abnormal problems and problems with nonoperator type equality constraints [9] using some extensions of the Dubovitskii-Milyutin method. In this paper, following the ideas in references [12,8,9], the Dubovitskii-Milyutin method as well as its generalization by Walczak [13] are applied to the above abstract optimal control problem and the local Maximum Principle is formulated for the cases without and with nonoperator equality constraints.

The classical sufficiency theorems which apply field theories (Weierstrass type sufficiency theorems) [1,10,15] or dynamic programming as well as (optimal) feedback control require strong assumptions imposed not only on the data of the problem, but also on objects which are constructed during the study of these problems. Unfortunately, such strong assumptions are very rarely satisfied by problems in optimal control theory. This is so because the approach is classical and problems are new - constraints appear (see [15]). Actually, the constraints do not allow for the regularity of the constructed objects (field of extremals, a feedback control, a value function). A dual approach to these objects allows us to overcome all these difficulties (compare [11]). Now the whole construction is done in the dual space (in the space of multipliers) - the natural space for problems with constraints. Some regularity is required too, however, only for objects which are dual to the classical one and so the last one need not be regular. In the paper the main ideas of the dual field theory and the dual feedback control are described.

2. Necessary conditions for optimality

Let $X \subset H$ be a real separable reflexive Banach space embedded continuously and densely in a Hilbert space H. Let U be a separable reflexive Banach space modelling the control space. Furthermore, let $||\cdot||$, $|\cdot|$, and $||\cdot||_*$ denote the norms in the spaces X, H, and X^* respectively where X^* denotes the dual space to X. We denote the inner product in H by $(\cdot,\cdot)$ and the duality pairings from X^* on X and from U^* on U by $\langle\cdot,\cdot\rangle$. Then we have for every $x \in X \subset H$ and $h \in H \subset X^*$ that $(x,h) = \langle h,x\rangle$. Let $A(t) : X \to X^*$ and $f : [0,T] \times X \times U \to X$ be operators, $L : [0,T] \times X \times U \to R$ a functional, $U : [0,T] \to \mathcal{P}(U)$ a multifunction with nonempty, closed, convex subsets as images.

We make the following basic assumptions:

(H0) For each $(x,u) \in X \times U$, the functions $t \to L(t,x,u)$, $t \to f(t,x,u)$ are $\mathbf{L}$-measurable. For each $t \in [0,T]$, the functions $(x,u) \to L(t,x,u)$, $(x,u) \to f(t,x,u)$ are continuously differentiable. The set $\{(t,u) \in [0,T] \times U : u(t) \in U(t)\}$ is $\mathbf{L} \times \mathbf{B}$-measurable, i.e. belongs to the σ-algebra of subsets of $[0,T] \times U$ generated by products of Lebesgue measurable subsets of $[0,T]$ and Borel measurable subsets of U.

(H1) For each $x \in D(A(t))$ the function $t \to A(t)x$ is X-measurable in $[0,T]$. For $t \in [0,T]$, the linear operators $A(t) : X \to X^*$ are continuous.

In order to derive the necessary optimality conditions below, we also need some technical assumptions on the operators A and f and the function L which we list separately.

(A1) There exist functions $a^i(\cdot) \in L^2_+, i = 1,2,3$ and numbers $p \geq 2$, $q \geq 1$ such that $|L(t,x,u)| \leq a_1(t) + a_2(t)|x|^p + a_3(t)||u||^q$ a.e. in $[0,T]$.

(A2) $||f(t,x,u)||_* \leq b_1(t) + b_2(t)|x| + b_3(t)||u||$ with some functions $b_1, b_2 \in L^2_+$ and $b_3 \in L^\infty_+$.

(A3) For $V = \{u \in L^2(U) : u(t) \in U(t)\, a.e.\}$ we have $intV \neq \emptyset$.

Let us call the problem (1.1)-(1.4) under assumptions (H0)-(H1) and (A1)-(A3) Problem I. If $u(t)$ is a control function subject to (1.3) and $x(t)$ is an absolutely continuous function corresponding to $u(t)$ (by (1.2)) and $L(t,x(t),u(t))$ is integrable, then the pair $(x(t),u(t))$ will be called admissible and $x(t)$ is an admissible trajectory. Along an admissible pair $(x(t),u(t))$ we need the following assumptions:

(B1) For $t \in [0,T]$, the linear operators $B(t) = A(t) - f_x(t,x(t),u(t)) : X \to X^*$ are coercive uniformly in $[0,T]$ i.e. there are $\lambda > 0, \alpha > 0$ such that $\langle B(t)h,h\rangle + \lambda|h|^2 \geq \alpha||h||^2$.

(B2) There exists a constant $c > 0$ such that for $t \in [0,T]$ we have

$$||B(t)||_{L(X,X^*)} \leq c$$

(B3) There exists a constant $\eta > 0$ such that for $t \in [0,T]$ we have

$$||f_u(t,x(t),u(t))||_{L(X,X^*)} \leq \eta$$

2.A Extremum principle

Denote by $W([0,T])$ the Banach space

$$W([0,T]) = \{x \in L^2(X) : \dot{x} \in L^2(X^*)\}$$

with norm

$$||x||^2_{W([0,T])} = \int_0^T ||x(t)||^2 dt + \int_0^T ||\dot{x}(t)||^2_* dt. \tag{2.1}$$

Since the space of absolutely continuous functions $x : [0,T] \to X$ is dense in $W([0,T])$ and $J(x,u)$ is continuous in $W([0,T])$ by (A3), Problem I considered in both spaces has the same value. However, to derive the necessary conditions it is more convenient to work in the space $W([0,T])$. Thus we seek a solution of the above problem in the subspaces of periodic or antiperiodic solutions, $WP([0,T]) = \{x \in W([0,T]) : x(0) = x(T)\}$, respectively $WA([0,T]) = \{x \in W([0,T]) : x(0) = -x(T)\}$. We simply use the notation $\tilde{W}([0,T])$ to denote either of them depending on the circumstances. We also denote the space of square-integrable Lebesgue measurable functions mapping $[0,T]$ into U by $L^2(U)$. Using the Dubovitskii-Milyutin method necessary conditions for optimality for Problem I can be derived in the form of the extremum principle given below. In its proof we need the following auxiliary result (see [6] Chapter 3 Vol. 1):

Lemma 2.1. *Suppose $(H0)$ and $(H1)$, and let $(x(t), u(t))$ be an admissible pair such that assumptions $(B1)$, and $(B2)$ hold. Furthermore suppose the mapping $t \to g(t)$ belongs to $L^2(X^*)$. Then there exist a unique solution $h \in \tilde{W}([0,T])$ to the variational equation*

$$\dot{h}(t) + A(t)h(t) - f_x(t, x(t), u(t))h(t) = g(t) \tag{2.2}$$

and a unique solution $y \in \tilde{W}([0,T])$ to the covariational or adjoint equation

$$-\dot{y}(t) + A^*(t)y(t) - f_x^*(t, x(t), u(t))y(t) = g(t) \qquad \text{a.e..} \tag{2.3}$$

For the Dubovitskii-Milyutin formalism we introduce the operator $F : \tilde{W}([0,T]) \times L^2(U) \to L^2(X^*)$ given by

$$F(x,u) = \dot{x}(t) + A(t)x(t) - f(t, x(t), u(t)). \tag{2.4}$$

In terms from optimization theory, our optimal control problem can be formulated as minimizing a functional $I(x,u)$ under the constraints $(x,u) \in Z_1 \cap Z_2$ where

$$Z_1 = \{(x,u) : F(x,u) = 0\} \tag{2.5}$$

$$Z_2 = \{(x, u) : u \in V\}. \tag{2.6}$$

We say the pair $(x, u) \in \tilde{W}([0, T]) \times L^2(U)$ is admissible if $(x, u) \in Z_1 \cap Z_2$. Note that the set Z_1 is an equality constraint, i.e. $intZ_1 = \emptyset$, and Z_2 is an inequality constraint, i.e. $intZ_2 \neq \emptyset$. Using the Dubovitskii-Milyutin method the following necessary conditions for optimality can be proven:

THEOREM 2.1. **(Extremum Principle)** *Assume $H(0), (H1)$, and $(A1)$ to $(A3)$ hold and let the admissible process (x, u) be optimal for problem I. Furthermore suppose that conditions (B1) to (B3) are satisfied along (x, u). Then there exist $y^0 \leq 0$ and $y \in \tilde{W}([0, T])$ with the property that*

$$\|y(t)\|_{X^*} + |y^0| \tag{2.7}$$

does not vanish and such that they satisfy the adjoint equation

$$\dot{y}(t) = -f_x^*(t, x(t), u(t))y(t) + A^*(t)y(t) - y^0 L_x(t, x(t), u(t)) \qquad \text{a.e.} \tag{2.8}$$

and the following maximum condition

$$\int_0^T \langle y^0 L_u(t, x(t), u(t)) + f_u^*(t, x(t), u(t))y(t), u - u(t)\rangle dt \leq 0 \tag{2.9}$$

for all $u \in V$.

PROOF: In the Dubovitskii-Milyutin approach we need to determine the cone of decrease for the functional I at (x, u), $DC(I, (x, u))$, the tangent cone to the set Z_1 at (x, u), $TC(Z_1, (x, u))$ and the feasible cone to the set Z_2 at (x, u), $FC(Z_2, (x, u))$ as well as the corresponding dual cones.

Repeating arguments from [12 and 8] and applying Theorem 7.4 from [3], it follows that

$$\begin{aligned} DC(I, (x, u)) = \{(h, v) \in \tilde{W}([0, T]) \times L^2(U) : \\ \int_0^T L_x(t, x(t), u(t))h(t) + L_u(t, x(t), u(t))v(t)dt < 0\} \end{aligned} \tag{2.10}$$

and

$$\begin{aligned} DC(I, (x, u))^* = \{f_0(h, v) : \text{ there exists a } y^0 \leq 0 \text{ such that} \\ f_0(h, v) = y^0 \int_0^T L_x(t, x(t), u(t))h(t) + L_u(t, x(t), u(t))v(t)dt\} \end{aligned} \tag{2.11}$$

In view of the assumptions of the theorem the operator F defined in (2.4) is continuously Fréchet differentiable at (x,u) with derivative given by

$$F'(x,u)(h,v)(t) = \dot{h}(t) + A(t)h(t) - f_x(t,x(t),u(t))h(t) - f_u(t,x(t),u(t))v(t). \tag{2.12}$$

It follows from Lemma 2.1 that in fact for arbitrary $v \in L^2(U)$ and $z \in L^2(X^*)$, the equation

$$\dot{h}(t) + A(t)h(t) - f_x(t,x(t),u(t))h(t) - f_u(t,x(t),u(t))v(t) = z(t) \tag{2.13}$$

has a unique solution $h \in \tilde{W}([0,T])$. Hence $F'(x,u)$ maps $\tilde{W}([0,T]) \times L^2(U)$ onto $L^2(X^*)$ and the classical Lusternik Theorem applies. We obtain that

$$\begin{aligned} TC(Z_1,(x,u)) = \{&(h,v) \in \tilde{W}([0,T]) \times L^2(U) : \\ &\dot{h}(t) + A(t)h(t) - f_x(t,x(t),u(t))h(t) - f_u(t,x(t),u(t))v(t) = 0.\} \end{aligned} \tag{2.14}$$

The structure of the corresponding dual cone will not be required in the remainder of the proof.

We now analyze the inequality constraint Z_2 given by (2.5). Theorem 10.5 from [3] implies that

$$\begin{aligned} FC(Z_2,(x,u)) = \{&(h,v) \in \tilde{W}([0,T]) \times L^2(U) : \\ &v = \lambda(w-u), w \in V\} \end{aligned} \tag{2.15}$$

$$\begin{aligned} FC(Z_2,(x,u))^* = \{&f_2(h,v) = (0, f_2'(v)) \text{ where } f_2' \text{ is a functional} \\ &\text{supporting the set } V \text{at } u\}. \end{aligned} \tag{2.16}$$

The Dubovitskii-Milyutin lemma [3] implies that there exist linear functionals $f_0 \in DC(I,(x,u))^*$, $f_1 \in TC(Z_1,(x,u))^*$ and $f_2 \in FC(Z_2,(x,u))^*$, not all zero, such that

$$f_0(h,v) + f_1(h,v) + f_2(h,v) \equiv 0 \qquad \text{for all } (h,v) \in \tilde{W}([0,T]) \times L^2(U).$$

Let $(h,v) \in TC(Z_1,(x,u))$. Since $TC(Z_1,(x,u))$ is a subspace, we have $f_1(h,v) = 0$. Hence, for $(h,v) \in TC(Z_1,(x,u))$ and using (2.11) and (2.16), the Euler-Lagrange equation takes the form

$$y^0 \int_0^T L_x(t,x(t),u(t))h(t) + L_u(t,x(t),u(t))v(t)dt + f_2'(v) = 0. \tag{2.17}$$

By Lemma 2.1 the adjoint equation (2.8) has a solution $y \in \tilde{W}([0,T])$. Using the adjoint equation in (2.17) we obtain

$$-y^0 \int_0^T L_x(t, x(t), u(t))h(t)dt =$$
$$\int_0^T \langle \dot{y}(t), h(t)\rangle dt - \int_0^T \langle A^*(t)y(t), h(t)\rangle dt + \int_0^T \langle f_x^*(t, x(t), u(t))y(t), h(t)\rangle dt. \tag{2.18}$$

Using Lemma 5.1 from [14] about integration by parts and the fact that $h \in \tilde{W}([0,T])$ we obtain

$$-y^0 \int_0^T L_x(t, x(t), u(t))h(t)dt = \int_0^T \langle y(t), -\dot{h}(t) - A(t)h(t) + f_x(t, x(t), u(t))h(t)\rangle dt. \tag{2.19}$$

Since $(h, v) \in TC(Z_1, (x, u))$, using (2.14) in (2.19) we have that

$$y^0 \int_0^T L_x(t, x(t), u(t))h(t) = \int_0^T \langle y(t), f_u(t, x(t), u(t))v(t)\rangle dt. \tag{2.20}$$

Combining equations (2.17) and (2.20) it follows therefore that

$$f_2'(v) = -\int_0^T \langle y^0 L_u(t, x(t), u(t)) + f_u^*(t, x(t), u(t))y(t), v(t)\rangle dt.$$

The maximum condition (2.9) of the theorem follows now directly from the definition of a supporting functional to V at (x, u).

This proves the theorem for the case when the cone of decrease for I at (x, u), $DC(I, (x, u))$, given by (2.10) is nonempty. If it is empty, then actually

$$\int_0^T L_x(t, x(t), u(t))h(t) + L_u(t, x(t), u(t))v(t)dt = 0$$

for all $(h, v) \in \tilde{W}([0,T]) \times L^2(U)$. This equation takes the form of the Euler-Lagrange equation with multipliers equal to $y^0 = 0$ and $f_2' \equiv 0$. Now proceeding as above, the theorem follows.

REMARK 2.1. *If we assume in addition that*

(A6) *the mapping* $t \to |U(t)| = sup\{||u|| : u \in U(t)\}$ *belongs to* L^2_+, *then, proceeding like in Theorem 3.1 in [12], we can express the maximum condition in the following simpler form:*

$$\langle y^0 L_u(t, x(t), u(t)) + f_u^*(t, x(t), u(t))y(t), v - u(t)\rangle \leq 0 \tag{2.21}$$

for all $v \in U(t)$ and almost everywhere in $[0, T]$.

REMARK 2.2. *If $U(t) = L^2(U)$ for all $t \in [0, T]$, i.e. there are no constraints on the control functions, then the maximum condition (2.21) becomes*

$$y^0 L_u(t, x(t), u(t)) = -f_u^*(t, x(t), u(t))y(t) \qquad \text{a.e.}$$

2B. AN OPTIMAL CONTROL PROBLEM WITH NONOPERATOR EQUALITY CONSTRAINTS

We now consider Problem I without the assumption (A3) that the set V has a nonempty interior in $L^2(U)$ and call this problem II. The geometric model for the Dubovitskii-Milyutin method changes here since we now have two equality constraints and one of them is in the nonoperator form. In order to prove the extremum principle, we need a generalization of the Dubovitskii-Milyutin method due to Walczak [13] which deals with such a situation.

Let us consider the problem of minimizing a functional $I(x)$ in a Banach space under n inequality constraints $Z_1, \ldots, Z_n$ and m equality constraints $Z_{n+1}, \ldots,$ Z_{n+m}. In [13] necessary conditions for optimality for this problem were formulated using properties of *cones of the same sense* (CSS) and *cones of the opposite sense* (COS).

THEOREM 2.2. *(Walczak, [13]) Suppose x_0 is a solution to problem II. Let C_0 denote a nonempty open and convex cone contained in the cone of decrease of the functional I at a point x_0 and let C_i, $i = 1, \ldots, n$ denote open convex cones contained in the feasible cones for the set Q_i at x_0. Let C_i, $i = n+1, \ldots, n+m$ denote closed convex cones contained in the tangent cones to the sets Q_i, respectively, and suppose that the cone $\tilde{C}$ is contained in the tangent cone to the set $V = \bigcap_{i=n+1}^{n+m} Q_i$ at x_0. If the cones C_i^*, $i = n+1, \ldots, n+m$, are either of* the same sense *or of* the opposite sense, *then there exist linear functionals $f_i^* \in C_i^*$, $i = n+1, \ldots, n+m$, such that*

$$\sum_{i=0}^{n+m} f_i \equiv 0.$$

Furthermore, if the cones C_i^*, $i = n+1, \ldots, n+m$, are of the same sense, then $\sum_{i=0}^{n} ||f_i|| > 0$ and if these cones are of the opposite sense, then $\sum_{i=n+1}^{n+m} ||f_i|| > 0$.

For our purposes here the following criterion for when cones are of the same sense will suffice.

LEMMA 2.2. *(Walczak [13]) Suppose E_1 and E_2 are normed spaces and $A : E_2 \to E_1$ is a bounded linear operator with dual operator A^*. Set $E = E_1 \times E_2$ and let*

$$C_1 = \{(x, y) \in E : x = Ay\}, \qquad C_2 = E_1 \times \bar{C}_2$$

where $\bar{C}_2$ is a cone in E_2. Then we have

$$C_1^* = \{(x^*, y^*) \in E^* : y^* = -A^* x^*\}$$
$$C_2^* = \{(0, y^*) \in E^* : y^* \in \bar{C}_2^*\}$$

and the cones C_1^* and C_2^* are of the same sense.

THEOREM 2.3. **(Extremum Principle)** *Assume (H0), (H1) and (A1), (A2) hold. Let (x, u) be an optimal process for problem II and suppose assumptions (B1)-(B3) are satisfied along (x, u). Then there exist a $y^0 \leq 0$ and $y(\cdot) \in \tilde{W}([0, T])$, with the property that $\|y(t)\|_{X^*} + |y^0|$ does not vanish and such that they satisfy the adjoint equation (2.8) and the maximum condition (2.9).*

PROOF: First assume that the cone of decrease for the functional I at (x, u), $DC(I, (x, u))$, is nonempty. In this case we can use the formula (2.11) for the corresponding dual cone. For the first equality constraint in the operator form (2.5) with the operator F given in (2.4) the form of the tangent cone $TC(Z_1, (x, u))$ will be as before, given in (2.14). For the equality constraint Z_2 in nonoperator form we obtain from Theorem 10.5 in [3] that

$$TC(Z_2, (x, u)) = \tilde{W}([0, T]) \times \bar{C}_2$$

where $\bar{C}_2 \subset L^2(U)$ denotes the tangent cone to the set V at u. In order to apply Theorem 2.2 we need to verify that the dual cones $TC(Z_1, (x, u))^*$ and $TC(Z_2, (x, u))^*$ are of the same sense. The operator F is continously Fréchet differentiable and it follows from Lemma 2.1 that the equation

$$\dot{h} + A(t)h - f_x(t, x(t), u(t))h - f_u(t, x(t), u(t))v = 0, \tag{2.22}$$

has a unique periodic (respectively antiperiodic) solution $h \in \tilde{W}([0, T])$ for every $v \in L^2(U)$. Hence the operator F_x is continous, one-to-one and onto and therefore has a continuous inverse by the open mapping theorem. Thus in a neighborhood V of (x, u) the assumptions for the implicit function theorem are satisfied for the operator F and the set Z_1 can be described in the form $Z_1 = \{(\tilde{x}, w) : \tilde{x} = S(w)\}$ where $S : L^2(U) \to \tilde{W}([0, T])$ is a C^1 operator which satisfies $F(S(w), w) = 0$ for every w with the property that $(S(w), w) \in V$. Differentiating we obtain that

$$TC(Z_1, (x, u)) = \{(h, v) \in \tilde{W}([0, T]) \times L^2(U) : h = S'(u)v\}.$$

Hence by Lemma 2.2 the cones $TC(Z_1,(x,u))^*$ and $TC(Z_2,(x,u))^*$ are of the same sense.

It is easily seen that the cone of decrease for I at x_0, $DC(I,(x,u))$, and the tangent cones $TC(Z_1,(x,u))$ and $TC(Z_2,(x,u))$ are convex. Hence, in order to apply Theorem 2.2, we only have to show that

$$TC(Z_1,(x,u)) \cap TC(Z_2,(x,u)) \subset TC(Z_1 \cap Z_2,(x,u)).$$

The proof of this relation is similar to the one given in [9]. Take an arbitrary $(h,v) \in TC(Z_1,(x,u)) \cap TC(Z_2,(x,u))$. Since Z_2 only involves the control variable, by the definition of tangent direction, there exists an operator $r_u^2(t)$ such that

$$\frac{||r_u^2(t)||}{t} \to 0 \qquad \text{as } t \to 0^+$$

and

$$(x,u) + t(h,v) + (r_x^2(t), r_u^2(t)) \in Z_2$$

for sufficiently small t and arbitrary $r_x^2(t)$. Furthermore, we also have for sufficiently small t that

$$(S(u + tv + r_u^2(t)), u + tv + r_u^2(t)) \in Z_1. \tag{2.23}$$

Since S is Fréchet differentiable, there exists an operator $r_x^1(t)$ such that

$$\frac{||r_x^1(t)||}{t} \to 0 \qquad \text{as } t \to 0^+$$

and

$$S(u + tv + r_u^2(t)) = S(u) + tS'(u)v + r_x^1(t).$$

Now using (2.23) we get that

$$(x,u) + t(h,v) + (r_x^1(t), r_u^1(t)) \in Z_1.$$

It is therefore enough to choose $r_x^1(t) = r_x^2(t)$ and it follows that $(h,v) \in TC(Z_1 \cap Z_2,(x,u))$. Thus all the assumptions of Theorem 2.2 are fulfilled. This theorem then implies that there exist linear functionals $f_0 \in DC(I,(x,u))^*$ and $f_i \in TC(Z_i,(x,u))^*$ for $i = 1,2$, not all zero, such that

$$f_0(h,v) + f_1(h,v) + f_2(h,v) \equiv 0.$$

Proceeding analogously as before, we obtain the result. In the case when the cone of decrease is empty, the result follows as a degenerate case by an argument given in Theorem 2.1. This completes the proof.

REMARK 2.3. *The Maximum principles formulated in Theorems 2.1 and 2.3 give the local form of the maximum condition. The stronger global versions can be derived following the arguments given in [3].*

3. Canonical spray of flights

We say that an admissible pair $(x(t), u(t))$, $t \in [0, T]$ is a *line of flight* (briefly l.f.) if it satisfies the following conditions (the maximum principle): along $x(t)$ there exist a conjugate vector function $y(t)$, absolutely continuous in $[0, T]$ with values in X^*, and a number $y^0 \leq 0$ such that $\|y(t)\|_{X^*} + |y^0|$ is nonvanishing and

$$\dot{y}(t) = -f_x^*(t, x(t), u(t))y(t) - y^0 L_x(t, x(t), u(t)) + A^*(t)y(t), \qquad \text{a.e.} \tag{3.1}$$

$$\begin{aligned} &\langle y(t), f(t, x(t), u(t))\rangle + y^0 L(t, x(t), u(t)) \\ &= \sup\{\langle y(t), f(t, x(t), u)\rangle + y^0 L(t, x(t), u) : u \in U(t)\} \end{aligned} \tag{3.2}$$

$$l(x(0), x(T)) = l(y(0), y(T)) = 0 \tag{3.3}$$

or

$$\text{condition (3.3) is not satisfied.} \tag{3.3'}$$

We call a triple $(x(t), y(t), u(t))$ of functions and a number y^0 such that $(x(t), u(t))$ define a line of flight and $(y(t), y^0)$ are the corresponding conjugate function satisfying conditions (3.1)–(3.3) or (3.3') a *canonical line of flight* (briefly c.l.f). In the usual way we define an open arc line of flight or canonical line of flight.

In order to construct a field of extremals from canonical lines of flights defined above, we must choose a certain family of canonical lines of flight which satisfy extra regularity hypotheses. In this section we describe such a family of canonical lines of flight Let W be a separable reflexive Banach space and on an open set $G \subset W$ define a pair of differentiable continuous functions

$$t^-(\sigma), t^+(\sigma), \qquad t^-(\sigma) < t^+(\sigma), \qquad \sigma \in G,$$

with values in [0,T]. Let

$$\begin{aligned} S^- &= \{(t, \sigma) : t = t^-(\sigma) \geq 0, \ \sigma \in G\}, \\ S &= \{(t, \sigma) : t^-(\sigma) < t^+(\sigma), \ \sigma \in G\}, \\ S^+ &= \{(t, \sigma) : t = t^+(\sigma) \leq T, \ \sigma \in G\}, \end{aligned}$$

and set $[S] = S^- \cup S \cup S^+$. For a given line of flight $(x(t), u(t))$ we write $z(t) = (-x^0(t), x(t))$ where $x^0(t) = \int_t^T L(\tau, x(\tau), u(\tau))d\tau$, $t \in [0, T]$ and for a given canonical line of flight $(x(t), y(t), u(t), y^0)$ we write $p(t) = (y^0, y(t))$. Further, by a canonical line of flight we shall always mean the triple of functions $(z(t), p(t), u(t))$.

The family of canonical lines of flight is described by the functions

$$z(t, \sigma), p(t, \sigma), u(t, \sigma), \qquad (t, \sigma) \in [S] \tag{3.4}$$

and will be denoted by Σ. The sets of pairs (t,x) where $x = x(t,\sigma)$ with (t,σ) belonging to $S^-, S, S^+, [S]$ will be denoted by $E^-, E, E^+, [E]$, respectively, and the set of pairs $(t, z(t,\sigma))$ with (t,σ) in $S^-, S, S^+, [S]$ by $D^-, D, D^+, [D]$; E^{*-}, $E^*, E^{*+}, [E^*]$ will denote the sets of values of $(t, p(t,\sigma))$ with (t,σ) in S^-, S, $S^+, [S]$.

We shall make the following regularity hypotheses on Σ:

(H2) The functions $z(t,\sigma)$ and $p(t,\sigma)$ are C^1 in $[S]$ and $u(t,\sigma)$ is Borel measurable in $[S]$.

(H3) The functions $\tilde{L}(t,\sigma) = L(t, x(t,\sigma), u(t,\sigma))$, $\tilde{f}(t,\sigma) = f(t, x(t,\sigma), u(t,\sigma))$ are continuous in $[S]$; they have continuous derivatives $\tilde{L}_\sigma(t,\sigma)$, $\tilde{f}_\sigma(t,\sigma)$ in $[S]$ and for each fixed (t,x) in $[E]$, the partial derivatives $\partial L(t,x,u(t,\sigma))/\partial\sigma$, $\partial f(t,x,u(t,\sigma))/\partial\sigma$ at $x = x(t,\sigma)$ satisfy the following relations:

$$\frac{\partial \tilde{L}}{\partial \sigma}(t,\sigma) = \frac{\partial L(t,x,u(t,\sigma))}{\partial \sigma} + L_x(t,x,u(t,\sigma))x_\sigma(t,\sigma),$$
$$\frac{\partial \tilde{f}}{\partial \sigma}(t,\sigma) = \frac{\partial L(t,x,u(t,\sigma))}{\partial \sigma} + f_x(t,x,u(t,\sigma))x_\sigma(t,\sigma).$$

(H4) The maps $S^- \to D^-$, $S \to D$ defined by $(t,\sigma) \to (t, z(t,\sigma))$ have the following property: given any arc $C_z \subset D^-$ (or $C_z \subset D$) with the description $t_1 \le \tau \le t_2$, $(-x^0(\tau), x(\tau))$ where $x(t)$ is a trajectory of an admissible pair $(x(t), u(t))$, $t \in [0,T]$, $x^0(t) = \int_t^T L(\tau, x(\tau), u(\tau))d\tau$, issuing from $(t_1, z(t_1,\sigma_1))$, there exists a rectifiable curve $\Gamma \subset S^-$ (or $\Gamma \subset S$) issuing from (t_1,σ_1) such that every small arc of C_z issuing from $(t_1, z(t_1,\sigma_1))$ is the image under the map $(t,\sigma) \to (t, z(t,\sigma))$ of a small arc of Γ issuing from (t_1,σ_1).

(H5) There exists at least one $\sigma_0 \in G$ such that $t^-(\sigma_0) = 0$, $t^+(\sigma_0) = T$. If condition (3.3) is not satisfied for a canonical line of flight $z(t,\sigma_1)$, $p(t,\sigma_1)$, $u(t,\sigma_1)$, then for it $\langle y(t^+(\sigma_1),\sigma_1), x_{\sigma_1}(t^+(\sigma_1),\sigma_1)\rangle = 0$.

(H6) For each $(t_0, p_0) \in [E^*]$ the value of

$$\begin{aligned} V(t_0,p_0) &= y_0^0 \int_{t_0}^{t^+(\sigma)} L(t, x(t,\sigma), u(t,\sigma))dt - \langle x((t_0,\sigma), y(t_0,\sigma)\rangle \\ &= -\langle z(t_0,\sigma), p(t_0,\sigma)\rangle_z \end{aligned}$$

is the same for all σ which satisfy $p(t_0,\sigma) = p_0$.

Further we shall consider only those admissible states $x(t)$ which are defined in $[t^-(\sigma), T]$ for some $\sigma \in G$, whose graphs are contained in $X_g = E^- \cup \{(T,x) : x \in X\}$.

For $(t,p) \in [E^*]$ let $Z(t,p), U(t,p)$ stand for the sets of values of $z(t,\sigma)$ and $u(t,\sigma)$ at those $(t,\sigma) \in [S]$ for which $p(t,\sigma) = p$. For $(t,x) \in [E]$, $P(t,x), U(t,x)$ denote the sets of values of $p(t,\sigma)$ and $u(t,\sigma)$ at those $(t,\sigma) \in [S]$ for which $x(t,\sigma) = x$. By an admissible pair of functions

$$z(t,p) \in Z(t,p), \qquad u(t,p) \in U(t,p), \qquad (t,p) \in [E^*],$$

we shall mean single valued functions $(z(t,p), u(t,p))$ in $[E^*]$ such that for each $(t_0,p_0) \in [E^*]$ there exists a canonical line of flight $(z(t), p(t), u(t))$ for which $p(t_0) = p_0$, $z(t_0,p_0) = z(t_0)$, $u(t_0,p_0) = u(t_0)$. By an admissible pair of functions

$$p(t,x) \in P(t,x), \qquad u(t,x) \in U(t,x), \qquad (t,x) \in [E],$$

we mean single valued functions $(z(t,x), u(t,x))$ in $[E]$ such that for each $(t_0,x_0) \in [E]$ there is a canonical line of flight $z(t) = (x^0(t), x(t))$, $p(t) = (y^0, y(t)), u(t)$ for which $x(t_0) = x_0$, $p(t_0, x(t_0)) = p(t_0)$, $u(t_0, x(t_0)) = u(t_0)$.

REMARK 3.1. *Let C_z denote any small arc contained in D^- or D, with the description $t_1 \le \tau \le t_2$, $(-x^0(\tau), x(\tau))$ where $x(t)$ is a trajectory of an admissible pair $(x(t), u(t))$, $t \in [0,T]$, with $l(x(0), x(T)) = 0$, $x^0(t) = \int_t^T L(\tau, x(\tau), u(\tau))d\tau$, issuing from $(t_1, z(t_1,\sigma_1))$. We also represent C_z in terms of its arc length s as $t = t(s)$ $z = z(s) = (x^0(s), x(s))$, $s \in [0, s_{C_z}]$. Furthermore, let Γ denote a rectifiable curve in S^- or S such that small arcs of C_z issuing from $(t_1, z(t_1,\sigma_1))$ are, in accordance with (H4), the images under the map $(t,\sigma) \to (t, z(t,\sigma))$ of small arcs of Γ issuing from (t_1,σ_1). We represent Γ in terms of its arc length λ by functions $\bar{t}(\lambda), \bar{\sigma}(\lambda)$, so that the point (t_1,σ_1) corresponds to $\lambda = 0$. We can then define a continuous increasing function $s(\lambda)$ having inverse $\lambda(s)$, which satisfies the relation*

$$t(s(\lambda)) = \bar{t}(\lambda), \qquad z(s(\lambda)) = z(\bar{t}(\lambda), \bar{\sigma}(\lambda)).$$

In turn, let C_z be the image under the map $(t,\sigma) \to (t, p(t,\sigma))$ of Γ issuing from

$$(t_1, p(t_1,\sigma_1)) = (t_1, p_1) = (t_1, y_1^0, y_1).$$

We easily see that to small arcs of Γ issuing from from (t_1,σ_1) there correspond small arcs of C_p issuing from (t_1,p_1). Thus we can express the final points of the small arcs of C_p as a function of s, $(t(s), p(s))$. Denote by (t_2,p_2) the terminal point of C_p which corresponds to that of $C_z(t_2, -x^0(t_2), x(t_2))$.

LEMMA 3.1. *Let C_p be one of the arcs described in Remark 3.1. Then along C_p, the function $V(t,p)$ is bounded and there exists a Borel measurable admissible pair of functions $(z(t,p),\ u(t,p))$ along C_p. Moreover, the function $z(t,p)$, $L(t, x(t,p), u(t,p))$, $f(t, x(t,p), u(t,p))$ are bounded along it.*

PROOF: By the definition of C_p, it is the image under the map $(t,\sigma) \to (t, p(t,\sigma))$ of some $\Gamma \subset [S]$. Therefore we can treat $V(t,p)$ as a function $V(t, p(t,\sigma))$ along

Γ which by (H3) is continuous. The graph of Γ is a compact set in $[0,T] \times W$. Hence $V(t,p)$ is bounded along C_p. Applying the measurable selection theorem from [2] to the multifunction $(t,p) \to \{(t,\sigma) \in \Gamma : p(t,\sigma) = p\}$ defined on C_p and putting this selection into $z(t,\sigma), u(t,\sigma)$ we obtain the functions $z(t,p), u(t,p)$ as it is required in the assertion of the lemma. The proof of the last assertion is analogous to the first one.

LEMMA 3.2. *Let any $\sigma_0 \in G$ be given. Then for each $w \in W$ the function*

$$t \to \langle p(t,\sigma_0), z_\sigma(t,\sigma_0)w\rangle_z = -y^0(\sigma_0)x_\sigma^0(t,\sigma_0)w + -\langle z(t,\sigma_0), x_\sigma(t,\sigma_0)w\rangle \tag{3.5}$$

is constant in $(t^-(\sigma_0), t^+(\sigma_0))$.

PROOF: Let t' be any point in $(t^-(\sigma_0), t^+(\sigma_0))$ and let $z_0(t) = (-x_0^0(t), x_0(t))$, $p_0 = (y_0^0, y_0(t))$, $u_0(t)$ be the corresponding values of the functions $z(t,\sigma_0), p(t,\sigma_0)$, $u(t,\sigma_0)$ for $t \in [t', t^+(\sigma_0))$. Then from (1.2) we derive

$$\frac{\partial}{\partial t}x_\sigma(t,\sigma_0)w + A(t)x_\sigma(t,\sigma_0)w = \tilde{f}_\sigma(t,\sigma_0)w$$

and, analogously, at this point and $w \in W$ we find

$$-\frac{\partial}{\partial t}x_\sigma^0(t,\sigma_0)w = \tilde{L}_\sigma(t,\sigma_0)w;$$

from (3.1) we obtain for almost all $t \in [t', t^+(\sigma_0))$,

$$\begin{aligned}\langle \dot{y}(t), x_\sigma(t,\sigma_0)w\rangle &= -\langle y(t), f_x(t, x_0(t), u(t))x_\sigma(t,\sigma_0)w\rangle \\ &\quad - y_0^0\langle L_x(t, x_0(t), u_0(t)), x_\sigma(t,\sigma_0)w\rangle + \langle y(t), A(t)x_\sigma(t,\sigma_0)w\rangle\end{aligned}$$

and by the definition of y^0, we have at (t,σ_0) and $w \in W$

$$-\frac{\partial}{\partial t}y^0 x_\sigma^0(t,\sigma_0)w = 0.$$

We add both sides of the last four equalities with the first two multiplied by $y_0(t)$ and y_0^0, respectively. As a result we obtain at (t,σ_0) and $w \in W$

$$\begin{aligned}\frac{\partial}{\partial t}\langle p_0(t), z_\sigma(t,\sigma_0)w\rangle_z &= \langle y_0(t), \frac{\partial}{\partial \sigma}(\tilde{f}(t,\sigma_0) - f(t, x(t,\sigma_0), u(t,\sigma_0)))w\rangle \\ &\quad + y_0^0\frac{\partial}{\partial \sigma}(\tilde{L}(t,\sigma_0) - L(t, x(t,\sigma_0), u(t,\sigma_0)))w.\end{aligned}$$

Hence, using (H3) and next (3.2) we get

$$\frac{\partial}{\partial t}\langle p_0(t), z_\sigma(t,\sigma_0)w\rangle_z = 0 \qquad \text{for almost all } t \text{ in } [t', t^+(\sigma_0)).$$

Integrating this equality in the interval $[t', t^+(\sigma_0)]$ we get the assertion of the lemma.

LEMMA 3.3. *Let Γ denote any small rectifiable curve in $[S]$ with (t_0, σ_0) as the initial point and (t_1, σ_1) as the terminal one. Then*

$$\int_{\Gamma} \frac{d}{dt}\langle z(t,\sigma), p(t,\sigma)\rangle_z dt + \frac{d}{d\sigma}\langle z(t,\sigma), p(t,\sigma)\rangle_z d\sigma$$
$$= V(t_0, p(t_0, \sigma_0)) - V(t_1, p(t_1, \sigma_1)).$$

The proof follows directly from the definition of the function $V(t,p)$.

LEMMA 3.4. *In the set S^+ the quantity*

$$\langle p(t,\sigma), z_\sigma(t,\sigma)w\rangle_z$$

is identically zero, $w \in W$.

PROOF: First notice that the Fenchel conjugate of the function l defined in (1.4) is the same function (for $g : X \to \bar{R}$, $g^*(x^*) = \sup\{\langle x^*, x\rangle - g(x),\ x \in X\}$, $x^* \in X$), thus condition (3.3) may equivalently be written as

$$(y(0), -y(T)) \in \partial l(x(0), x(T)) \tag{3.6}$$

where ∂l denotes the subdifferential of the convex function l. In view of (H5) we can confine ourselves to the case when our canonical lines of flight satisfy condition (3.3). Since the function $\sigma \to \langle p(t^+(\sigma), \sigma), z_\sigma(t^+(\sigma), \sigma)w\rangle_z$ is continuous, the set G_0 of σ defining those canonical lines of flight for which condition (3.3) is satisfied is open. Hence and by (3.3) $l_\sigma(x(0,\sigma), x(T,\sigma)) = 0$. From this and (3.6) we infer the assertion of the lemma.

As a direct consequence of Lemmas 3.2 and 3.4 we have the following corollary.

COROLLARY 3.1. *The quantity (3.5) is identically zero in $[S]$.*

By the same arguments as in Lemma 3.1 we obtain the following lemma

LEMMA 3.5. *Along arc $t_1 \le \tau \le t_2$, $x(\tau)$ lying in E^- (or E), there exist Borel measurable functions $p(t,x), u(t,x)$, $(t,x) \in [E]$ with the property that the functions $p(t,x)$, $L(t,x,u(t,x))$, $f(t,x,u(t,x))$ are bounded along it.*

THEOREM 3.1. *Let C_z and C_p be as described in Remark 3.1. Then the following relation holds for some admissible pair $p(t,x), u(t,x)$, $(t,x) \in [E]$:*

$$\begin{aligned} V(t_1,p_1) - V(t_2,p_2) - \langle x(t_2), y_2\rangle + \langle x(t_1), y_1\rangle + x^0(t_2)y_2^0 - x^0(t_1)y_1^0 \\ = \int_{C_z} (y^0(t,x)L(t,x,u(t,x)) + \langle y(t,x), f(t,x,u(t,x)) \\ - A(t)x\rangle)dt - \langle p(t,x), dz\rangle_z \end{aligned} \tag{3.7}$$

PROOF: Let $e(s) = (\frac{dt}{ds}, \frac{dz}{ds})$ stand for the direction of the tangent to C_z defined for a.e. s in $[0, s_{C_z}]$. Let s_0 be any point in $[0, s_{C_z}]$ such that $e(s)$ and $p(s)$ are approximately continuous at it. We set $t_0 = t(s_0)$, $x_0 = x(s_0)$, $e_0 = e(s_0)$, $\dot{t}_0 = dt(s_0)/ds$, $\dot{z}_0 = dz(s_0)/ds$, $A_0 = A(t_0)x_0$. Let $p_0 = (y_0^0, y_0)$, u_0 be any admissible vectors from the sets $P(t_0, x_0), U(t_0, x_0)$ such that $p(t_0, \sigma_0) = p_0$, $u(t_0, \sigma_0) = u_0$ for any (t_0, σ_0) belonging to the graph of Γ. We also put $f_0 = f(t_0, x_0, u_0)$, $L_0 = L(t_0, x_0, u_0)$ and let λ_0 be such that $\sigma_0 = \bar{\sigma}(\lambda_0)$.

Denote by γ a sufficiently small arc Γ starting from (t_0, σ_0) defined over an interval $I = [\lambda_0, \lambda_2]$ of values of λ, i.e. the functions $\bar{t}(\lambda)$, $\bar{\sigma}(\lambda)$ are restricted now to the interval I. Denote by ΔV the difference in $V(t,p)$ at the ends of the small arc C_p starting from (t_0, p_0) which is the image of γ, and denote by Δs the corresponding difference in s. By Lemma 3.3, Corollary 3.1, and taking into account Remark 3.1, we obtain

$$\begin{aligned}
\Delta R = & -\Delta V - \langle x(\bar{t}(\lambda_2)), y(\bar{t}(\lambda_2)), \bar{\sigma}(\lambda_2))\rangle + \langle x(t_0), y_0\rangle \\
& + x^0(\bar{t}(\lambda_2))y^0(\bar{\sigma}(\lambda_2)) - x^0(t_0)y_0^0 \\
= & \int_\gamma (y^0(\sigma)\tilde{L}(t,\sigma) + \langle y(t,\sigma), \tilde{f}(t,\sigma) - A(t)x(t,\sigma)\rangle \\
& + \langle y_t(t,\sigma), x(t,\sigma)\rangle)dt + \langle z(t,\sigma), p_\sigma(t,\sigma)d\sigma\rangle_z \\
& - \int_\gamma (\langle z_t(t,\sigma), p(t,\sigma)\rangle_z + \langle z(t,\sigma), p_t(t,\sigma)\rangle_z dt \\
& + \langle p(t,\sigma), z_\sigma(t,\sigma)d\sigma\rangle_z + \langle z(t,\sigma), p_\sigma(t,\sigma)d\sigma\rangle_z \\
= & \int_I [(y^0(\bar{\sigma}(\lambda))\tilde{L}(\bar{t}(\lambda), \bar{\sigma}(\lambda)) + \langle y(\bar{t}(\lambda), \bar{\sigma}(\lambda)), \tilde{f}(\bar{t}(\lambda), \bar{\sigma}(\lambda)) \\
& - A(\bar{t}(\lambda))x(\bar{t}(\lambda), \bar{\sigma}(\lambda))\rangle)\frac{dt}{ds} - \langle p(\bar{t}(\lambda), \bar{\sigma}(\lambda)), \frac{dz}{ds}\rangle_z] ds(\lambda).
\end{aligned} \tag{3.8}$$

Since $p(t,\sigma)$, $\tilde{L}(t,\sigma)$, $\tilde{f}(t,\sigma)$, $z(t,\sigma)$, $A(t)x(t,\sigma)$ are continuous on γ, we deduce that they are bounded on I. This along with (3.8) implies the uniform boundedness of the ratio $\Delta R/\Delta s$ for all sufficiently small Δs. Thus the function $\bar{V}(s) = V(t(s), p(s))$ is locally Lipschitz (as $z(s)$ and $p(s)$ are locally Lipschitz). To prove the assertion of the theorem it is enough to show that

$$\lim \frac{\Delta R}{\Delta s} = \{y^0 L_0 + \langle y_0, f_0 - A_0\rangle\}\dot{t}_0 - \langle p_0, \dot{z}_0\rangle_z \qquad \text{as } \Delta s \to 0.$$

But this is quite analogous to the corresponding part of the proof of Lemma 25.3 in [15, Volume II] if we take there

$$\begin{aligned}
\varphi = \varphi(\lambda) = & \left(\left\{y^0\tilde{L} + \left\langle y, \tilde{f} - Ax\right\rangle\right\}\frac{dt}{ds} - \left\langle p, \frac{dz}{ds}\right\rangle_z\right) \\
& - \left(\left\{y_0^0 L_0 + \langle y_0, f_0 - A_0\rangle\right\}\dot{t}_0 - \left\langle p_0, \dot{z}_0\right\rangle_z\right).
\end{aligned}$$

REMARK 3.2. *Now let C_z be any trajectory contained in [D] with the description $0 \le t \le T$, $(-x^0(t), x(t))$ where $x(t)$ is a trajectory of an admissible pair $(x(t), u(t))$, $t \in [0,T]$, $l(x(0), x(T)) = 0$, $x^0(t) = \int_t^T L(\tau, x(\tau), u(\tau))d\tau$. Then we can divide C_z into a finite number of small arcs (see (H4)), which are described in Remark 3.1 and in the way presented there we are able to obtain an arc C_p with ends $(0, p(0)), (T, p(T))$, corresponding to C_z. Of course, for each such small arcs Theorem 3.1 holds. This implies the corollary below.*

COROLLARY 3.1. *Let C_z and C_p be as in Remark 3.2. Then the following relation holds for some admissible pair $(p(t,x), u(t,x))$, $(t,x) \in [E]$:*

$$\begin{aligned} V(0, p(0)) &+ \langle y(0), x(0)\rangle - x^0(0)y^0(0) \\ &= \int_{C_z} (y^0(t,x)L(t,x,u(t,x)) + \langle y(t,x), f(t,x,u(t,x)) \\ &\quad - A(t)x\rangle)dt - \langle p(t,x), dz\rangle_z \end{aligned} \tag{3.9}$$

The last corollary allows us to derive sufficient conditions for a relative minimum of J. Denote by $G_p \subset G$ the set of those σ for which $x(t,\sigma)$ satisfy (3.3) and let $E_p^- = \{(0,x) : x = x(0,\sigma),\ \sigma \in G_p\}$.

THEOREM 3.2. *Assume that a canonical spray of flights exists. Let a triple of functions $z^*(t) = (-x^{0*}(t), x^*(t))$, $p^*(t) = (y^{0*}, y^*(t))$, $u^*(t)$, $t \in [0,T]$, satisfying (3.3) be a member of our spray and suppose that*

$$J(x^*, u^*) = \min_{\sigma \in G_p} \int_0^T L(t, x(t,\sigma), u(t,\sigma))dt. \tag{3.10}$$

Then

$$-\bar{y}^0 J(x^*, u^*) \le -\bar{y}^0 J(x,u) \tag{3.11}$$

relative to all admissible pairs $(x(t), u(t))$, $t \in [0,T]$, with $x(t)$ satisfying (3.3), for which the graphs of $x(t)$ are contained in [E], $x(0) \in E_p^-$ and $\bar{y}^0 = y^0(\bar{\sigma})$ where $\bar{\sigma}$ is such that $x(0, \bar{\sigma}) = x(0)$.

PROOF: Let $(x(t), u(t))$, $t \in [0,T]$, be any admissible pair with $x(t)$ satisfying (3.3) and its graph contained in [E], $x(0) \in E_p^-$. Put $l(x(0), x(T)) = 0$, $x^0(t) = \int_t^T L(\tau, x(\tau), u(\tau))d\tau$ and let C_z denote the trajectory $z(t) = (-x^0(t), x(t))$ and C_p the corresponding (see Remark 3.2) trajectory in the (t,p)-space with the initial point $(\bar{y}^0, \bar{y}(0)) = p(0)$. Let $\bar{z}(t), \bar{p}(t) = (\bar{y}^0, \bar{y}(t))$, $\bar{u}(t)$, $t \in [0,T]$, $\bar{x}(0) = x(0)$, with $\bar{x}(t), \bar{y}(t)$ satisfying (3.3) be a member of our spray. Then by Corollary 3.1

and equation (3.2) we have

$$\begin{aligned}
&\bar{y}^0 \int_0^T L(t, \bar{x}(t), \bar{u}(t))dt - \bar{y}^0 \int_0^T L(t, x(t), u(t))dt \\
&\quad = \int_0^T [y^0(t, x(t))(L(t, x(t), u(t, x(t))) - L(t, x((t), u(t)))) \\
&\qquad + \langle y(t, x(t)), f(t, x, (t), u(t, x(t))) - f(t, x(t), u(t))\rangle]dt \geq 0.
\end{aligned} \tag{3.12}$$

Hence $-\bar{y}^0 J(\bar{x}, \bar{u}) \leq -y^0 J(x, u)$ and by (3.10) we get (3.11).

REMARK 3.3. *In the assertion of Theorem 3.2, the multiplier $\bar{y}^0$ depends on $x(t)$, i.e. $\bar{y}^0$ is determined by $x(0)$ (see Remarks 3.1 and 3.2). If $y^0(\sigma) \neq 0$ for all $\sigma \in G_p$, then (x^*, u^*) is a strong relative minimum for J.*

4. DUAL FEEDBACK CONTROL

In this section we assume all notations and assumptions of the previous sections. In Theorems 3.1 and 3.2 we used a Borel measurable selection $u(t, x)$ of the multifunction $U(t, x)$, $(t, x) \in [E]$. In applications this function is considered a feedback control or synthesis. In practice it often plays a more useful role than minimizers of functionals and it is very important to have an algorithm to determine feedback functions. In fact, in section 3 we gave a method to calculate the multifunction $U(t, x)$ whose existence is ensured by the existence of a canonical spray of flights. However, from the theorems in section 3 we cannot infer that there exists a selection of $U(t, x)$ which would be an optimal feedback control. That is so because relations (3.7) and (3.12) are satisfied only for some admissible pair $(p(t, x), u(t, x))$, $(t, x) \in [E]$, and hence we need the aditional requirement (3.10) in the sufficiency Theorem 3.2.

The aim of this section is to study properties of selections $u(t, p)$ of the multifunction $U(t, p)$ introduced at the beginning of Section 3 which we will call dual feedback controls. We give also sufficient conditions under which they become optimal feedback controls. To this effect we need one more hypothesis.

(H7) The maps $S^- \to E^{*-}$, $S \to E^*$ defined by $(t, \sigma) \to (t, p(t, \sigma))$ are descriptive: a map $S \to E^*$ is descriptive if for each $(t, \sigma) \in S$ and any given rectifiable curve $C \subset E^*$ starting from $(t, p(t, \sigma))$ there exists a rectifiable curve $\Gamma \subset S$ starting from (t, σ) such that every small arc of C starting from $(t, p(t, \sigma))$ is the image under the above map of a small arc of Γ starting from (t, σ) (see also [15]).

We put $\bar{V}(s) = V(t(s), p(s))$ along any rectifiable curve C in E^{*-} or E^* with the arc length description $t = t(s)$, $p = p(s)$, $0 \leq s \leq s_C$.

THEOREM 4.1. *The function $\bar{V}(s)$ is absolutely continuous along C and, for almost all s in $[0, s_C]$, and each admissible pair $(z(t,p), u(t,p))$, $(t,p) \in [E^*]$ we have:*

$$\begin{aligned}\frac{d}{ds}\bar{V}(s) = -(\{y^0(s)L(t(s), x(t(s),p(s)), u(t(s),p(s)))\\ + \langle y(s), f(t(s), x(t(s),p(s)), u(t(s),p(s)))\\ - A(t(s))x(t(s),p(s))\rangle\}\frac{dt}{ds}\\ + \langle z(t(s),p(s)), \frac{dp}{ds}\rangle_z\end{aligned} \tag{4.1}$$

PROOF: We limit ourselves to the case when C is contained in E^*. Let $e(s) = (\frac{dt}{ds}, \frac{dp}{ds})$ stand for the direction of the tangent to C defined for a.e. in $[0, s_C]$. Let s_0 be any point in $[0, s_C]$ such that $e(s)$ is approximately continuous at it. We set $t_0 = t(s_0)$, $(y_0^0, y^0) = p_0 = p(s_0) = (y^0(s_0), y(s_0))$, $e_0 = e(s_0)$, $\dot{t}_0 = dt(s_0)/ds$, $\dot{p}_0 = dp(s_0)/ds$. Let $z_0 = (-x_0^0, x_0), u_0$ be admissible vectors from the sets $Z(t,p), U(t,p)$ and let (t_0, σ_0) be any point in S for which $p(t_0, \sigma_0) = p_0$, $u(t_0, \sigma_0) = u_0$. We also put $f_0 = f(t_0, x_0, u_0)$, $L_0 = (t_0, x_0, u_0)$, $A_0 = A(t_0)x_0$.

Denote by Γ a rectifiable curve in S such that small arcs of the curve C starting from (t_0, p_0) are, in accordance with (H7), the image under the map $(t,\sigma) \to (t, p(t,\sigma))$ of small arcs γ of Γ starting from (t_0, σ_0). Let now

$$t = \bar{t}(\lambda), \qquad \sigma = \bar{\sigma}(\lambda), \qquad \lambda \in I = [0, h],$$

be the arc length parametric description of γ such that the point (t_0, σ_0) corresponds to 0. Next define a continuously increasing function $s = s(\lambda)$, $\lambda \in I$, such that $s(0) = s_0$, which satisfies in I the relation

$$t(s(\lambda)) = \bar{t}(\lambda), \qquad p(s(\lambda)) = p(\bar{t}(\lambda), \bar{\sigma}(\lambda)). \tag{4.2}$$

Let Δs and $\Delta \bar{V}$ be the corresponding difference in s and in $\bar{V}(s)$ at the ends of a small arc of C starting from (t_0, σ_0), being the image of γ. By Corollary 3.1 $\langle p(t,\sigma), z_\sigma(t,\sigma)w\rangle_z = 0$ along γ. Hence, taking into account (4.2) and Lemma 3.3, we conclude that

$$\begin{aligned}-\Delta\bar{V} &= \int_\gamma \frac{d}{dt}\langle z(t,\sigma), p(t,\sigma)\rangle_z + \frac{d}{d\sigma}\langle z(t,\sigma), p(t,\sigma)\rangle_z d\sigma\\ &\quad - \int_\gamma \langle p(t,\sigma), z_\sigma(t,\sigma)\rangle_z d\sigma\\ &= \int_I (\{y^0(\bar{\sigma}(\lambda))\tilde{L}(\bar{t}(\lambda), \bar{\sigma}(\lambda)) + \langle y(\bar{t}(\lambda), \bar{\sigma}(\lambda)), \tilde{f}(\bar{t}(\lambda), \bar{\sigma}(\lambda))\\ &\quad - A(\bar{t}(\lambda))x(\bar{t}(\lambda), \bar{\sigma}(\lambda))\rangle\}\frac{dt}{ds} - \langle p(\bar{t}(\lambda), \bar{\sigma}(\lambda))\rangle\}\frac{dt}{ds}\\ &\quad + \langle z(\bar{t}(\lambda), \bar{\sigma}(\lambda)), \frac{dp}{ds}\rangle_z)ds(\lambda).\end{aligned}$$

From this relation and (H2), (H3) we conclude that the ratio $\Delta\bar{V}/\Delta s$ is uniformly bounded for all suficiently small Δs. This proves that $\bar{V}$ is absolutely continuous. To verify (4.1), it is enough to show that

$$\lim \frac{\Delta\bar{V}}{\Delta s} = -(\{y^0 L_0 + \langle y_0, f_0 - A_0\rangle\}\dot{t}_0 - \langle z_0, \dot{p}_0\rangle_z \qquad \text{as} \quad \Delta s \to 0.$$

But this is quite analogous to the corresponding part of Lemma 25.3 in [15, Volume II] (compare also [11, Theorem 1']).

Integrating (4.1) along C, we obtain the following corollary:

COROLLARY 4.1. *For each admissible pair $z(t,p), u(t,p)$, $(t,p) \in [E^*]$,*

$$\begin{aligned} V(t_1,p_1) - V(t_2,p_2) = \int \{&y^0 L(t,x(t,p),u(t,p)) \\ &+ \langle y, f(t,x(t,p),u(t,p)) - A(t)x(t,p)\rangle\}dt + \langle z(t,p), dp\,\rangle_z, \end{aligned} \tag{4.3}$$

where $(t_1,p_1),(t_2,p_2)$ are the initial and final points of C.

Now we give a precise definition of a dual feedback control and we show that the selections $u(t,p)$ of $U(t,p)$ are really dual feedback controls.

Let a Borel measurable function $u = u(t,p)$ from a set $P \subset R^2 \times X^*$ of the points $(t,p) = (t,y^0,y)$, $t \in [0,T]$, $y^0 \leq 0$, into $U(t)$ be given. Then the differential equation

$$\dot{x} + A(t)x = f(t,x,u(t,p)), \tag{4.4}$$

has in general many solutions $x(t,p)$ in P. We say that $u = u(t,p)$ is a *dual feedback control* if we can choose any solution $x(t,p)$ of (4.4) such that for each admissible trajectory $x(t)$ lying in $T = \{(t,x) : x = x(t,p),\ (t,p) \in P\}$, there exists a function $p(t) = (y^0, y(t))$ of bounded variation lying in P which satisfies $x(t) = x(t,p(t))$.

PROPOSITION 4.1. *If $z(t,p), u(t,p)$, $(t,p) \in [E^*]$, is an admissible pair of functions, then $u(t,p)$ is a dual feedback control.*

PROOF: By the definition of an admissible pair of functions we easily see that $x(t,p)$ from $z(t,p) = (-x^0(t,p), x(t,p))$ is a solution to (4.4). In Remarks 3.1 and 3.2 for each admissible trajectory $z(t) = (-x^0(t), x(t))$ with $x(t)$ lying in $T = \{(t,x) : x = x(t,p),\ (t,p) \in [E^*]\} \subset [E]$, the construction of a function $p(t)$ is described. From that construction we see that $p(t)$ lies in $[E^*]$ and is of bounded variation and that $x(t,p(t)) = x(t)$.

For a given dual fedback $u(t,p)$ with corresponding trajectory $x(t,p)$, $(t,p) \in P$, let us define the dual value function $S_D(t,p)$ in the set P as

$$S_D(t,p) = \inf\{-y^0 \int_t^T L(\tau, x(\tau), u(\tau))d\tau\}$$

where the infimum is taken over all admissible pairs $(x(\tau), u(\tau))$ restricted to $[t,T]$ with $x(\tau)$ satisfying (3.3) and having graph in T.

A dual feedback $u(t,p)$ will be called optimal if for each $(t,p) \in P$ there exists an absolutely continuous function $\bar{p}(\tau) = (\bar{y}^0, \bar{y}(\tau))$, $\tau \in [0,T]$, with graph in P such that

$$S_D(t,p) = -y^0 \int_t^T L(\tau, x(\tau, \bar{p}(\tau)), u(\tau, p(\tau)))d\tau \tag{4.5}$$

($x(t,p)$ is a function corresponding to $u(t,p)$) and a function $V(t,p)$ in P such that the triple $V(t,p)$, $z(t,p) = (-x^0(t,p), x(t,p))$, $u(t,p)$ satisfies (4.3) for all rectifiable curves C lying in P.

The next theorem gives sufficient conditions for the existence of an optimal dual feedback control.

THEOREM 4.2. *Assume we are given a canonical spray of flights satisfying (H7) and that there exists an admissible pair of functions* $z(t,p) = (-x^0(t,p), x(t,p))$, $u(t,p)$, $(t,p) \in [E^*]$ *for it such that*

$$\begin{aligned} y^0 L(t, x(t,p), u(t,p)) &+ \langle y, f(t, x(t,p), u(t,p)) - A(t)x(t,p)\rangle \\ = \sup_{\tilde{u}(t,p)\in U(t,p)} &\{y^0 L(t, x(t,p), \tilde{u}(t,p)) \\ &+ \langle y, f(t, x(t,p), \tilde{u}(t,p)) - A(t)x(t,p)\rangle\} \end{aligned} \tag{4.6}$$

where $U(t,p)$ *is a multifunction corresponding to the canonical spray of flights described in Section 3. Moreover, suppose that for each* $(t,p) \in [E^*]$ *there exists a* $\bar{p}(\tau)$, $\tau \in [t,T]$, *such that* $\bar{x}(\tau) = x(\tau, \bar{p}(\tau))$, $\bar{p}(\tau), \bar{u}(\tau) = u(\tau, \bar{p}(\tau))$ *is a member of our spray. Then* $u(t,p)$ *is an optimal dual feedback control.*

PROOF: In view of Corollary 4.1 and Proposition 4.1, it is enough to prove equation (4.5). To this effect it suffices to follow the proof of Theorem 3.2. We obtain then for each $(t,p) \in [E^*]$

$$-y^0 \int_t^T L(\tau, \bar{x}(\tau), \bar{u}(\tau))d\tau \le -y^0 \int_t^T L(\tau, x(\tau), u(\tau))d\tau$$

for all admissible pairs $x(\tau), u(\tau)$ with $x(\tau)$ satisfying (3.3) and having their graph in T. That means just equality (4.5).

REMARK 4.1. *The existence of an optimal dual fedback control gives us more information about the problem under consideration then the sufficiency Theorem 3.2. However, to obtain a sufficiency thorem on the existence of an optimal feedback (Theorem 4.2) we need much stronger assumptions.*

REFERENCES

1. C. Carathéodory, "Variationsrechnung und Partielle Differential Gleichungen erster Ordnung," Teubner, Leipzig, 1936.
2. C. Castain, M. Valadier, "Convex analysis and measurable multifunctions," Lecture Notes in Math., 580, Springer, 1977.
3. I.V. Girsanov, "Lectures on the mathematical theory of extremum problems," Springer, Berlin, 1972.
4. A. Friedman, *Optimal control in Banach space with fixed end-points*, Journal of Mathematical Analysis and Applications **24** (1968), 161-181.
5. J.L.Lions, "Optimal control of systems governed by partial differential equations," Springer, New York, 1971.
6. J.L.Lions, E.Magenes, "Non-homogeneous boundary value problems and applications, Vols. 1-3," Springer, New York, 1972.
7. J.L.Lions, *Optimal control of non well-posed distributed systems*, Mathematical Control Theory, Banach Center Publications **14** (1985), 299-311.
8. U. Ledzewicz-Kowalewska, *The extremum principle for some types of distributed parameter systems*, Applicable analysis **48** (1993), 1-21.
9. U. Ledzewicz, *On distributed parameter control systems in the abnormal case and in the case of nonoperator equality constraints*, Journal of Applied Mathematics and Stochastic Analysis **6** (1993), 137-152.
10. A. Nowakowski, *Sufficient conditions for a strong relative minimum in an optimal control problem*, Journal of Optimization Theory and Aplications **50** (1986), 129–147.
11. —, *Field theories in the modern calculus of variations*, Transactions of the American Mathematical Society **309** (1988), 725–752.
12. N.S. Papageorgiou, *Optimality conditions for systems with insufficient data*, Bulletin of the Australian Mathematical Society **41** (1990), 45-55.
13. S. Walczak, *On some properties of cones in normed spaces and their application to investigating extremal problems*, Journal of optimization theory and Applications **42** (1984), 561-582.
14. H. Tanabe, "Equations of evolution," Pitman, London, 1979.
15. L.C. Young, "Lectures on the calculus of variations and optimal control theory," Saunders, Philadelphia, 1969.

14 Pareto Optimality Conditions for Abnormal Optimization and Optimal Control Problems

Urszula Ledzewicz[1] Southern Illinois University at Edwardsville, Edwardsville, Illinois

Heinz Schaettler[2] Washington University, St. Louis, Missouri

Abstract. We formulate first and second order necessary conditions for Pareto optimality problems in a Banach space which have a nontrivial form in the case of equality constraints given by nonregular operators. The results are then applied to a Pareto type optimal control problem.

1. Introduction

We consider the problem of minimizing a multi-functional in a Banach space in the Pareto sense under constraints given by sets with both empty and nonempty interiors (so-called equality and inequality constraints). The equality constraints will be given as kernels of some operators (operator equality constraints). We formulate first and second order necessary conditions for optimality using the Dubovitskii-Milyutin method [see, for instance, Girsanov, 7]. This method has been used earlier by Censor to obtain first order necessary conditions with several inequality and one equality constraint [Censor, 6]. Censor's results have been generalized for problems with several equality constraints given by operators [Kotarski, 10] and, under additional assumptions, to problems with multi equality constraints with a general structure [Kotarski, 9] extending respectively results in [Ledzewicz-Kowalewska, 11] and [Walczak, 16] to the Pareto case. In [Kotarski, 9] also an application of the results to a Pareto optimal control problem for distributed parameter systems is given.

----

[1] *Research is supported in part by NSF Grant DMS – 91009324, SIUE Research Scholar Award and Fourth Quarter Fellowship, Summer 1992.*

[2] *Research is supported in part by NSF Grant DMS – 9100043*

Here we consider the Pareto optimization problem with inequality and equality constraints given in operator form in the so-called abnormal case when the necessary conditions for optimality are satisfied in a trivial way independent of the minimized multi-index. This case typically occurs when the operator which describes the equality constraints is not regular, i.e. when the Lusternik theorem is not applicable to describe the tangent space. Avakov has given several generalizations of Lusternik's theorem to nonregular operators in [Avakov, 1,2,4] and applied them to obtain optimality conditions for the smooth and smooth-convex problems. The necessary conditions of optimality from [Avakov, 2] have been used in [Avakov, 2,3] to derive a Maximum principle for optimal control problems under smooth-convex assumptions in the sense of [Ioffe, Tikhomirov, 8]. We used Avakov's idea to extend the Dubovitskii-Milyutin formalism [Girsanov, 7 and Ledzewicz-Kowalewska, 11] to the abnormal case [Ledzewicz and Schaettler, 13 & 14]. In this paper the Pareto optimization problem under several equality constraints is considered like in [Kotarski, 10] and the results from [Ledzewicz and Schaettler, 13 & 14] are extended to the Pareto case. Second order approximating cones are defined following the idea of [Ben-Tal and Zowe, 5] but allowing for reparametrization of the approximating curves. First and second order conditions for optimality are formulated which are nontrivial also in the abnormal case. Then these results are applied to a Pareto optimal control problem and an extended version of the local Maximum principle is derived. For the case of a single performance function, this result reduces to the local Maximum principle of [Ledzewicz, 12].

2. First and Second Order Conditions for Pareto Optimality

Let X and Y be Banach spaces, let $F: X \to Y$ be an operator and let $I_k: X \to \mathbb{R}$, $k=1,2,\ldots,s$ be functionals. We call the following optimization problem Problem 1:

Minimize the vector cost-functional $I(x)=(I_1(x),\ldots,I_s(x))$ in the sense of Pareto

subject to $x \in Z = \bigcap_{i=1}^{n+1} Z_i$ where $x \in Z_i$, $i=1,\ldots,n$ are inequality constraints

(i.e. int $Z_i \neq \emptyset$) and $x \in Z_{n+1}=\{x \in X: F(x)=0\}$ is an equality constraint.

Definition 2.1. A point $x_0 \in X$ is local *Pareto optimal* if $x_0 \in \bigcap_{i=1}^{n+1} Z_i$ and if there exists a neighborhood $U(x_0)$ of x_0 with the property that there is no $x \neq x_0$, $x \in \bigcap_{i=1}^{n+1} Z_i \cap U(x_0)$ which satisfies $I_k(x) \leq I_k(x_0)$, for all $k=1,2,\ldots,s$ and with strict inequality for at least one index k, $1 \leq k \leq s$.

We briefly recall some required definitions about second order approximations from [Ledzewicz and Schaettler, 14, this volume].

Definition 2.2. A vector h is said to be a *feasible* direction to the inequality constraint $Z \subseteq X$ at x_0 if there exists an $\epsilon_0 > 0$ and a neighborhood V of h such that for all $0 < \epsilon < \epsilon_0$ we have $x_0 + \epsilon V \subseteq Z$. The set of all feasible directions forms an open cone, the cone of feasible directions to Z at x_0, and is denoted by $FC(Z,x_0)$.

Definition 2.3. We call the pair (v,γ) a *second order feasible* direction to the set Z at the point $x_0 \in Z$ in direction of $h \in X$ if there exist an $\epsilon_0 > 0$ and a neighborhood V of v_0 such that for all ϵ, $0 < \epsilon < \epsilon_0$, we have

$$x_0 + \epsilon \sqrt{\gamma}\, h + \epsilon^2 V \subseteq Z$$

Again the second order feasible directions form an open cone, the second order feasible cone to Z at x_0 in direction of h, denoted by $FC^{(2)}(Z;x_0,h)$.

Definition 2.4. The first and second order feasible cones (in direction of h) to the set $Z=\{x \in X: f(x)<f(x_0)\}$ at x_0 are called the first and second order *cones of decrease* (in direction of h) for the functional f at x_0.

Definition 2.5. [see also Kotarski, 10] The first and second order feasible cones (in direction of h) to the set $Z=\{x \in X: f(x) \leq f(x_0)\}$ at x_0 are called the first and second order *cones of nonincrease* (in direction of h) for the functional f at x_0.

Definition 2.6. We call the pair $(v,\gamma) \in E \times \mathbb{R}_+$ a *second order tangent direction* to the set Z at the point $x_0 \in Z$ in direction of $h \in X$ if there exists an $\epsilon_0 > 0$ such that for all ϵ, $0<\epsilon<\epsilon_0$, there exists a point $r(\epsilon) \in X$ of order $o(\epsilon^2)$ which has the property that

$$x(\epsilon) = x_0 + \epsilon \sqrt{\gamma}\, h + \epsilon^2 v + r(\epsilon) \in Z.$$

The set of all second order tangent directions to Z at x_0 in direction of h forms a cone, the second order tangent cone to Z at x_0 in direction h, denoted by $TC^{(2)}(Z;x_0,h)$.

We are now ready to state necessary conditions for Pareto optimality for Problem 1. Assume that the vector $h \in X$ is such that

(1) for all k=1,2,...,s, the second order cones of decrease and nonincrease for the functionals I_k, at x_0 in the direction of h, $DC^{(2)}(I_k;x_0,h)$ and $NC^{(2)}(I_k;x_0,h)$, are nonempty, (open) and convex.

(2) for all i=1,2,...,n, the second order feasible cones for the inequality constraints Z_i,

$FC^{(2)}(Z_i;x_0,h)$, are nonempty, (open) and convex.

(3) the second order tangent cone to the set Z_{n+1} at the point x_0 in direction h, $TC^{(2)}(Z_{n+1};x_0,h)$, is nonempty, (closed) and convex.

Theorem 2.1. Suppose $x_0 \in Z=\bigcap_{i=1}^{n+1} Z_i$ is a local Pareto minimum on Z and assumptions (1) - (3) for Problem 1 are satisfied. Then there exist linear functionals

$$q_k(h) = (a_k(h), b_k(h)) \in DC^{(2)}(I_k;x_0,h)^* , \qquad \text{for } k=1,2,\dots,s,$$

$$q_{jk}(h)=(a_{jk}(h),b_{jk}(h)) \in NC^{(2)}(I_j;x_0,h)^*, \qquad \text{for } j,k=1,2,\dots,s,\ j \neq k,$$

$$g_{ik}(h)=(\lambda_{ik}(h),\mu_{ik}(h)) \in FC^{(2)}(Z_i;x_0,h)^*, \qquad \text{for } i=1,2,\dots,n, \text{ and}$$

$$g_{n+1,k}(h) = (\lambda_{n+1,k}(h), \mu_{n+1,k}(h)) \in TC^{(2)}(Z_{n+1};x_0,h)^*,$$

such that for k=1,2,...,s we have

$$a_k(h) +\sum_{\substack{j=1 \\ j\neq k}}^{s} a_{jk}(h) +\sum_{i=1}^{n}\lambda_{ik}(h) + \lambda_{n+1,k}(h) \equiv 0, \qquad \text{(Euler-Lagrange equation)} \quad (1)$$

$$b_k(h) +\sum_{\substack{j=1 \\ j\neq k}}^{s} b_{jk}(h) +\sum_{i=1}^{n}\mu_{ik}(h) + \mu_{n+1,k}(h) \equiv 0, \qquad \text{(Second order condition)} \quad (2)$$

and for each k=1,...,s, not all of $a_k(h)$, $a_{jk}(h)$, j=1,...,s, $j \neq k$, and $\lambda_{ik}(h)$, i=1,...,n+1, vanish identically.

Proof. We first show that since x_0 is a Pareto optimum, we have for each fixed k=1,...,s, that

$$DC^{(2)}(I_k;x_0,h) \cap \bigcap_{\substack{j=1 \\ j\neq k}}^{s} NC^{(2)}(I_j;x_0,h) \cap \bigcap_{i=1}^{n} FC^{(2)}(Z_i;x_0,h) \cap TC^{(2)}(Z_{n+1};x_0,h) = \emptyset$$

Assume to the contrary that (v,γ) lies in this intersection and let $x_t = x+t\sqrt{\gamma}h+t^2v$. Then, by definition of second order tangent directions there exists a function $r(t)=o(t^2)$ such that $x_t+r(t)$ is an admissible direction for the equality constraint. Since $(v,\gamma) \in \bigcap_{i=1}^{n} FC^{(2)}(Z_i;x_0,h)$, this curve is feasible. Furthermore, since also $(v,\gamma) \in DC^{(2)}(I_k;x_0,h)$, the functional I_k is strictly smaller than at x_0 and because $(v,\gamma) \in$

$\bigcap_{j=1}^{k-1} NC^{(2)}(I_j;x_0,h) \cap \bigcap_{j=k+1}^{s} NC^{(2)}(I_j;x_0,h)$, the functionals I_j, $j \neq k$, do not increase along this direction. But this contradicts the Pareto minimality of x_0.

The cones $DC^{(2)}(I_k;x_0,h)$, $NC^{(2)}(I_j;x_0,h)$ for $j \neq k$, and $FC^{(2)}(Z_i;x_0,h)$ for $i=1,2,\ldots,n$, are nonempty, open and convex and $TC^{(2)}_{n+1}(Z_{n+1};x_0,h)$ is a nonempty, closed and convex cone. So the conditions of the Dubovitskii-Milyutin Lemma [Girsanov, 7, Lemma 5.11] are satisfied and hence there exist linear functionals

$$q_k(h) \in DC^{(2)}(I_k;x_0,h)^*, \quad q_{jk}(h) \in NC^{(2)}(I_j;x_0,h)^* \text{ for } j \neq k,$$
$$g_{i,k}(h) \in FC^{(2)}(Z_i;x_0,h)^*, \qquad \text{for } i=1,2,\ldots,n,$$

and

$$g_{n+1,k}(h) \in TC^{(2)}(Z_{n+1};x_0,h)^*,$$

not all zero, such that

$$q_k(h) + \sum_{j=1}^{k-1} q_{jk}(h) + \sum_{j=k+1}^{s} q_{jk}(h) + \sum_{i=1}^{n+1} g_{ik}(h) \equiv 0.$$

Equations (1) and (2) follow if we split the functionals into its components. The argument that the X^* components of these multipliers cannot all vanish identically is like in the proof of Theorem 3.1. in [Ledzewicz and Schaettler, 14, this volume] and is omitted here. This completes the proof. Q.E.D.

3. Cones of Nonincrease and Decrease

For Pareto optimization problems in addition to tangent cones, feasible cones and cones of decrease also cones of nonincrease and their duals need to be considered. A particularly simple situation arises when the duals to the cones of decrease and nonincrease coincide. One sufficient condition for this to hold is related to Ponstein convexity.

Definition 3.1. [Ponstein, 15] A functional $I: X \to \mathbb{R}$ is called Ponstein convex if for every $x_1 \neq x_2$ and $\alpha,\beta>0$ such that $\alpha+\beta=1$ we have that $I(x_1) \leq I(x_2)$ implies $I(\alpha x_1+\beta x_2)< I(x_1)$.

Theorem 3.1. [Censor, 6] Let $I: X \to \mathbb{R}$ be continuous and Ponstein convex. If $x_0 \in \text{dom } I$ and $\inf I(x) < I(x_0)$, then $DC(I,x_0)^* = NC(I,x_0)^*$.

The drawback of this result is that in general it is rather difficult to assert that a given functional is Ponstein convex. In general, there is no relation between convexity and Ponstein convexity. It is clear that Ponstein convex functions are strictly quasi-convex (i.e. functions which have the property that if $I(x_2) < I(x_1)$ then $I(\alpha x_1+\beta x_2) < I(x_1)$ with $\alpha,\beta > 0$ and $\alpha+\beta=1$), but strictly quasi-convex functions need not be Ponstein convex [Censor, 6].

The situation becomes much simpler for differentiable functionals where often not only the duals, but the cones of decrease and nonincrease themselves coincide. Let $Z = \{x \in X: I(x) \leq I(x_0)\}$ and suppose f is twice Fréchet differentiable at x_0. The following result is an easy extension of the corresponding result in [Ledzewicz and Schaettler, 13].

Theorem 3.2. If $I'(x_0) \neq 0$, then $FC(Z,x_0) = \{h \in X: I'(x_0)h<0\}$ and for every $h \in \partial FC(Z,x_0) = \{h \in X: I'(x_0)h = 0\}$ we have

$$\begin{aligned} FC^{(2)}(Z;x_0,h) &= DC^{(2)}(I;x_0,h) = NC^{(2)}(I;x_0,h) \\ &= \{(v,\gamma) \in X \times \mathbb{R}_+ : I'(x_0)v + \tfrac{\gamma}{2} I''(x_0)(h,h) < 0 \} \end{aligned} \tag{3}$$

Remark 3.1 Let $x_0 \in Z$ and suppose I is twice Fréchet differentiable at x_0 with $I'(x_0) = 0$. Then

$$FC^{(2)}(Z;x_0,h) = DC^{(2)}(I;x_0,h) = NC^{(2)}(I;x_0,h) = \emptyset \qquad \text{if } I''(x_0)(h,h) > 0$$

and

$$FC^{(2)}(Z;x_0,h) = DC^{(2)}(I;x_0,h) = NC^{(2)}(I;x_0,h) = X \times (0,\infty) \quad \text{if } I''(x_0)(h,h) < 0.$$

Corollary 3.1. If either $I'(x_0) \neq 0$ or $I''(x_0)(h,h) \neq 0$, then

$$DC^{(2)}(I;x_0,h) = NC^{(2)}(I;x_0,h).$$

These results will suffice for our purpose here.

4. Convex Inequality Constraints

For a later application of Theorem 2.1. we need to know whether the X^* component of a functional in the dual to the second order feasible cone also lies in the dual to the feasible cone. Clearly, if $(\lambda,\mu) \in FC^{(2)}(Z;x_0,h)^*$, then $\lambda \in FC(Z;x_0)^*$ provided

$$FC(Z,x_0) \times \{0\} \subseteq \text{Clos } FC^{(2)}(Z;x_0,h) . \tag{4}$$

We show that convexity of Z is sufficient for (4) to hold. More generally, suppose $v \in FC(Z,x_0)$ and $(w,\gamma) \in FC^{(2)}(Z;x_0,h)$. Then there exist a $t_0>0$, a neighborhood V of v, and a neighborhood W of w such that for all t, $0<t<t_0$, we have $x_0+t^2V \subseteq \text{int } Z$ and $x_0+t\sqrt{\gamma}\, h+t^2W \subseteq \text{int } Z$. If for all $\tilde{w} \in W$ and $\tilde{v} \in V$ the convex combinations lie in the interior of Z,

$$\begin{aligned} x(t,\lambda) &= \lambda(x_0 + t\sqrt{\gamma}\, h + t^2 \tilde{w}) + (1-\lambda)(x_0 + t^2 \tilde{v}) \\ &= x_0 + t\sqrt{\lambda^2\gamma}\, h + t^2\left(\lambda\tilde{w} + (1-\lambda)\tilde{v}\right) \in \text{int } Z, \end{aligned}$$

then $C_\lambda = (\lambda w+(1-\lambda)v,\ \lambda^2\gamma) \in FC^{(2)}(Z;x_0,h)$ for $0 \leq \lambda \leq 1$. In particular, $C_0 = (v,0) \in FC^{(2)}(Z;x_0,h)$. Thus we have proven the following result.

Theorem 4.1. Suppose that there exists a neighborhood U of x_0 such that $U \cap \text{int } Z$ is convex and $FC^{(2)}(Z;x_0,h)$ is nonempty. Then

$$FC(Z,x_0) \times \{0\} \subseteq FC^{(2)}(Z;x_0,h).$$

Convexity of the constraint set near x_0 is a simple sufficient condition for (4) to hold, but in general it is not necessary as the following example shows.

Example: Let $Z= \left\{(x_1,x_2) \in \mathbb{R}^2:\ x_1 \geq 0,\ x_2 \leq x_1^2\right\}$, $x_0= (0,0)$ and take $h = \binom{1}{0}$. Then

$$FC(Z,x_0) = \{(v_1,v_2) \in \mathbb{R}^2:\ v_1>0,\ v_2<0\}$$

and

$$FC^{(2)}(Z;0,h) = \{(v,\gamma) \in \mathbb{R}^3:\ \gamma>0,\ v_2<\gamma\}.$$

The constraint set is not convex near x_0, but clearly any convex combination of curves corresponding to first and second order feasible directions lies in the interior of Z. Hence the condition given above would apply.

5. The Smooth Convex Problem with Several Equality Constraints

We now consider a Pareto optimization problem with several inequality constraints. Suppose x_0 is a local Pareto minimum for the problem to

$$\text{minimize } I(x)=(I_1(x),...,I_s(x)) \quad \text{subject to } x \in Z = \bigcap_{i=1}^{n+m} Z_i$$

where $x \in Z_i$, $i=1,...,n$, are inequality constraints (i.e. int $Z_i \neq \emptyset$),

and $x \in Z_{n+j} = \{ x \in X: F_j(x) = 0 \}$, $j=1,...,m$, are equality constraints.

We assume that the functionals I_k, $k=1,...,s$, are twice Fréchet differentiable at x_0. Furthermore, for $i=1,2,...,p$, the inequality constraints have the form $Z_i=\{x \in X: f_i(x) \leq 0\}$, and the functionals f_i: $X \to \mathbb{R}$ are twice Fréchet differentiable at x_0. For $i=p+1,...,n$, the inequality constraints Z_i are convex sets. The equality constraints are given in operator form as $Z_{n+j}=\{x \in X: F_j(x)=0\}$ for $j=1,2,...,m$, where the operators $F_j: X \to Y_j$ take values in Banach spaces $Y_1,Y_2,...,Y_m$. Define the operator F, $F:X \to Y$, with image in the product space $Y = Y_1 \times Y_2 \times \ldots \times Y_m$, as

$$F(x) = (F_1(x), F_2(x), \ldots, F_m(x)). \tag{5}$$

Hence $\{x \in X: F(x)=0\} = \bigcap_{i=n+1}^{n+m} Z_i = Z$. Define a new operator $G(x_0,h)$, depending on a parameter h, as

$$G(x_0,h): \quad X \to Y \times Y/\operatorname{Im} F'(x_0)$$

$$v \to G(x_0,h)v = (F'(x_0)v, \pi F''(x_0)(h,v))$$

where $\pi:Y \to Y/\operatorname{Im} F'(x_0)$ is the quotient map. Like in [Ledzewicz and Schaettler, 14, this volume], the following assumption allows to describe the second order tangent cone to the equality constraint Z.

(A) (i) $x_0 \in Z$ and F is three times Fréchet differentiable at x_0.
(ii) Im $F'(x_0)$ is closed.
(iii) $G(x_0,h)h=0$, i.e. $F'(x_0)h = 0$ and $F''(x_0)(h, h) \in \operatorname{Im} F'(x_0)$.
(iv) Im $G(x_0,h)$ is closed in Im $F'(x_0) \times Y/\operatorname{Im} F'(x_0)$.

Theorem 5.1. Let $J(x_0)=\{i: 1 \leq i \leq p, f_i(x_0)=0\}$ denote the set of active indices and define

$N_1(x_0) = \{ h \in X: I'_k(x_0)h \leq 0$ for $k=1,...,s$, and $f'_i(x_0)h \leq 0$ for $i \in J(x_0) \}$,

$N_2(x_0) = \{ h \in X: FC^{(2)}(Z_i;x_0,h)$ is nonempty and convex for $i=p+1,...,n \}$,

$N_3(x_0) = \{ h \in X$: assumption (A) is satisfied at x_0 for the operator F describing the equality constraints $Z = \bigcap_{j=n+1}^{n+m} Z_j \}$.

Under the assumptions made above, for every

$$h \in N(x_0) := N_1(x_0) \cap N_2(x_0) \cap N_3(x_0) \tag{6}$$

there exist Lagrange multipliers

$a_k(h) \le 0$ for $k=1,\dots,s$, $\quad b_i(h) \le 0$ for $i \in J(x_0)$,

$g_i(h) = (\lambda_i(h), \mu_i(h)) \in FC^{(2)}(Z_i;x_0,h)^*$, $\quad$ for $i=p+1,\dots,n$,

$q_j^*(h),\ s_j^*(h) \in Y_j^*$, $j=1,2\dots,m$,

such that

$$(s_1^*(h),\dots, s_m{}^*(h)) \in \operatorname{Im} F'(x_0)^{\perp}, \tag{7}$$

$$\sum_{k=1}^{s} a_k(h)\, I_k'(x_0) + \sum_{i \in J(x_0)} b_i(h)\, f_i'(x_0) + \sum_{i=p+1}^{n} \lambda_i(h)$$

$$+ \sum_{j=1}^{m} F_j'^*(x_0)\, q_j^*(h) + \sum_{j=1}^{m} (F_j''(x_0)(h))^*\, s_j^*(h) \equiv 0 \tag{8}$$

$$\frac{1}{2} \sum_{k=1}^{s} a_k(h)\, I_k''(x_0)(h,h) + \frac{1}{2} \sum_{i \in J(x_0)} b_i(h)\, f_i''(x_0)(h,h) + \sum_{i=p+1}^{n} \mu_i(h)$$

$$+ \frac{1}{2}\sum_{j=1}^{m} \langle q_j^*(h), F_j''(x_0)(h,h)\rangle + \frac{1}{6} \sum_{j=1}^{m} \langle s_j^*(h), F_j'''(x_0)(h,h,h)\rangle \le 0. \tag{9}$$

Furthermore, the following second order complementary slackness conditions hold:

$$a_k(h)\langle I_k'(x_0),h\rangle=0 \qquad \text{for} \quad k=1,\dots,s, \quad \text{and} \tag{10}$$

$$b_i(h)\langle f_i'(x_0),h\rangle=0 \qquad \text{for} \quad i \in J(x_0) \tag{11}$$

In addition, if $\operatorname{Im} G(x_0,h) = \operatorname{Im} F'(x_0) \times Y/\operatorname{Im} F'(x_0)$, then not all the multipliers $a_k(h)$ for $k=1,\dots,s$, $b_i(h)$ for $i \in J(x_0)$, and $g_i(h)$ for $i=p+1,\dots,n$, vanish identically.

Proof: If there exists an index $k \in \{1,\dots,s\}$ such that $\langle I_k'(x_0),h\rangle<0$, then h is a direction of decrease for the functional I_k and so $DC^{(2)}(I_k;x_0,h) = X \times (0,\infty)$. Similarly, if $i \in J(x_0)$ and $\langle f_i'(x_0), h\rangle<0$, then h is a feasible direction and so $FC^{(2)}(Z_i;x_0,h) =$

$X\times(0,\infty)$. Also, for $i\notin J(x_0)$ every direction h is feasible and so the second order cone is also $X\times(0,\infty)$. These cones therefore do not contribute to the empty intersection property in the Dubovitskii-Milyutin Lemma and the second order dual cones are $\{0\}$. This already verifies the second order complementary slackness conditions (10) and (11). For the rest of the argument, we may simply delete these cones.

For $h\in N(x_0)$ define $K_1(h)=\{k\in\{1,...,s\}: \langle I_k'(x_0),h\rangle=0\}$, $K_2(h)=\{i\in J(x_0): \langle f_i'(x_0), h\rangle=0\}$, and let $K(h)=K_1(h)\cup K_2(h)$ denote the set of second order active indices for h. If there exists an index $k\in K_1(h)$ such that $I_k'(x_0)=0$ and $I_k''(x_0)(h,h)\geq 0$, then the statement of the theorem holds trivially by choosing as corresponding multiplier $a_k(h)=-1$ and setting all other multipliers 0. Analogously, if there exists an index $i\in K_2(h)$ such that $f_i'(x_0)=0$ and $f_i''(x_0)(h,h)\geq 0$, then the statement of the theorem also holds trivially by choosing as corresponding multiplier $b_i(h)=-1$ and all other multipliers as 0. We may therefore assume that for all $k\in K_1(h)$ we have $I_k''(x_0)(h,h)<0$ whenever $I_k'(x_0)=0$ and for all $i\in K(h)$ we have $f_i''(x_0)(h,h)<0$ whenever $f_i'(x_0)=0$.

This has two consequences. First, by Corollary 3.1. the second order cones of decrease and of nonincrease for the functionals I_k are the same. But then all the Euler-Lagrange equations (1) and second order conditions (2) in Theorem 2.1. for different indices $k\in\{1,...,s\}$ state the same conditions. Consequently all these conditions reduce to just one Euler-Lagrange equation and one second order condition. Second, under these conditions the second order cones of decrease for the functionals I_k, $DC^{(2)}(I_k;x_0,h)$, $k\in K_1(h)$, and the second order feasible cones for the inequality constraints Z_i, $FC^{(2)}(Z_i;x_0,h)$, $i\in K_2(h)$, are given in Theorem 3.2. (for nonzero derivatives) or in Remark 3.1. (for zero derivatives). These sets are open half spaces and so in particular nonempty, convex cones. Furthermore, under assumption (A) the second order tangent cone $TC^{(2)}(Z=\bigcap_{j=n+1}^{n+m} Z_j;\ x_0,h)$ to the equality constraint defined by (5) is given by [Ledzewicz and Schaettler, 14, this volume, Corollary 4.1.] and in particular is a nonempty, closed, linear subspace. Therefore, by Theorem 2.1., there exist linear functionals

$$g_k(h) = (\alpha_k(h),\beta_k(h)) \in DC^{(2)}(I_k;x_0,h)^* \qquad \text{for } k\in K_1(h),$$

$$g_{s+i}(h) = (\lambda_i(h),\mu_i(h)) \in FC^{(2)}(Z_i;x_0,h)^* \qquad \text{for } i\in K_2(h) \text{ and } i=p+1,...,n,$$

and

$$g_{s+n+1}(h) = (\lambda_{n+1}(h),\ \mu_{n+1}(h)) \in TC^{(2)}(Z;\ x_0,\ h)^*,$$

such that

$$\sum_{k \in K_1(h)} \alpha_k(h) + \sum_{i \in K_2(h)} \lambda_i(h) + \sum_{i=p+1}^{n} \lambda_i(h) + \lambda_{n+1}(h) \equiv 0 \tag{12}$$

and

$$\sum_{k \in K_1(h)} \beta_k(h) + \sum_{i \in K_2(h)} \mu_i(h) + \sum_{i=p+1}^{n} \mu_i(h) + \mu_{n+1}(h) \equiv 0, \tag{13}$$

while not all of the $\alpha_k(h)$, $k \in K_1(h)$ and $\lambda_i(h)$, $i \in \{K_2(h),p+1,...,n+1\}$ vanish identically.

It remains to establish the form of the functionals claimed in the theorem. No further specifications were made for the functionals corresponding to the convex inequality constraints. The basic structure of the functionals is the same for the cones of decrease for the functionals I_k, $k \in K_1(h)$, and the feasible cones to the inequality constraints Z_i, $i \in K_2(h)$ and we will only consider the latter. So let $i \in K_2(h)$. The required dual cones are given in [Ledzewicz and Schaettler, 13 and 14, this volume]. If $f_i'(x_0) \neq 0$, then by Theorem 5.2. in [Ledzewicz and Schaettler, 14, this volume] there exist real numbers $b_i(h) \leq 0$ and $c_i(h) \geq 0$ such that

$$\lambda_i(h) = b_i(h) f_i'(x_0)^*, \quad \mu_i(h) = b_i(h) \tfrac{1}{2} f_i''(x_0)(h,h) + c_i(h) . \tag{14}$$

If $f_i'(x_0)=0$ (and $f_i''(x_0)(h,h)<0$), then $FC^{(2)}(Z_i;x_0,h) = X \times (0,\infty)$ and so we get the same form with $b_i(h)=0$ and $c_i(h)=0$. The dual cone to the second order tangent cone to Z in direction of h is given by Theorem 4.2. in [Ledzewicz and Schaettler, 14, this volume] which states that there exist elements $y_1^*(h) \in \mathrm{Im}F'(x_0)^*$, $y_2^*(h) \in \mathrm{Im}F'(x_0)^{\perp}$ and a number $r(h) \geq 0$ such that

$$\lambda_{n+1}(h) = F'(x_0)^* y_1^*(h) + F''(x_0)(h,\cdot)^* y_2^*(h) \tag{15}$$

and

$$\mu_{n+1}(h) = \tfrac{1}{2} \langle y_1^*(h), F''(x_0)(h,h)\rangle + \tfrac{1}{6} \langle y_2^*(h), F'''(x_0)(h,h,h)\rangle + r(h). \tag{16}$$

Extend $y_1^*(h)$ and $y_2^*(h)$ to the full space Y^* and write

$$y_1^*(h) = \left(q_1^*(h),..., q_m^*(h)\right) \quad \text{and} \quad y_2^*(h) = \left(s_1^*(h),..., s_m^*(h)\right).$$

Then using the decomposition of a continuous linear functional on the product space as well as formula (5) we have for $v \in X$ and $\gamma \in \mathbb{R}$ that

$$\begin{aligned}
&< \left(\lambda_{n+1}(h), \mu_{n+1}(h)\right), (v,\gamma) > \\
&= \langle y_1^*(h), F'(x_0)v + \tfrac{\gamma}{2} F''(x_0)(h,h)\rangle + \langle y_2^*(h), F''(x_0)(h,v) + \tfrac{\gamma}{6} F'''(x_0)(h,h,h)\rangle \\
&= \sum_{j=1}^{m} \langle q_j^*(h), F_j'(x_0)v + \tfrac{\gamma}{2} F_j''(x_0)(h,h) \rangle
\end{aligned}$$

$$+ \sum_{j=1}^{m} < s_j^*(h), F_j''(x_0)(h,v) + \frac{\gamma}{6} F_j'''(x_0)(h,h,h)>$$

$$= < \sum_{j=1}^{m} F_j'^*(x_0)\, q_j^*(h) + \sum_{j=1}^{m} F_j''(x_0)(h,\cdot)^* s_j^*(h), v> +$$

$$+ \left(\frac{1}{2} \sum_{j=1}^{m} <q_j^*(h), F_j''(x_0)(h,h)> + \frac{1}{6} \sum_{j=1}^{m} <s_j^*(h), F_j'''(x_0)(h,h,h)> \right) \gamma$$

Dropping the nonnegative summands in the μ-terms, equations (8) and (9) follow. Finally, if $G(x_0,h)$ is onto, then $y_1^*(h)$ and $y_2^*(h)$ vanish which implies that not all the multipliers $a_k(h)$ for $k \in K_1(h)$, $b_i(h)$ for $i \in K_2(h)$, and $g_i(h)$ for $i=p+1,\ldots,n$, are nontrivial. Q.E.D.

Remark 5.1 In order to specify the structure of the functionals in the dual to the second order feasible cones for the convex inequalities further, it is necessary to specify the convex sets. Recall, however, that we have in general that $\lambda_i(h) \in FC(Z_i,x_0)^*$ for convex inequality constraints.

6. Pareto Optimal Control Problem in the Abnormal Case

We now apply Theorem 5.1. to derive first order necessary conditions for Pareto optimization in an optimal control problem. Second order conditions follow in principle in the same way except that the structure of the convex inequality constraints need to be specified further (like a polyhedron, ball, etc.) if one wants to calculate the second order feasible cone and its dual. We consider the problem to minimize in the Pareto sense the multi-index cost functional

$$I(x,u) = (I_1(x,u), I_2(x,u),\ldots,I_s(x,u))$$

$$= \left(\int_0^1 f^1(t,x(t),u(t))dt, \ldots, \int_0^1 f^s(t,x(t),u(t))dt \right) \tag{17}$$

subject to the constraints

$$\dot{x}(t) = f(t,x(t),u(t)), \qquad x(0) = 0, \tag{18}$$

$$g(x(1)) = 0, \tag{19}$$

$$u(\cdot) \in U, \quad U = \{ u(\cdot) \in L_\infty^r(0,1): u(t) \in M \text{ for } t \in [0,1] \text{ a.e.}\}. \tag{20}$$

Here $x(\cdot)$ lies in the space $\overline{W}_{11}^n(0,1)=\{x:[0,1]\mapsto\mathbb{R}^n: x(\cdot)$ is absolutely continuous and $\dot{x}(\cdot) \in L_1^n(0,1)$, $x(0)=0\}$, and f^k, $k=1,2,\ldots,s$, and f are functions from $\mathbb{R}\times\mathbb{R}^n\times\mathbb{R}^r$ into $\mathbb{R}$

and $\mathbb{R}^n$ respectively; g maps $\mathbb{R}^n$ into $\mathbb{R}^k$. We assume that

(A1) The functions f and f^k are Lebesgue measurable in t for every (x,u) fixed; for $t \in [0,1]$ fixed the functions f^k, k=1,2,...,s are once and f and g are twice continuously differentiable in (x,u). These derivatives are Lebesgue measurable in t for (x,u) fixed. All functions are bounded on bounded subsets of (x,u)-space.

(A2) The control set $M \subseteq \mathbb{R}^r$ is closed and convex with nonempty interior.

We call the problem defined by (17)-(20) under assumptions (A1) and (A2) Problem 2.

Definition 6.1. An admissible process $(x_0(\cdot),u_0(\cdot)) \in \overline{W}^n_{11}(0,1) \times L^r_\infty(0,1)$ for Problem 2 is called an *abnormal process* (abnormal mode) if there exist a function $p(\cdot)$: $[0,1] \to \mathbb{R}^n$ and a vector $a \in \mathbb{R}^k$ such that for $t \in [0,1]$ a.e. and all $u \in M$

$$\dot{p} = -f^*_x(t,x_0(t),u_0(t))p, \qquad p(1) = -g^*_x(x_0(1))a \tag{21}$$

$$< f^*_u(t,x_0(t),u_0(t))p(t),u_0(t)> = < f^*_u(t,x_0(t),u_0(t))p(t),u> \tag{22}$$

Hence abnormal processes satisfy the Pontryagin Maximum Principle with $\lambda_0=0$.

We now formulate Problem 2 in the framework used in Theorem 5.1. Define operators F_1 and F_2 given by the following formulas:

$$F_1: \overline{W}^n_{11}(0,1) \times L^r_\infty(0,1) \to \overline{W}^n_{11}(0,1)$$

$$(x,u) \mapsto F_1(x,u)(t) = x(t) - \int_0^t f(s,x(s),u(s))ds; \tag{23}$$

$$F_2: \overline{W}^n_{11}(0,1) \times L^r_\infty(0,1) \to \mathbb{R}^k$$

$$(x,u) \mapsto F_2(x,u) = g(x(1)). \tag{24}$$

Then let F: $\overline{W}^n_{11}(0,1) \times L^r_\infty(0,1) \to \overline{W}^n_{11}(0,1) \times \mathbb{R}^k$ be given by $F(x,u)=(F_1(x,u), F_2(x,u))$. The first and second order Fréchet derivatives of F can easily be calculated as

$$F'(x_0,u_0)(\bar{h},\bar{v})(t) = \Big(\bar{h}(t) - \int_0^t f_x(s,x_0(s),u_0(s))\bar{h}(s) + f_u(s,x_0(s),u_0(s))\bar{v}(s)\, ds, \\ g_x(x_0(1))\bar{h}(1)\Big); \tag{25}$$

and

$$F''(x_0,u_0,h,v)(\bar{h},\bar{v})(t) = \Big(- \int_0^t f_{xx}(s,x_0(s),u_0(s))h(s)\bar{h}(s) + f_{xu}(s,x_0(s),u_0(s))h(s)\bar{v}(s) \\ + f_{ux}(s,x_0(s),u_0(s))v(s)\bar{h}(s) + f_{uu}(s,x_0(s),u_0(s))v(s)\bar{v}(s)\, ds,$$

$$g_{xx}(x_0(1))h(1)\bar{h}(1)\Big) ; \tag{26}$$

where (h,v), $(\bar{h},\bar{v}) \in \overline{W}^n_{11}(0,1) \times L^r_\infty(0,1)$. For arbitrary $(h,v) \in \overline{W}^n_{11}(0,1) \times L^r_\infty(0,1)$ define the operator

$$G(x_0,u_0,h,v):\ \overline{W}^n_{11}(0,1) \times L^r_\infty(0,1) \to \overline{W}^n_{11}(0,1) \times \mathbb{R}^k \times (\overline{W}^n_{11}(0,1) \times \mathbb{R}^k)/\mathrm{Im}F'(x_0,u_0)$$

given by

$$G(x_0,u_0,h,v)(\bar{h},\bar{v}) = \big(F'(x_0,u_0)(\bar{h},\bar{v}),\ \pi(F''(x_0,u_0,h,v)(\bar{h},\bar{v}))\big), \tag{27}$$

where π is the canonical projection from $\overline{W}^n_{11}(0,1) \times \mathbb{R}^k$ onto $\overline{W}^n_{11}(0,1) \times \mathbb{R}^k/\mathrm{Im}F'(x_0,u_0)$. Now introduce the following set $N(x_0,u_0)$ of parameters:
$N(x_0,u_0) = \{(h,v) \in \overline{W}^n_{11}(0,1) \times L^r_\infty(0,1)$:

(i) $dh/dt - f_x(t,x_0(t),u_0(t))\, h(t) - f_u(t,x_0(t),u_0(t))\, v(t) = 0$,

(ii) $g_x(x_0(1))\, h(1) = 0$,

(iii) $$\Big(-\int_0^t f_{xx}(t,x_0(t),u_0(t))\, h(t)h(t) + 2\, f_{xu}(t,x_0(t),u_0(t))\, h(t)v(t) + f_{uu}(t,x_0(t),u_0(t))\, v(t)v(t)\, dt,\ g_{xx}(x_0(1))\, h(1)h(1)\Big) \in \mathrm{Im}\, F'(x_0,u_0)\ ,$$

(iv) $\int_0^1 f^k_x\,(t,x_0(t),u_0(t))\, h(t) + f^k_u(t,x_0(t),u_0(t))\, v(t)\, dt \le 0$ for $k=1,2,\dots,s$

(v) the second order feasible cone to the set U at the point u_0 in the direction of v is nonempty and convex. }

Theorem 6.1. (Extended Local Maximum Principle) Suppose (x_0,u_0) is a Pareto optimal process for Problem 2. Then for every $(h,v) \in N(x_0,u_0)$, there exist multipliers $\lambda_k=\lambda_k(h,v) \ge 0$, for $k=1,2,\dots,s$, $a=a(h,v) \in \mathbb{R}^k$, $b=b(h,v) \in \mathbb{R}^k$, absolutely continuous functions $p(\cdot)= p(h,v)(\cdot)$ and $\psi(\cdot)=\psi(h,v)(\cdot)$: $[0;1] \to \mathbb{R}^n$, all depending on (h,v) and not all (identically) zero, which satisfy the following conditions for $t \in [0,1]$ a.e.:

- $$\dot{\psi}(t)= -\ f^*_x(t,x_0(t),u_0(t))\psi(t) \tag{28}$$

with terminal condition

$$\psi(1)= -\ g^*_x(x_0(1))b\ ; \tag{29}$$

$$f^*_u(t,x_0(t),u_0(t))\psi(t) = 0 \tag{30}$$

- $$\dot{p}(t) = -\sum_{k=1}^{s} \lambda_k f_x^{k*}(t,x_0(t),u_0(t)) - f_x^*(t,x_0(t),u_0(t))\, p(t)$$
$$- f_{xx}^*(t,x_0(t),u_0(t))(h(t))\, \psi(t) - f_{ux}^*(t,x_0(t),u_0(t))(v(t))\, \psi(t)\,, \tag{31}$$

with terminal condition

$$p(1) = -\, g_x^*(x_0(1))a - (g_{xx}(x_0(1))h(1))^*\, b; \tag{32}$$

- for every $w \in M$ we have

$$\Big(\sum_{k=1}^{s} \lambda_k f_u^{k*}(t,x_0(t),u_0(t)) + f_u^*(t,x_0(t),u_0(t))\, p(t) + f_{xu}^*(t,x_0(t),u_0(t))(h(t))\, \psi(t)$$
$$+ f_{uu}^*(t,x_0(t),u_0(t))(v(t))\, \psi(t),\ u_0(t) - w \Big) \geq 0 \tag{33}$$

Furthermore, $\lambda_k = 0$ if

$$\int_0^1 f_x^k\,(t,x_0(t),u_0(t))\, h(t) + f_u^k(t,x_0(t),u_0(t))\, v(t)\, dt < 0\,. \tag{34}$$

In addition, if $\lambda_k=0$ for $k=1,2...,s$ and $b=0$, then there exists a subset $D \subseteq [0,1]$ of positive measure such that

$$f_u^*(t,x_0(t),u_0(t))p(t) \neq 0 \quad \text{for every } t \in D. \tag{35}$$

Remark 6.1. Notice that this Extremum Principle has a parametrized form in the sense that the multipliers as well as equations and conditions from the formulation of the theorem depend on the pair of parameters $(h,v) \in N(x_0,u_0)$. Notice also the extra terms in the adjoint equation (31) and in the maximum condition (33) which involve the second derivatives. Because of these two facts we can say that we have an "extended" adjoint equation with an "extended" maximum condition. The two new multipliers $\psi(\cdot)$ and b are associated with these extra terms. They are determined by the additional equation (28) with terminal condition (29) and the stationary point condition (30). These appear due to the possible nonregularity of the operator F given by (23) and (24). In this case by the Hahn-Banach theorem there exist nonvanishing multipliers $\psi(\cdot)$ and b which satisfy (see equation (40) below)

$$\int_0^1 \Big(\dot{\bar{h}}(t) - f_x(t,x_0(t),u_0(t))\bar{h}(t) - f_u(t,x_0(t),u_0(t))\bar{v}(t) \Big)\, \psi(t)dt + (b,\, g_x(x_0(1))\bar{h}(1)) = 0\,.$$

In the regular case, i.e. when $\operatorname{Im} F'(x_0,u_0) = \overline{W}_{11}^n(0,1) \times \mathbb{R}^k$, then this condition holds with trivial multipliers $\psi=0$, $b=0$ and we have the following changes in the formulation of Theorem 6.1:

— the additional equation (28) with terminal condition (29) and the stationary point condition (30) are satisfied trivially with $\psi \equiv 0$;

— the "extended" adjoint equation (31) reduces to the classical form of the adjoint equation :

$$\dot{p}(t) = -\sum_{k=1}^{s} \lambda_k f_x^k(t,x_0(t),u_0(t)) - f_x^*(t,x_0(t),u_0(t))p(t);$$

—the "extended" maximum condition (32) reduces to the classical form: for every $u \in M$ and $t \in [0,1]$ a.e. we have

$$\Big(\sum_{k=1}^{s} \lambda_k f_u^k(t,x_0(t),u_0(t)) + f_u^*(t,x_0(t),u_0(t))p(t),\ u_0(t) - u \Big) \geq 0$$

Remark 6.2. The proof is an application of Theorem 5.1. However, here we only derive the first order necessary conditions corresponding to equation (8). It is therefore enough to assume that the operator F is twice Fréchet differentiable rather than three times as it is required in assumption (A).

Proof. Let $Z_1=\{(x,u):\ u \in U\}$ and $Z_{1+i}=\{(x,u) \in \overline{W}_{11}^n(0,1) \times L_\infty^r(0,1):\ F_i(x,u)=0\}$, for $i=1,2$. The set U and the operators F_1 and F_2 are given by (20), (23) and (24) respectively. With this notation Problem 2 can be written in the form of the Dubovitskii-Milyutin formalism as the problem of minimizing the multi-index functional $I(x,u)$ under the constraints $(x,u) \in Z_i$, $i=1,2,3$, where Z_1 is a convex inequality constraint and Z_2 and Z_3 are equality constraints.

We first show that for $(h,v) \in N(x_0,u_0)$ the assumptions of Theorem 5.1. are satisfied. We need to analyse the operator $F'(x_0,u_0)$ given by (25). It is easy to see that for every $a(t) \in \overline{W}_{11}^n(0,1)$ the equation

$$\bar{h}(t) - \int_0^t f_x(s,x_0(s),u_0(s))\,\bar{h}(s) + f_u(s,x_0(s),u_0(s))\,\bar{v}(s)\,ds = a(t) \tag{36}$$

has a solution $(\bar{h},\bar{v})$. For instance, take $\bar{v}=0$ and solve the resulting linear integral equation. Therefore $\mathrm{Im}F'(x_0,u_0)$ has finite codimension in $\overline{W}_{11}^n(0,1) \times \mathbb{R}^k$. But then $\mathrm{Im}F'(x_0,u_0)$ is closed and the operator $G(x_0,u_0,h,v)$ given by (27) has a closed image in $\overline{W}_{11}^n(0,1) \times \mathbb{R}^k \times (\overline{W}_{11}^n(0,1) \times \mathbb{R}^k)/\mathrm{Im}F'(x_0,u_0)$ as well. Hence assumptions (Aii) and (Aiv) of section 5 are satisfied. All other conditions which enter the definition of the set $N(x_0)$ in (6) in Theorem 5.1. are part of the definition of the set $N(x_0,u_0)$.

Hence (and taking into account Remark 5.1.) there exist numbers $\lambda_k=\lambda_k(h,v) \geq 0$, a linear functional $f_{s+1}(h,v)=\tilde{f}_{s+1}(v)$ supporting the set U at the point u_0, functionals $q_1^*(h,v)$, $s_1^*(h,v) \in \overline{W}_{11}^n(0,1)^*$, and vectors $q_2^*(h,v)$, $s_2^*(h, v) \in \mathbb{R}^k$, not all zero, such that

$$-\sum_{k=1}^{s} \lambda_k(h,v)I_k'(x_0,u_0) + f_{s+1}(h,v) + F_1'^*(x_0,u_0)\, q_1^*(h,v) + F_1'^*(x_0,u_0)\, q_2^*(h,v)$$

$$+ (F_1''(x_0,u_0)(h,v))^* s_1^*(h,v) + (F_2''(x_0,u_0)(h,v))^* s_2^*(h,v) \equiv 0\,, \tag{37}$$

and

$$(s_1^*(h,v),\, s_2^*(h,v)) \in \mathrm{Im}F'(x_0,u_0)^{\perp}\,. \tag{38}$$

(Recall that the dual to the feasible cone to a closed convex set with nonempty interior at a point u_0 consists of all functionals supporting the set at u_0. [Girsanov, 7, Theorem 10.5.]) Using equations (25) and (26) and the definition of the adjoint operator as well as the general form of linear functionals in the spaces $\overline{W}_{11}^n(0,1)$ and $\mathbb{R}^k$, it follows that there exist functions $p(\cdot)= p(h,v)(\cdot)$ and $\psi(\cdot)=\psi(h,v)(\cdot)$ in $L_\infty^n(0,1)$, depending on (h,v), and vectors $a=a(h,v)$, $b=b(h,v) \in \mathbb{R}^k$ such that we have for all $(\bar{h},\bar{v}) \in \overline{W}_{11}^n(0,1) \times L_\infty^r(0,1)$:

$$-\sum_{k=1}^{s} \lambda_k \int_0^1 f_x^k(t,x_0(t),u_0(t))\bar{h}(t)+f_u^k(t,x_0(t),u_0(t))\bar{v}(t)\, dt + f_1'(\bar{v})$$

$$+\int_0^1 \Big(\dot{\bar{h}}(t) - f_x(t,x_0(t),u_0(t))\bar{h}(t) - f_u(t,x_0(t),u_0(t))\bar{v}(t)\Big)\, p(t)\, dt$$

$$+ (a,\, g_x(x_0(1))\bar{h}(1))$$

$$-\int_0^1 \Big(f_{xx}(t,x_0(t),u_0(t))h(t)\bar{h}(t) + f_{xu}(t,x_0(t),u_0(t))h(t)\bar{v}(t)$$

$$+ f_{ux}(t,x_0(t),u_0(t))v(t)\bar{h}(t) + f_{uu}(t,x_0(t),u_0(t))v(t)\bar{v}(t)\Big)\, \psi(t)\, dt$$

$$+ \big(b,\, g_{xx}(x_0(1))h(1)\bar{h}(1)\big) = 0\,. \tag{39}$$

And equation (38) is equivalent to

$$\int_0^1 \Big(\dot{\bar{h}}(t) - f_x(t,x_0(t),u_0(t))\bar{h}(t) - f_u(t,x_0(t),u_0(t))\bar{v}(t)\Big)\, \psi(t)\, dt + (b,\, g_x(x_0(1))\bar{h}(1)) = 0. \tag{40}$$

Using the definition of the adjoint operator, integrating the second term by parts and rearranging terms yields

$$\int_0^1 \dot{\bar{h}}(t)\Big(\psi(t)+\int_0^t f_x^*(s,x_0(s),u_0(s))\psi(s)ds\Big)\,dt - \int_0^1 \bar{v}(t)\, f_u^*(t,x_0(t),u_0(t))\psi(t)\,dt$$

$$\Big(-\int_0^1 f_x^*(s,x_0(s),u_0(s))\psi(s)ds+ g_x^*(x_0(1))b\Big)\,\bar{h}(1) = 0 \tag{41}$$

for all $\bar{h}\in \overline{W}_{11}^n(0,1)$ and $\bar{v}\in L_\infty^r(0,1)$. Conditions (28)-(30) follow directly from this. (If necessary, change $\psi(\cdot)$ on a set of measure zero to make it absolutely continuous.) Conditions (31) and (32) follow from (39) in the same way if we set $\bar{v}\equiv 0$. If we set $\bar{h}\equiv 0$ in (39), then we get

$$\tilde{f}_{s+1}(\bar{v}) = \int_0^1 \Big(\sum_{k=1}^{s} \lambda_k f_u^k(t,x_0(t),u_0(t)) + f_u^*(t,x_0(t),u_0(t))\, p(t) \tag{42}$$
$$+ \big(f_{xu}(t,x_0(t),u_0(t))h(t)+f_{uu}(t,x_0(t),u_0(t))v(t)\big)^*\psi(t)\Big)\, \bar{v}(t)\, dt\, .$$

But $\tilde{f}_{s+1}(\bar{v})$ is a supporting functional for the set U at u_0 and therefore also

$$\int_0^1 \Big(\sum_{k=1}^{s} \lambda_k f_u^k(t,x_0(t),u_0(t)) + f_u^*(t,x_0(t),u_0(t))\, p(t) \tag{43}$$
$$+ \big(f_{xu}(t,x_0(t),u_0(t))h(t)+f_{uu}(t,x_0(t),u_0(t))v(t)\big)^*\psi(t),\ u(t)-u_0(t)\Big)\, dt \le 0$$

for all $u(\cdot)\in U$. Hence, proceeding as in the proof of the Maximum principle [Girsanov, 7], we obtain the maximum condition (33). Condition (34) is the second order complementary slackness condition (10) of Theorem 5.1. Finally, note that $\psi(\cdot)\equiv 0$ if b=0. Therefore, if also λ_k=0 for all k, then (42) gives that

$$\tilde{f}_{s+1}(\bar{v}) = \int_0^1 f_u^*(t,x_0(t),u_0(t))\, p(t)\, \bar{v}(t)\, dt\, . \tag{44}$$

It follows from the proof of Theorem 5.1. that the functional $q^*(h,v)=(q_1^*(h,v),q_2^*(h,v))$ is an extension of a linear functional in $\mathrm{Im}F'(x_0,u_0)^*$. Since F' is not onto, there exists a pair $(\bar{h},\bar{v})$ such that $\big(q^*(h,v),\ F'(x_0,u_0)(\bar{h},\bar{v})\big)>0$. But then $\tilde{f}_{s+1}$ cannot vanish identically and so (35) follows. Q.E.D.

REFERENCES

1. Avakov, E.R. (1985), Extremum conditions for smooth problems with equality type constraints. U.S.S.R. Computational Mathematics and Mathematical Physics, Vol. 25, No.5, pp. 680-693.

2. Avakov, E.R. (1988) Necessary conditions for the minimum for the nonregular problems in Banach spaces. Maximum Principle for abnormal problems of optimal

control. Trudy Matematiceskovo Instituta AN – SSSR, Vol. 185, pp. 3-29 (in Russian).

3. Avakov, E.R. (1988) The Maximum principle for abnormal optimal control problems. Soviet Math. Dokl., Vol.37, No.1, pp.231-234.

4. Avakov, E.R. (1989) Necessary extremum conditions for smooth abnormal problems with equality and inequality-type constraints, Mathematicheskie Zametki, Vol. 45, No.6, pp.3-11.

5. Ben-Tal, A. and Zowe, J. (1982) A unified theory of first and second order conditions for extremum problems in topological vector spaces, Mathematical Programming Study, Vol. 19, pp. 39-76.

6. Censor, Y. (1977) Pareto optimality in multiobjective problems, Applied Mathematics and Optimization, Vol.4, No.1, pp.41 - 59

7. Girsanov, I.V. (1972) Lectures on Mathematical Theory of Extremum Problems, Lecture Notes in Economics and Mathematical Systems, Berlin, Germany, Springer Verlag.

8. Ioffe, A.D. and Tikhomirov, W.M. (1979) Theory of Extremal Problems, North Holland, Amsterdam, Holland.

9. Kotarski, W. (1989) Characterization of Pareto optimal points in problems with multi-equality constraints, Optimization, Vol.20, No.1, pp. 93-106.

10. Kotarski, W. (1990) On some specification of the Dubovicki-Milutin theorem for Pareto optimal problems, Nonlinear Analysis, Theory, Methods & Applications,Vol.14, No.3, pp. 287-291.

11. Ledzewicz-Kowalewska, U. (1986) On some specification of the Dubovitskii-Milyutin method, Nonlinear Analysis, Vol.10, No.12, pp. 1367-71.

12. Ledzewicz, U. (1993) Extension of the local Maximum Principle to abnormal optimal control problems, Journal of Optimization Theory and Applications, Vol.77, No.3, pp.661-681.

13. Ledzewicz, U. and Schaettler, H. (1993) Second order conditions for extremum problems with nonregular equality constraints, to appear.

14. Ledzewicz, U. and Schaettler, H. (1993) A theory of first and second order conditions for nonregular extremum problems, this volume.

15. Ponstein J. (1967) Seven kinds of convexity, SIAM Review ,9, pp.115-119.

16. Walczak S. (1984) On some properties of cones in in normed spaces and their applications to investigating extremal problems, Journal of Optimization Theory and Applications, Vol. 42, No. 4, pp. 561-581.

15 A Theory of First and Second Order Conditions for Nonregular Extremum Problems

Urszula Ledzewicz Southern Illinois University at Edwardsville, Edwardsville, Illinois

Heinz Schaettler Washington University, St. Louis, Missouri

Abstract. Using results of Avakov about tangent directions to equality constraints given by smooth operators, we formulate a theory of first and second order conditions for optimality in the sense of Dubovitskii-Milyutin which is nontrivial also in the case of equality constraints given by nonregular operators. Second order feasible and tangent directions are defined to construct conical approximations to inequality and equality constraints. In particular, the result generalizes Avakov's result for the smooth case.

1. Introduction

We consider the problem of minimizing a functional in a Banach space under constraints given by sets with both empty and nonempty interiors (so-called equality and inequality constraints). The equality constraints will be given as kernels of some operators (operator equality constraints). More specifically, let X and Y be Banach spaces, let $F: X \to Y$ be an operator and let $I: X \to \mathbb{R}$ be a functional. We consider the following optimization problem, which we will call Problem 1:

Minimize I(x) subject to $x \in Z = \bigcap_{i=1}^{n+1} Z_i$

where $x \in Z_i$, i=1,...,n are inequality constraints (i.e. int $Z_i \neq \emptyset$)

and $x \in Z_{n+1} = \{ x \in X: F(x) = 0\}$ is an equality constraint.

---

[1] *Research is supported in part by NSF Grant DMS – 91009324, SIUE Research Scholar Award and Fourth Quarter Fellowship, Summer 1992.*

[2] *Research is supported in part by NSF Grant DMS – 9100043*

In the Dubovitskii-Milyutin approach necessary conditions for optimality at a point x_0 are derived from separate conical approximations to the constraint sets Z_i, $i=1,\dots,n+1$, and to the set $\{x \in X: I(x) < I(x_0)\}$ (see [Girsanov, 5]). The definitions are made in such a way that, if these approximating cones are all nonempty and convex, then x_0 is a local minimum for Problem 1 if and only if there exists no point common to all approximating cones. By the Dubovitskii-Milyutin lemma (see [Girsanov, 5, Lemma 5.11] this can then equivalently be described in terms of linear functionals from the corresponding polar or dual cones. The monograph [Girsanov, 5] entirely deals with an exposition of the Dubovitskii-Milyutin approach. As can be seen there many first order conditions for optimality can be derived in this way ranging from the Kuhn-Tucker theorem for nonlinear programming to the Pontryagin maximum principle in optimal control theory. A second order approach along the lines of the Dubovitskii-Milyutin lemma was introduced in [Ben-Tal and Zowe, 4]. In our paper we give a slightly modified definition of second order approximations. Second order feasible and tangent cones are defined in $X \times \mathbb{R}_+$ rather than in X as it is done in [Ben-Tal and Zowe, 4] to allow for reparametrizations of the approximating curves. This is motivated by results of Avakov [Avakov, 1-3] about tangent directions to equality constraints given by nonregular operators and has the advantage that these so-defined second order directions form cones, not just sets as in [Ben-Tal and Zowe, 4]. Therefore our definition reduces the required calculations of the polar sets to rather direct applications of known results. In addition, generalizations of our definitions to approximations of arbitrary order are straightforward and yield always conical structures.

The main focus in our paper is on the structure of tangent directions to the equality constraint. The classical answer to this question is given by the Lusternik theorem and requires that the operators are continuously Fréchet differentiable and regular in the sense that the Fréchet derivative at the point x_0 is onto. Optimization problems involving nonregular operators belong to the class of abnormal problems and there are relatively few results concerning this type of problems. In a series of papers, Avakov [Avakov, 1-3] presented several generalizations of the Lusternik theorem to the case of nonregular operators. The regularity assumption is weakened at the expense of stronger Fréchet differentiability assumptions. His results were already applied to optimization problems like the smooth [Avakov, 1,3] and the smooth-convex problem [Avakov, 2] to obtain extended necessary conditions of optimality. They have a structure of first order conditions though they involve second order derivatives in the abnormal cases.

In this paper we combine the results of Avakov concerning the optimization problem with mixed equality and inequality constraints given by smooth operators [Avakov, 3] with those of [Ben-Tal and Zowe, 4] to formulate a second order theory in the sense of Dubovitskii-Milyutin. Our results give nontrivial optimality conditions also in the case of equality constraints given by nonregular operators. Here we present only a brief exposition of the theory, more details and complete proofs can be found in [Ledzewicz and Schaettler, 9]. The main result is a theorem which states first and second order necessary conditions of optimality for Problem 1. It extends classical results to abnormal cases caused by nonregularity of the operator. Avakov's results for the smooth case [Avakov, 3] are a direct corollary of the general theorem.

These results already found applications to optimal control problems and extended forms of the maximum principle have been formulated which give nontrivial conditions also in abnormal cases [see Avakov, 2 and Ledzewicz 7,8]. In [Ledzewicz and Schaettler, 10] in this volume we derive the corresponding extended conditions for Pareto optimality in abnormal cases and apply them to optimal control problems.

2. Second Order Conical Approximations

Definition 2.1. [Girsanov, 5] A vector h is said to be a *feasible direction* to the inequality constraint $Z \subseteq X$ at x_0 if there exists an $\epsilon_0 > 0$ and a neighborhood V of h such that for all $0 < \epsilon < \epsilon_0$ we have $x_0 + \epsilon V \subseteq Z$. The set of feasible directions to Z at x_0 forms an open cone with apex at 0 called cone of feasible directions or, in short, the feasible cone and is denoted by $FC(Z,x_0)$.

Definition 2.2. [Ledzewicz and Schaettler, 9] We call the pair (v,γ) a *second-order feasible direction* to the set Z at the point $x_0 \in Z$ in direction of $h \in X$ if there exist an $\epsilon_0 > 0$ and a neighborhood V of v_0 such that for all ϵ, $0<\epsilon<\epsilon_0$, we have $x_0 + \epsilon\sqrt{\gamma}h + \epsilon^2 V \subseteq Z$. We denote the set of all second-order feasible directions to Z at x_0 in direction of h by $FC^{(2)}(Z;x_0,h)$.

It is easy to see, that $FC^{(2)}(Z;x_0,h)$ is a cone. For, if $(v,\gamma) \in FC^{(2)}(Z;x_0,h)$ and $\lambda > 0$, then, by reparametrizing ϵ as $\epsilon = \sqrt{\lambda}\tilde{\epsilon}$ and defining $\tilde{r}(\tilde{\epsilon}) = r(\sqrt{\lambda}\tilde{\epsilon}) = r(\epsilon)$ we have $x(\tilde{\epsilon}) = x_0 + \tilde{\epsilon}\sqrt{\lambda\gamma}\,h + \tilde{\epsilon}^2\lambda\, v + \tilde{r}(\tilde{\epsilon}) \in Z$ and $\tilde{r}(\tilde{\epsilon})$ is still of order $o(\tilde{\epsilon}^2)$. We call $FC^{(2)}(Z;x_0,h)$ the second order feasible cone to Z at x_0 in direction of h. Furthermore, it is easily seen that $FC^{(2)}(Z;x_0,h)$ is open [Ledzewicz and Schaettler, 9].

Remark 2.1. This definition of second-order feasible directions (in $X \times \mathbb{R}$ instead of in X) was precisely chosen to allow for reparametrizations which makes the set of second order feasible directions a cone. It differs from the one given by [Ben-Tal and Zowe, 4] where second order tangent directions are defined in X and form sets, not necessarily cones. Consequently our approach has the advantage that second order necessary conditions for optimality can be derived using classical results about dual cones [Girsanov, 5].

Note that, if $h \in FC(Z,x_0)$, then $FC^{(2)}(Z;x_0,h) = X \times (0,\infty)$. Typically h will lie in $\partial FC(Z,x_0)$, but even if $h \notin \text{Clos } FC(Z,x_0)$, then $FC^{(2)}(Z;x_0,h)$ need not be empty. Besides from giving additional information for directions h in the boundary of the feasible cone, second order feasible cones may also give information when the first order feasible cone is empty. Indeed, quadratic approximations around a direction h may lie in the interior of the set even when no cones around h do as the following simple example shows.

Example 2.1. Let $Z = \{(x_1,x_2)\colon\ x_1 \geq 0,\ x_1^2 \leq x_2 \leq 2x_1^2\}$ and consider $x_0 = (0,0)$. Since feasible cones are necessarily open, we have $FC(Z,0) = \emptyset$ and no boundary directions exist. But clearly the direction $h = \binom{1}{0}$ plays a special role and

$$FC^{(2)}(Z;0,h) = \left\{ (v,\gamma) \in \mathbb{R}^3\colon\ \gamma > 0,\ \ \gamma < v_2 < 2\gamma \right\}.$$

Of course, in general a second order approximation need not give additional information over first order approximations (see [Ledzewicz and Schaettler, 9]).

Definition 2.3. Let $f\colon X \to \mathbb{R}$ be a functional and let $x_0 \in X$. The first respectively second order feasible cones (in direction of h) to the set $Z=\{x \in X\colon f(x) < f(x_0)\}$ at x_0 are called the first and *second order cones of decrease* (in direction of h) for the functional f at x_0.

Definition 2.4. [Girsanov, 5] Let $Z \subseteq X$ be an equality constraint, i.e. $\text{int } Z = \emptyset$. A vector h is said to be a *tangent direction* to Z at $x_0 \in Z$ if for every sufficiently small $\epsilon > 0$ there exists a vector $r(\epsilon) \in X$ of order $o(\epsilon)$ with the property that $x(\epsilon) = x_0 + \epsilon h + r(\epsilon) \in Z$.

It is well-known that the tangent directions generate a cone with apex at 0, called the cone of tangent directions to the set Z at the point x_0, in short the tangent cone, denoted by $TC(Z,x_0)$.

Definition 2.5. [Ledzewicz and Schaettler, 9] We call the pair $(v,\gamma) \in X \times \mathbb{R}_+$ a *second order tangent direction* to the set Z at the point $x_0 \in Z$ in direction of $h \in E$ if there exists an $\epsilon_0 > 0$ such that for all ϵ, $0 < \epsilon < \epsilon_0$, there exists a point $r(\epsilon) \in E$ of order $o(\epsilon^2)$ which has the property that

$$x(\epsilon) = x_0 + \epsilon \sqrt{\gamma}\, h + \epsilon^2 v + r(\epsilon) \in Z.$$

We denote the set of all second order tangent directions to Z at x_0 in direction of h by $TC^{(2)}(Z;x_0,h)$. It follows like for the second order feasible cone that $TC^{(2)}(Z;x_0,h)$ is a cone and we call it the second order tangent cone to Z at x_0 in direction h.

Remark 2.2. Our definitions of second order feasible and tangent directions can be extended in a straightforward way to approximations of higher order. For instance, given $(h,v) \in X^2$, call a triple $(w,\gamma_1,\gamma_2) \in X \times \mathbb{R}_+^2$ a *third order tangent direction* to the set Z at the point $x_0 \in Z$ in direction of the (ordered) pair $(h,v) \in X^2$ if there exists an $\epsilon_0 > 0$ such that for all ϵ, $0 < \epsilon < \epsilon_0$, there exists a point $r(\epsilon) \in X$ of order $o(\epsilon^3)$ which has the property that

$$x(\epsilon) = x_0 + \epsilon \sqrt[3]{\gamma_1}\, h + \epsilon^2 \sqrt[3]{\gamma_2^2}\, v + \epsilon^3 w + r(\epsilon) \in Z.$$

Here h stands for a first and v for a second order approximating direction. The coefficients γ_1 and γ_2 in this definition again allow for reparametrizations and therefore the approximating sets become cones.

3. First and Second Order Conditions for Optimization Problems in Banach Spaces

We now return to Problem 1: Minimize I(x) subject to $x \in Z = \bigcap_{i=1}^{n+1} Z_i$ where $x \in Z_i$, i=1,...,n are inequality constraints and $x \in Z_{n+1}=\{x \in X\colon F(x)=0\}$ is an equality constraint. We assume that the vector $h \in X$ is such that

(1) the second order cone of decrease for the functional I at x_0 in the direction of h, $DC^{(2)}(I;x_0,h)$, is nonempty, (open) and convex;

(2) the second order feasible cones for the inequality constraints Z_i, $FC^{(2)}(Z_i;x_0,h)$, i=1,2,...,n, are nonempty, (open) and convex;

(3) the second order tangent cone to the set Z_{n+1} at the point x_0 in direction h, $TC^{(2)}(Z_{n+1};x_0,h)$, is nonempty, (closed) and convex.

Theorem 3.1. Suppose I attains a local minimum on $Z = \bigcap_{i=1}^{n+1} Z_i$ at x_0 and assumptions (1)-(3) for Problem 1 are satisfied. Then there exist linear functionals

$$g_0(h) = (\lambda_0(h),\mu_0(h)) \in DC^{(2)}(I;x_0,h)^*, \; g_i(h) = (\lambda_i(h),\mu_i(h)) \in FC^{(2)}(Z_i;x_0,h)^*,$$

for i=1,2,...,n, and

$$g_{n+1}(h) = (\lambda_{n+1}(h),\mu_{n+1}(h)) \in TC^{(2)}(Z_{n+1};x_0,h)^*,$$

such that

$$\lambda_0(h) + \lambda_1(h) + \ldots + \lambda_n(h) + \lambda_{n+1}(h) \equiv 0, \quad \text{(Euler-Lagrange equation)} \quad (1)$$

$$\mu_0(h) + \mu_1(h) + \ldots + \mu_n(h) + \mu_{n+1}(h) \equiv 0, \quad \text{(Second order condition)} \quad (2)$$

while not all of the $\lambda_i(h)$, i=0,...,n+1, vanish identically.

Proof. We show first that

$$DC^{(2)}(I;x_0,h) \cap \bigcap_{i=1}^{n} FC^{(2)}(Z_i;x_0,h) \cap TC^{(2)}(Z_{n+1};x_0,h) = \emptyset. \quad (3)$$

Suppose (v,γ) lies in this intersection and let $x_t = x_0 + t\sqrt{\gamma}h + t^2v$. By the definition of tangent directions there exists a function $r(t) = o(t^2)$ such that $x_t+r(t)$ is an admissible direction for the equality constraint. Since $(v,\gamma) \in \bigcap_{i=1}^{n} FC^{(2)}(Z_i;x_0,h)$ this curve is feasible and since $(v,\gamma) \in DC^{(2)}(I;x_0,h)$ the functional I is strictly smaller than at x_0. This contradicts the minimality of x_0.

By our assumptions $DC^{(2)}(I;x_0,h)$ and $FC^{(2)}(Z_i;x_0,h)$ for i=1,2,...,n, are nonempty, open, convex cones and $TC^{(2)}_{n+1}(Z_{n+1};x_0,h)$ is a nonempty, closed, convex cone. So the conditions of the Dubovitskii-Milyutin Lemma [Girsanov, 5, Lemma 5.11] are satisfied and there exist linear functionals $g_0(h) \in DC^{(2)}(I;x_0,h)^*$, $g_i(h) \in FC^{(2)}(Z_i;x_0,h)^*$, for i=1,2,...,n, and $g_{n+1}(h) \in TC^{(2)}(Z_{n+1};x_0,h)^*$, not all zero, such that

$$g_0(h) + g_1(h) + g_2(h) + \ldots + g_n(h) + g_{n+1}(h) \equiv 0. \quad (4)$$

Writing $g_i(h)$ as $(\lambda_i(h), \mu_i(h))$ equations (1) and (2) follow. Furthermore, if all $\lambda_i(h)$ vanish identically, then we also have that all $\mu_i(h)$ vanish identically. For, since $g_i(h)=(\lambda_i(h), \mu_i(h))$ lies in the dual to the corresponding second order cone, we have then that $\mu_i(h) \geq 0$. But

$$\mu_0(h) + \mu_1(h) + \mu_2(h) + \ldots + \mu_n(h) + \mu_{n+1}(h) \equiv 0$$

and so all $\mu_i(h)$ vanish contradicting the nontriviality of the multipliers $g_i(h)$. Q.E.D.

In order to put this abstract result into more definite forms we need to specify the constraints and calculate the corresponding approximating cones and their duals. As an example, we will derive first and second order conditions of optimality for the so-called smooth nonlinear programming problem. We concentrate on equality constraints given by nonregular operators, only briefly mentioning the results for inequality constraints.

4. Second Order Tangent Cones for Nonregular Operators

Recall that we assume that the equality constraint is given in operator form, i.e. X and Y are Banach spaces, $F: X \to Y$ is an operator and $Z = \{x \in X: F(x) = 0\}$. If F is continuously Fréchet differentiable in a neighborhood of the point x_0 and if the derivative $F'(x_0)$ is onto (it is then customary to call F *regular* at x_0), then by the classical Lusternik theorem, the tangent cone to Z at x_0 is given as

$$TC(Z,x_0) = \{h \in X: F'(x_0)h = 0\} = \ker F'(x_0). \tag{5}$$

Avakov generalized the Lusternik theorem to the case of nonregular operators which are twice Fréchet differentiable at x_0 [Avakov, 1,2]. Assuming that $\operatorname{Im} F'(x_0)$ is closed in Y, for $h \in X$ introduce a linear mapping $G(x_0,h)$: $X \to Y \times Y/\operatorname{Im} F'(x_0)$ given by the formula

$$v \to G(x_0,h)v = (F'(x_0)v,\ \pi F''(x_0)(h,v)) \tag{6}$$

where π: $Y \to Y/\operatorname{Im} F'(x_0)$ is the quotient map from Y into $Y/\operatorname{Im} F'(x_0)$. Suppose that

$$G(x_0,h)h = 0 \qquad \text{and} \qquad \operatorname{Im} G(x_0,h) = \operatorname{Im} F'(x_0) \times Y/\operatorname{Im} F'(x_0).$$

Then $h \in TC(Z,x_0)$ [Avakov, 1,2]. Furthermore, if $\operatorname{Im} G(x_0,h) = \operatorname{Im} F'(x_0) \times Y/\operatorname{Im} F'(x_0)$ for every $h \neq 0$ such that $G(x_0,h)h = 0$, then

$$TC(Z,x_0) = \left\{ h \in X:\ F'(x_0)h = 0,\ F''(x_0)(h,h) \in \operatorname{Im} F'(x_0) \right\}. \tag{7}$$

Our aim is to develop second-order necessary conditions for optimality in the presence of nonregular equality constraints generalizing corresponding results of Dubovitskii-Milyutin [Girsanov, 5]. For this we calculate the second order tangent cone to Z at x_0 in direction of a vector h based on Avakov's results [Avakov, 1-3]. We make the following assumptions:

(A) (i) $x_0 \in Z$ and F is three times Fréchet differentiable at x_0.

(ii) Im $F'(x_0)$ is closed.

(iii) h satisfies $G(x_0,h)h = 0$, i.e. $F'(x_0)h = 0$ and $F''(x_0)(h,h) \in \mathrm{Im}\, F'(x_0)$.

(iv) the operator G is onto: $\mathrm{Im}\, G(x_0,h) = \mathrm{Im}\, F'(x_0) \times Y/\mathrm{Im}\, F'(x_0)$.

The following result allows to calculate second-order tangent directions.

Theorem 4.1. [Avakov, 3] Under assumptions (A), for every $h \in X$, $||h|| = 1$, there exist numbers $\epsilon = \epsilon(h)$ and $K = K(h)$ and a neighborhood U of x_0 such that for any $x \in U$ which satisfies

$$|| \frac{x - x_0}{||x - x_0||} - h || < \epsilon$$

and arbitrary elements $y \in \mathrm{Im}\, F'(x_0)$, there exists a mapping r: $U \to X$, $x \mapsto r(x)$, such that

$$F(x + r(x)) = F(x_0) \tag{8}$$

and

$$|| r(x) || \le K \Big(\frac{||F(x) - F(x_0) - y||}{||x - x_0||} + || y || \Big). \tag{9}$$

Lemma 4.1. Suppose conditions (A) are satisfied. If

$$F'(x_0)v + \frac{\gamma}{2} F''(x_0)(h,h) = 0 \tag{10}$$

and

$$F''(x_0)(h,v) + \frac{\gamma}{6} F'''(x_0)(h,h,h) \in \mathrm{Im}\, F'(x_0), \tag{11}$$

then

$$(v,\gamma) \in TC^{(2)}(Z;x_0,h).$$

Proof. Let $x_\epsilon = x_0 + \epsilon\sqrt{\gamma}\, h + \epsilon^2 v$. Then we have by Taylor's theorem for sufficiently small ϵ

$$F(x_\epsilon) - F(x_0) = \epsilon\sqrt{\gamma}\, F'(x_0)h + \epsilon^2 \Big(F'(x_0)v + \frac{\gamma}{2} F''(x_0)(h,h)\Big) + \tag{12}$$

$$+ \sqrt{\gamma}\, \epsilon^3 \Big(F''(x_0)(h,v) + \frac{\gamma}{6} F'''(x_0)(h,h,h)\Big) + o(\epsilon^3).$$

By assumption $F'(x_0)h = 0$, $F'(x_0)v + \frac{\gamma}{2} F''(x_0)(h,h) = 0$ and

$$F''(x_0)(h,v)+ \frac{\gamma}{6} F'''(x_0)(h,h,h) \in \text{Im } F'(x_0)$$

Choosing

$$y = \sqrt{\gamma}\, \epsilon^3 \left(F''(x_0)(h,v) + \frac{\gamma}{6} F'''(x_0)(h,h,h)\right) \in \text{Im } F'(x_0).$$

in Theorem 4.1., it follows that there exists a function $r(\epsilon)$ such that

$$F(x_\epsilon + r(\epsilon)) = F(x_0)$$

and

$$||r(\epsilon)|| \leq K\left(\frac{||F(x_\epsilon) - F(x_0) - y||}{||x_\epsilon - x_0||} + ||y|| \right) \leq K \left(\frac{o(t^3)}{C\, t} + o(t^3)\right) = o(t^2).$$

Hence (v,γ) is a second-order tangent direction to Z at x_0. Q.E.D.

Conversely, if $(v,\gamma) \in TC^{(2)}(Z;x_0,h)$, then there exists a function $r(\epsilon)$ of order $o(\epsilon^2)$ such that

$$\begin{aligned} F(x_0) &= F(x_\epsilon + r(\epsilon)) = F(x_0 + \epsilon\sqrt{\gamma}\, h + \epsilon^2 v + r(\epsilon)) \\ &= F(x_0) + F'(x_0)(\epsilon\sqrt{\gamma}\, h + \epsilon^2 v + r(\epsilon)) \\ &\quad + \frac{1}{2} F''(x_0)\, (\epsilon\sqrt{\gamma}\, h + \epsilon^2 v + r(\epsilon)\, ,\, \epsilon\sqrt{\gamma}\, h + \epsilon^2 v + r(\epsilon)) \\ &\quad + \frac{1}{6} F'''(x_0)\, (\epsilon\sqrt{\gamma}\, h + \ldots\, ,\, \epsilon\sqrt{\gamma}\, h + \ldots\, ,\, \epsilon\sqrt{\gamma}\, h + \ldots\,) + \ldots \\ &= F(x_0) + \epsilon\sqrt{\gamma}\, F'(x_0)h + \epsilon^2 \left(F'(x_0)v + \frac{\gamma}{2} F''(x_0)(h,h)\right) + \\ &\quad + \sqrt{\gamma}\epsilon^3 \left(F''(x_0)(h,v) + \frac{\gamma}{6} F'''(x_0)(h,h,h)\right) + F'(x_0)\, r(\epsilon) + o(\epsilon^3)\, . \end{aligned}$$

Furthermore, by (12)

$$\begin{aligned} F(x_\epsilon) = F(x_0) &+ \epsilon\sqrt{\gamma}\, F'(x_0)\, h + \epsilon^2 \left(F'(x_0)v + \frac{\gamma}{2} F''(x_0)(h,h)\right) + \\ &+ \sqrt{\gamma}\epsilon^3 \left(F''(x_0)(h,v) + \frac{\gamma}{6} F'''(x_0)(h,h,h)\right) + o(\epsilon^3) \end{aligned} \tag{13}$$

and thus

$$F(x_\epsilon) = F(x_0) - F'(x_0)r(\epsilon) + o(\epsilon^3). \tag{14}$$

Comparing this with the Taylor formula (13) implies

$$F'(x_0)h=0, \quad F'(x_0)v + \frac{\gamma}{2} F''(x_0)(h,h)=0$$

and

$$F''(x_0)\, (h,v) + \frac{\gamma}{6} F'''(x_0)(h,h,h) \in \text{Im } F'(x_0\,)\, .$$

Thus we have the following result:

Corollary 4.1 Under assumptions (A) we have

$$TC^{(2)}(Z;x_0,h) = \Big\{(v,\gamma) \in X \times \mathbb{R}_+ : \; F'(x_0)v + \tfrac{\gamma}{2} F''(x_0)(h,h) = 0 \, , \; F''(x_0)(h,v) + \tfrac{\gamma}{6} F'''(x_0)(h,h,h) \in \operatorname{Im} F'(x_0) \Big\} .$$

Note that the second order tangent cone is a nonempty linear subspace and thus in particular it is convex. To apply Theorem 3.1. the dual cone is needed.

Theorem 4.2. Assuming conditions (A), the dual cone to $TC^{(2)}(Z;x_0,h)$ is given by

$$\begin{aligned} TC^{(2)}(Z;x_0,h)^* = \Big\{ (\lambda,\mu) \in X^* \times \mathbb{R}^* : \; & \text{there exist } \; y_1^* \in \operatorname{Im} F'(x_0)^* \, , \\ & y_2^* \in \big(\operatorname{Im} F'(x_0)\big)^{\perp} \text{ and } \; r \geq 0 \; \text{ such that} \\ & \lambda = F'(x_0)^* \, y_1^* + (F''(x_0)(h,\cdot))^* \; y_2^* \\ & \mu = < y_1^*, \, \tfrac{1}{2} F''(x_0)(h,h)> + <y_2^*, \, \tfrac{1}{6} F'''(x_0)(h,h,h)> + r \Big\} \end{aligned}$$

Proof: Without the sign restriction on γ, the second order tangent cone would be the kernel of a bounded linear operator. Define a map

$$\mathfrak{G} : \; X \times \mathbb{R} \to \operatorname{Im} F'(x_0) \; \times \; Y / \operatorname{Im} F'(x_0) \tag{15}$$

$$(v,\gamma) \mapsto \mathfrak{G}(x_0,h)(v,\gamma) = \Big(F'(x_0)v + \tfrac{\gamma}{2} F''(x_0)(h,h), \; \pi\big(F''(x_0)(h,v) + \tfrac{\gamma}{6} F'''(x_0)(h,h,h)\big)\Big).$$

Since $F''(x_0)(h,h) \in \operatorname{Im} F'(x_0)$, the first component lies in $\operatorname{Im} F'(x_0)$. The second order tangent cone can then be described as

$$TC^{(2)}(Z;x_0,h) = \operatorname{Ker} \mathfrak{G}(x_0,h) \cap \{\gamma \in \mathbb{R} : \; \gamma > 0\}. \tag{16}$$

The dual cone can therefore be expressed as (see for instance [Girsanov, 5, Lemma 5.9])

$$TC^{(2)}(Z;x_0,h)^* = \big(\operatorname{Ker} \mathfrak{G}(x_0,h) \big)^* + \{0\} \times \{\mu \in \mathbb{R} : \; \mu \geq 0 \} .$$

By assumption (A iv) the operator $\mathfrak{G}(x_0,h)$ restricted to $X \times \{0\}$ is onto and therefore by the annihilator lemma or closed-range theorem (see, for instance, [Ioffe, Tikhomirov, 6, p.16]), the dual to $\operatorname{Ker} \mathfrak{G}(x_0,h)$ is given by

$$\big(\operatorname{Ker} \mathfrak{G}(x_0,h)\big)^* = \operatorname{Ker} \mathfrak{G} \, (x_0,h)^{\perp} = \operatorname{Im} \mathfrak{G}(x_0,h)^*$$

where

$$\mathfrak{G}(x_0,h)^*: \left(\text{Im } F'(x_0)\right)^* \times \left(Y/\text{Im } F'(x_0)\right)^* \to X^* \times \mathbb{R}$$

denotes the adjoint map. Let $(\lambda,\mu) \in \text{Im } \mathfrak{G}(x_0,h)^*$, say $(\lambda,\mu) = \mathfrak{G}(x_0,h)^*\left(y_1^*,\tilde{y}_2^*\right)$ where $y_1^* \in \text{Im } F'(x_0)^*$ and $\tilde{y}_2^* \in \left(Y/\text{Im } F'(x_0)\right)^*$. Then we have for all $(v,\gamma) \in TC^{(2)}(Z;x_0,h)$ that

$$0 = <(\lambda,\mu), (v,\gamma)>=<\mathfrak{G}(x_0,h)^*(y_1^*, \tilde{y}_2^*), (v,\gamma)>=< (y_1^*, \tilde{y}_2^*), \mathfrak{G}(x_0,h)(v,\gamma)>=$$
$$= < y_1^*, F'(x_0)v+\frac{\gamma}{2} F''(x_0)(h,h)> + < \tilde{y}_2^*, \pi \left(F''(x_0)(h,v) + \frac{\gamma}{6} F'''(x_0)(h,h,h)\right) >$$

Let $\tau : \left(Y/ \text{ Im } F'(x_0)\right)^* \to \text{Im } F'(x_0)^{\perp}$ be the canonical isometric isomorphism. Then we have for all $v \in \left(Y/\text{Im } F'(x_0)\right)^*$ that $\tau \circ v^* = v^* \circ \pi$ and therefore, setting $y_2^*(h) = \tau \circ \tilde{y}_2^*(h)$, we get

$$0 =< y_1^*, (F'(x_0), \tfrac{1}{2}F''(x_0)(h,h)) (v,\gamma) >+ < y_2^*, (F''(x_0) (h,\cdot) , \tfrac{1}{6} F'''(x_0)(h,h,h)) (v,\gamma) >$$
$$= < F'(x_0)^* y_1^*+F''(x_0)(h,\cdot)^* y_2^*, v > + \Big(< y_1^*, \tfrac{1}{2} F''(x_0)(h,h)>$$
$$+ < y_2^*, \tfrac{1}{6} F'''(x_0)(h,h,h) > , \gamma\Big). \tag{17}$$

This gives the desired form of λ and μ. All the steps can be retraced and therefore a pair (λ,μ) which satisfies the last equation also lies in the image of $\text{Im}\mathfrak{G}(x_0,h)^*$. Q.E.D.

Remark 4.1. Under further differentiability assumptions on the operator F, higher order tangent directions (see Remark 2.2.) and corresponding degrees of abnormality for the problem can be defined analogously. Following Avakov's idea [Avakov, 1,2] if the operator F is p-times Fréchet differentiable, replace the operator G by the following mapping

$$G_p(x_0,h): \quad X \to Y \times V_1 \times V_2 \times \dots \times V_p$$
$$G_p(x_0,h)x = (F'(x_0)x, \pi_1(F''(x_0)(h)x), \pi_2\pi_1(F'''(x_0)(h,h)x),$$
$$\pi_p \dots \pi_1 F^{(p+1)}(x_0)(h,h,\dots,h)x))$$

where $V_1 = Y/\text{Im}F'(x_0)$ and $V_i = V_{i-1}/\text{Im}\pi_{i-1} \cdot\cdot\cdot \pi_1(F^{(i)}(x_0)(h,h,..,h))$ for $i = 2,3,\dots,p$. Let $W_k = \text{Im } F'(x_0) \times V_1 \times V_2 \times \dots \times V_k$, $k=1,\dots,p$. Accordingly we call the operator F *p-regular* at x_0 if and only if $\text{Im}G_k(x_0,h)$ is closed, but properly contained in W_k for $k<p$ and $\text{Im}G_p(x_0,h)=W_p$. In this sense 0-regularity corresponds to regularity in the classical sense. Abnormal problems are split further into several categories of abnormality in the way that we call a problem p-abnormal if the operator F is p-regular.

5. Second Order Feasible Cones for Smooth Constraints

We now consider the second order feasible cone in the so-called smooth case, i.e. when the inequality constraint is given by a smooth functional f on X in the form $Z = \{x \in X: f(x) \leq 0\}$. Here we only state the results and refer the reader to [Ledzewicz and Schaettler, 9] for the proofs.

Theorem 5.1 [Ledzewicz and Schaettler, 9] Let $x_0 \in Z$ and suppose f is twice Fréchet differentiable at x_0 with $f'(x_0) \neq 0$. Then $FC(Z,x_0) = \{h \in X: f'(x_0)h < 0\} \neq \emptyset$ and for every $h \in \partial\, FC(Z,x_0) = \{ h \in X: f'(x_0)h = 0\}$ we have

$$FC^{(2)}(Z;x_0,h) = \left\{(v,\gamma) \in X \times \mathbb{R}_+ : \; f'(x_0)v + \frac{\gamma}{2} f''(x_0)(h,h) < 0 \right\}. \tag{18}$$

Remark 5.1. Analogous to [Ben-Tal and Zowe, 4] a second order directional derivative of f at x_0 in direction of $h \in X$ can be defined as

$$\nabla^2 f(x_0,h)(v,\gamma) = \lim_{t \to 0^+} \frac{f(x_0 + t\sqrt{\gamma}h + t^2 v) - f(x_0) - tf'(x_0)\sqrt{\gamma}h}{t^2} .$$

By Theorem 5.1, if f is twice Fréchet differentiable at x_0, then this limit exists and is given by

$$\nabla^2 f(x_0,h)(v,\gamma) = f'(x_0)v + \frac{\gamma}{2} f''(x_0)(h,h). \tag{19}$$

Notice that h is a parameter in which the expression is quadratic, but $\nabla^2 f(x_0,h)$ acts linearly on (v,γ). With this notation, the second order feasible cone can be described as

$$FC^{(2)}(Z;x_0,h) = \{(v,\gamma) \in X \times (0, \infty): \; \nabla^2 f(x_0,h)(v,\gamma) < 0\} .$$

Using the Minkowski-Farkas lemma [Girsanov, 5], the dual cone can easily be calculated.

Theorem 5.2. [Ledzewicz and Schaettler, 9] Let $x_0 \in Z$ and suppose f is twice Fréchet differentiable at x_0 with $f'(x_0) \neq 0$. Then

$$FC^{(2)}(Z;x_0,h)^* = \left\{(\lambda, \mu) \in X^* \times \mathbb{R} : \text{there exist } x_1 \leq 0 \text{ and } x_2 \geq 0 \text{ such that} \right.$$
$$\left. \begin{pmatrix} \lambda \\ \mu \end{pmatrix} = \begin{pmatrix} f'(x_0)^* & 0 \\ \frac{1}{2} f''(x_0)\,(h,h) & 1 \end{pmatrix} \begin{pmatrix} x_1 \\ x_2 \end{pmatrix} \right\}. \tag{20}$$

Remark 5.2. Let $x_0 \in Z$ and suppose f is twice Fréchet differentiable at x_0 with $f'(x_0)=0$. Then

$$FC^{(2)}(Z;x_0,h) = \emptyset \qquad \text{if } f''(x_0)(h,h) > 0$$

and

$$FC^{(2)}(Z;x_0,h) = X \times (0,\infty) \qquad \text{if } f''(x_0)(h,h) < 0.$$

6. Second Order Necessary Conditions for Optimality in the Nonregular Case

Avakov's results for the smooth case [Avakov, 3] can easily be derived from Theorem 3.1. In this case the inequality constraints Z_i, $i = 1,2,...,n$, are of the form $Z_i=\{x \in X: f_i(x) \leq 0\}$ for differentiable functionals f_i: $X \mapsto \mathbb{R}$, $i=1,...,n$, on X. The equality constraint is as in Problem 1 and assumption (3) will be guaranteed by

(3′) Suppose assumptions (A) of Section are satisfied for the equality constraint, i.e.

- (i) F is three times Fréchet differentiable at x_0,
- (ii) Im $F'(x_0)$ is closed,
- (iii) h satisfies $G(x_0,h)h = 0$, i.e. $F'(x_0)h = 0$ and $F''(x_0)(h,h) \in \text{Im } F'(x_0)$,
- (iv) the operator G is onto: $\text{Im } G(x_0,h) = \text{Im } F'(x_0) \times Y/\text{Im } F'(x_0)$.

We call this Problem 1′.

Corollary 5.1. [Avakov, 3] Suppose x_0 is a local minimum for Problem 1′ and assume that the functionals I and f_i, $i=1,...,n$ are twice Fréchet differentiable at x_0. Furthermore, let $J(x_0)=\{i: 1 \leq i \leq n, f_i(x_0)=0\}$ denote the active indices and define

$$N(x_0)=\Big\{ h \in E: \; I'(x_0)\, h \leq 0 \;\text{ and }\; f_i'(x_0)\, h \leq 0 \;\text{ for } i \in J(x_0) \Big\}. \tag{21}$$

Then for every $h \in N(x_0)$ for which assumption (2′) for the equality constraint Z_{n+1} is satisfied at x_0, there exist Lagrange multipliers $\nu_0(h) \leq 0$ and $\nu_i(h) \leq 0$ for $i \in J(x_0)$, $y_1^*(h) \in \text{Im } F'(x_0)^*$, and $y_2^*(h) \in \big(\text{Im } F'(x_0)\big)^{\perp}$, which do not all vanish, such that

$$\nu_0(h)\, I'(x_0) + \sum_{i \in J(x_0)} \nu_i(h) f_i'(x_0) + F'^*(x_0)\, y_1^*(h) + (F''(x_0)(h,\cdot))^*\, y_2^*(h) \equiv 0, \tag{22}$$

$$\nu_0(h)\, I''(x_0)(h,h) + \sum_{i \in J(x_0)} \nu_i(h) f_i''(x_0)(h,h) +$$

$$< y_1^*(h), F''(x_0)(h,h)> + \tfrac{1}{3} < y_2^*(h), F'''(x_0)(h,h,h)> \; \leq 0. \tag{23}$$

Furthermore, the following second order complementary slackness conditions hold:

$$\nu_0(h)<I'(x_0), h>=0 \quad \text{and} \quad \nu_i(h)< f_i'(x_0), h>=0 \qquad \text{for } i \in J(x_0). \tag{24}$$

REFERENCES

1. Avakov, E.R. (1985), Extremum conditions for smooth problems with equality-type constraints. U.S.S.R. Computational Mathematics and Mathematical Physics, Vol. 25, No.5, pp. 680-693.

2. Avakov, E.R. (1988) Necessary conditions for the minimum for the nonregular problems in Banach spaces. Maximum principle for abnormal problems of optimal control. Trudy Matematiceskovo Instituta AN – SSSR, Vol. 185, pp. 3-29.

3. Avakov, E.R. (1989) Necessary extremum conditions for smooth abnormal problems with equality and inequality-type constraints, Mathematicheskie Zametki, Vol. 45, No.6, pp.3-11.

4. Ben-Tal, A. and Zowe, J. (1982) A unified theory of first and second order conditions for extremum problems in topological vector spaces, Mathematical Programming Study, Vol. 19, pp. 39-76.

5. Girsanov, I.V. (1972) Lectures on Mathematical Theory of Extremum Problems, Lecture Notes in Economics and Mathematical Systems, Berlin,Germany, Springer Verlag.

6. Ioffe, A.D. and Tikhomirov, W.M. (1979) Theory of Extremal Problems, North Holland, Amsterdam, Holland.

7. Ledzewicz, U., (1993) On abnormal optimal control problems with mixed equality and inequality constraints, Journal of Mathematical Analysis and Applications, Vol.173, No.1, pp.18-42.

8. Ledzewicz, U. (1993) Extension of the local Maximum Principle to abnormal optimal control problems, Journal of Optimization Theory and Applications, Vol.77, No.3, pp. 661-681.

9. Ledzewicz, U. and Schaettler, H. (1993) Second order conditions for extremum problems with nonregular equality constraints, to appear.

10. Ledzewicz, U. and Schaettler, H. (1993) Pareto optimality conditions for abnormal optimization and optimal control problems, this volume.

16 Existence, Approximation, and Suboptimality Conditions for Minimax Control of Heat Transfer Systems with State Constraints

Boris S. Mordukhovich Wayne State University, Detroit, Michigan

Kaixia Zhang Wayne State University, Detroit, Michigan

Abstract. In this paper, we consider a minimax control problem for heat transfer systems with uncertain perturbations and pointwise state constraints. We study properties of mild solutions to such parabolic systems with Dirichlet boundary controls and prove an existence theorem for optimal solutions to the minimax problem. We develop penalized procedures to approximate state constraints and then establish convergence results for these approximations. Using a variational analysis, we obtain necessary optimality conditions for approximating solutions which ensure suboptimality conditions in the original minimax problem with state constraints.

Keywords: Heat transfer equations, uncertain disturbances, Dirichlet boundary controls, minimax criterion, state constraints, approximation, and suboptimality.

1991 AMS Subject Classification:

Primary:49K20, 49K35; Secondary: 49J20, 35K50

This paper is concerned with a minimax control problem for multivariate heat transfer equations with distributed uncertain perturbations (disturbances) and Dirichlet boundary controls. We study such a problem under hard (pointwise) control and state constraints. Parts of our motivations come from Mordukhovich [10] and Mordukhovich and Zhang [11] where some feedback control problems have been considered in connection with practical applications.

We define an optimal solution to the minimax problem as a *saddle point* of the cost functional (relative to differential games) and prove an existence theorem in the class of measurable controls and disturbances under natural assumptions. To overcome difficulties connected with discontinuous Dirichlet boundary conditions, we use a concept of *mild solutions* in the framework of the semigroup model developed by Washburn [15] and Lasiecka and Triggiani [7].

To study necessary conditions for optimal solutions to the minimax problem, we split the original problem into two interrelated optimal control problems with separated distributed perturbations and Dirichlet boundary controllers under moving state constraints. To relax state constraints, we develop penalization procedures involving C^∞-approximations of multi-valued maximal monotone operators. In this way, we obtain convergence results and necessary optimality conditions for approximating disturbances and controls which appear to be suboptimal solutions to the original state-constrained problem.

We refer the reader to Barbu [1, 2], Barbu and Precupanu [3], Bonnans and Tiba [4], Friedman [5], He [6], Tiba [14], and bibliographies therein for related approaches and results for the cases of distributed (domain) controls as well as Neumann boundary control problems in the framework of variational inequalities. The case of Dirichlet boundary controls offers the lowest regularity properties and still appear to be the most difficult and challenging. Some results for such problems are available in the recent book of Barbu [2] in the settings with continuous Dirichlet boundary controllers. More detailed considerations and generalizations of the results presented in this paper can be found in [12].

Now let us consider the heat transfer system

$$\begin{cases} \dfrac{\partial y}{\partial t} = \Delta y + Bw + f \ \text{ in } \ Q := (0,T) \times \Omega, \\ y(0,x) = y_0(x), \ x \in \Omega, \\ y(t,\xi) = u(t,\xi) \text{ a.e. } (t,\xi) \in \Sigma := (0,T] \times \Gamma \end{cases} \tag{1}$$

under the following pointwise constraints:

$$a \le y(t,x) \le b \text{ a.e. } (t,x) \in \Omega, \tag{2}$$

$$c \le w(t,x) \le d \text{ a.e. } (t,x) \in Q,$$

$$\nu \le u(t,\xi) \le \mu \text{ a.e. } (t,\xi) \in \Sigma$$

where each of the intervals $[a,b]$, $[c,d]$, and $[\nu,\mu]$ contains 0 and $\Omega \subset \mathbf{R}^N$ is a bounded open set with smooth enough boundary Γ.

Let $X := L^2(\Omega;\mathbf{R})$, $U := L^2(\Gamma;\mathbf{R})$, and $W := L^2(\Omega;\mathbf{R})$ be, respectively, spaces of states, controls, and disturbances. (In what follows, we remove $\mathbf{R}$ from the latter and similar space notation for real-valued functions). Denote by

$$U_{ad} := \{u = u(\cdot) \in L^p(0,T;U) \mid \nu \le u(t,\xi) \le \mu \text{ a.e. } (t,\xi) \in \Sigma\}$$

the set of all admissible controls, and by

$$W_{ad} := \{w = w(\cdot) \in L^p(0,T;W) \mid c \le w(t,x) \le d \text{ a.e. } (t,x) \in Q, \ \iint_Q w(t,x)dtdx = \beta \ne 0\}$$

the set of all admissible disturbances.

Throughout the paper we impose the following hypotheses:

(H1) Δ is a Laplacian in Ω.

(H2) One has $a \le y_0(x) \le b$ for $x \in \Omega$, $\Delta y_0 \in X$, and $f \in L^\infty(Q)$.

(H3) The linear operator $B : L^p(0,T;W) \to L^p(0,T;X)$ is bounded.

We always assume that $p > (N+2)/2$, $p \ge 2$. Later we make more restrictions on number p.

We are going to study system (1) in the framework of mild solutions. Let us consider the *Dirichlet operator* D defined by $z = Du$ where z satisfies

$$\Delta z = 0 \text{ in } Q,$$

$$z(t,\xi) = u(t,\xi), \quad (t,\xi) \in \Sigma.$$

By a *mild solution* to system (1) corresponding to $(u,w) \in L^p(0,T;U) \times L^p(0,T;W)$ we mean a continuous $y : [0,T] \to X$ such that

$$\begin{aligned} &y(t) = S(t)y_0 + \int_0^t S(t-\tau)((Bw)(\tau) + f(\tau))d\tau + \int_0^t \Delta S(t-\tau)(Du)(\tau)d\tau = \\ &S(t)y_0 + \int_0^t S(t-\tau)((Bw)(\tau) + f(\tau))d\tau + \\ &\int_0^t \Delta^{3/4+\delta} S(t-\tau)\Delta^{1/4-\delta}(Du)(\tau)d\tau \quad \forall t \in [0,T] \end{aligned} \tag{3}$$

where $S(t)$ is a strongly continuous analytic semigroup generated by Δ and $\delta \in (0, 1/4]$.

We refer the reader to [2, 7, 12, 15] and bibliographies therein for various properties of mild solutions. It is essential for this paper that the assumptions imposed ensure the existence and uniqueness of a mild solution to (1) for any admissible u and w. We assume that system (1), (2) has at least one feasible triple (u,w,y) in the sense defined.

On the set of such feasible solutions, we consider the cost functional

$$J(u,w) := \iint_Q g(t,x,y(t,x))dtdx + \iint_Q \varphi(t,x,w(t,x))dtdx + \iint_\Sigma h(t,\xi,u(t,\xi))dtd\xi. \quad (4)$$

Hereafter we use the following assumptions on the integrands in (4) where $1/q + 1/p = 1$.

(H4a) For any bounded set $V \subset \mathbf{R}$ there exists a nonnegative function $\eta \in L^1(0,T)$ such that $g(t,x,y)$ is measurable in $(t,x) \in Q$, continuous in $y \in V$, and

$$|g(t,x,y)| \leq \eta(t) \ \ \forall y \in V \ \text{ a.e. } (t,x) \in Q.$$

(H4b) $g(t,x,y)$ is differentiable in y and there exists a nonnegative function $\eta_1(t) \in L^q(0,T)$ such that

$$|\frac{\partial g}{\partial y}(t,x,y)| \leq \eta_1(t) \ \ \forall y \in V \ \text{ a.e. } (t,x) \in Q.$$

(H5a) $\varphi(t,x,w)$ is measurable in $(t,x) \in Q$, continuous and *concave* in $w \in [c,d]$, and

$$|\varphi(t,x,w)| \leq \kappa(t) \ \ \forall w \in [c,d] \ \text{ a.e. } (t,x) \in Q$$

for some nonnegative function $\kappa \in L^1(0,T)$.

(H5b) $\varphi(t,x,w)$ is differentiable in w and there exists a nonnegative function $\kappa_1(t) \in L^q(0,T)$ such that

$$|\frac{\partial \varphi}{\partial w}(t,x,w)| \leq \kappa_1(t) \ \ \forall w \in [c,d] \ \text{ a.e. } (t,x) \in Q.$$

(H6a) $h(t,\xi,u)$ is measurable in $(t,\xi) \in \Sigma$, continuous and *convex* in $u \in [\nu,\mu]$, and

$$|h(t,\xi,u)| \leq \gamma(t) \ \ \forall u \in [\nu,\mu] \ \text{ a.e. } (t,\xi) \in \Sigma$$

for some nonnegative function $\gamma \in L^1(0,T)$.

(H6b) $h(t,\xi,u)$ is differentiable in u and there exists a nonnegative function $\gamma_1(t) \in L^q(0,T)$

such that

$$|\frac{\partial h}{\partial u}(t,\xi,u)| \le \gamma_1(t) \ \ \forall u \in [\nu,\mu] \ \text{ a.e. } (t,\xi) \in \Sigma.$$

The main concern of this paper is the following minimax control problem:

(P) Find an admissible control $\bar{u} \in U_{ad}$ and an admissible disturbance $\bar{w} \in W_{ad}$ such that $(\bar{u}, \bar{w})$ is a *saddle point* for the functional $J(u,w)$ subject to system (1) and state constraints (2). This means that

$$J(\bar{u},w) \le J(\bar{u},\bar{w}) \le J(u,\bar{w}) \ \ \forall u \in U_{ad} \ \& \ w \in W_{ad}$$

under relations (1) and (2). Such a pair $(\bar{u},\bar{w})$ is called an *optimal solution* to (P).

Let us define an operator L from $L^p(0,T;U)$ into $L^r(0,T;H^{1/2-\varepsilon}(\Omega))$ by the formula

$$Lu = (Lu)(t) := \Delta \int_0^t S(t-\tau)Du(\tau)d\tau = \int_0^t \Delta^{3/4+\delta}S(t-\tau)\Delta^{1/4-\delta}Du(\tau)d\tau \tag{5}$$

where $p,\ r \in [1,\infty]$, $\delta \in (0,1/4]$, and $\varepsilon \in (0,1/2]$. Here $H^{1/2-\varepsilon}(\Omega) \subset L^2(\Omega) = X$ is the Sobolev space which norm $\|y\|_{1/2-\varepsilon}$, being stronger than $\|y\|_X$, can be defined as $\|y\|_{1/2-\varepsilon} := \|\Delta^{1/4-\varepsilon/2}y\|_X$; see [9, p. 21].

It is well known that the operator L in (5) may be *unbounded* for some p and r. However, this operator enjoys nice regularity (continuity) properties for p big enough, as one can see from the following assertion. Similar but somewhat different results are proved in Lasiecka and Triggiani [8, Theorem 2.5].

Proposition 1. *Let $p > 4/\varepsilon$ for any $\varepsilon \in (0,1/2]$. Then one has $Lu \in C(0,T;H^{1/2-\varepsilon}(\Omega))$ for any $u \in L^p(0,T;U)$. Moreover, the operator $L : L^p(0,T;U) \to C(0,T;H^{1/2-\varepsilon}(\Omega))$ is linear and continuous.*

Proof. Obviously L is linear. To show that L is continuous, we use the following estimates (see [7, 15]):

$$\|\Delta^{\delta} D\|_U \leq M, \ \|\Delta^{3/4+\delta} S(t)\| \leq \frac{M_1}{t^{3/4+\delta}} \ \forall \delta \in (0, 1/4], \ t \in [0, T].$$

It is clear that for any $t \in [0, T]$ one has

$$\|(Lu)(t)\|_{1/2-\varepsilon} = \| \int_0^t \Delta^{1/4-\varepsilon/2} \Delta S(t-\tau) Du(\tau) d\tau \|_X =$$
$$\| \int_0^t \Delta^{1-\varepsilon/4} S(t-\tau) \Delta^{1/4-\varepsilon/4} Du(\tau) d\tau \|_X \leq M M_1 \int_0^t (t-\tau)^{-(1-\varepsilon/4)} \|u\|_U d\tau \leq$$
$$M M_1 (\int_0^t (t-\tau)^{-(1-\varepsilon/4)q} d\tau)^{1/q} \|u\|_{L^p(0,T;U)}$$

with $1/p + 1/q = 1$. Since $p > 4/\varepsilon$ infers $q < \frac{4}{4-\varepsilon}$, we get

$$\|(Lu)(t)\|_{1/2-\varepsilon} \leq M M_1 (\frac{1}{1-(1-\varepsilon/4)q})^{1/q} t^{\frac{1-(1-\varepsilon/4q}{q}} \|u\|_{L^p(0,T;U)}. \tag{6}$$

Let us prove that $Lu \in C(0, T; H^{1/2-\varepsilon}(\Omega))$, i.e., $(Lu)(t)$ is continuous at any point $t_0 \in [0, T]$ as a $H^{1/2-\varepsilon}(\Omega)$-valued function. Indeed, taking for definiteness $t \geq t_0$, one has

$$(Lu)(t) - (Lu)(t_0) = \int_{t_0}^t \Delta S(t-\tau) Du(\tau) d\tau +$$
$$(S(t-t_0) - I) \int_0^{t_0} \Delta S(t-\tau) Du(t) d\tau.$$

The latter implies that

$$\|(Lu)(t) - (Lu)(t_0)\|_{1/2-\varepsilon} \to 0 \ \text{ as } \ t \to t_0$$

by virtue of (6) and the strong continuity of $S(\cdot)$.

Moreover, from (6) and the definition of the norm in $C(0, T; H^{1/2-\varepsilon}(\Omega))$ we immediately get inequality

$$\|Lu\|_{C(0,T;H^{1/2-\varepsilon}(\Omega))} \leq K \|u\|_{L^p(0,T;U)} \ \text{ for } \ K := M M_1 (\frac{1}{1-(1-\varepsilon/4)q})^{1/q} T^{1-(1-\varepsilon/4)q} q$$

which ensures the required continuity of L. This ends the proof. □

We now establish the following existence theorem for optimal solutions to problem (P).

Theorem 2. *Let hypotheses* (H1)–(H3) *and* (H4a)–(H6a) *hold and let, in addition, the integrand g be linear in y and p satisfy the assumption in Proposition* 1. *Then the functional $J(u,w)$ has a saddle point $(\bar{u},\bar{w}) \in U_{ad} \times W_{ad}$ subject to system* (1). *Moreover, if the corresponding trajectory for* (1) *satisfies state constraints* (2), *then $(\bar{u},\bar{w})$ is an optimal solution to the original minimax problem* (P).

Proof. Let us consider the functional $J(u,w)$ defined on the set $U_{ad} \times W_{ad} \subset L^p(0,T;U) \times L^p(0,T;W)$. Note that both U_{ad} and W_{ad} are convex and weakly compact in $L^p(0,T;U)$ and $L^p(0,T;W)$, respectively. Moreover, one can always use the sequential weak topologies on these spaces by virtue of their reflexivity.

Further, one can see that J is convex-concave on $U_{ad} \times W_{ad}$ due to convexity of h in u, concavity of φ in w, and linearity of g in y which is linearly dependent on (u,w). Let us show that J is weakly lower semicontinuous with respect to u for any fixed w and it is weakly upper semicontinuous with respect to w for any fixed u. For definiteness, we consider only the first case; the second one is symmetric.

Indeed, let $u_n \to \tilde{u}$ weakly in $L^p(0,T;U)$ as $k \to \infty$. According to the classical Masur theorem, there is a sequence of convex combinations of u_n which converges to $\tilde{u}$ strongly in $L^p(0,T;U)$. It is clear that this sequence also converges to $\tilde{u}$ strongly in $L^p(\Sigma)$. Now employing (H6a), we get

$$\iint_\Sigma h(t,\xi,\tilde{u}(t,\xi))dtd\xi \le \liminf_{n\to\infty} \iint_\Sigma h(t,\xi,u_n(t,\xi))dtd\xi.$$

Due to the linearity of operator (3) and a classical functional result (see, e.g., [3, Proposition 2.8 on p. 37], we can conclude from Proposition 1 that the mild solutions y_n corresponding to u_n (under fixed w) weakly in $C(0,T;H^{1/2-\varepsilon}(\Omega))$ converge to the solution $\tilde{y}$ corresponding to $\tilde{u}$. Moreover, there exists a subsequence of $\{y_n\}$ which converges to $\tilde{y}(t,x)$ *almost everywhere* in Q; see [12] for more details. This implies

$$\int\int_Q g(t,x,\tilde{y}(t,x))dtdx = \lim_{n\to\infty} \int\int_Q g(t,x,y_n(t,x))dtdx$$

by virtue of Lebesque's limiting integral theorem under assumption (H4a). Therefore, J is weakly lower semicontinuous with respect to u (and weakly upper semicontinuous with respect to w).

Now the existence of a saddle point $(\bar{u},\bar{w})$ for J subject to system (1) follows from the classical minimax theorem in infinite dimensions (see, e.g., [3, Theorem 3.5 on p. 162]). Obviously, $(\bar{u},\bar{w})$ is an optimal solution to the original minimax problem (P) if the corresponding trajectory $\bar{y}$ satisfies state constraints (2). □

Next we consider effective procedures which allow us to approximate a given solution $(\bar{u},\bar{w})$ to the original minimax problem with state constraints by families of optimal control problems without state constraints. Developing this line, we are going to obtain necessary optimility conditions for approximating controls and disturbances which appear to be *suboptimal solutions* to the original problem. Note that for these purposes we don't need to assume any more that g is linear in y.

Let us split the original system (1) into two subsystems with separated disturbances and controls. The first system

$$\begin{cases} \dfrac{\partial y_1}{\partial t} = \Delta y_1 + Bw + f \ \text{ in } \ Q = (0,T) \times \Omega, \\ y_1(0,x) = y_0(x) \in H_0^1(\Omega) \cap H^2(\Omega), \\ y_1(t,\xi) = 0, \ (t,\xi) \in \Sigma \end{cases} \tag{7}$$

has zero boundary conditions and depends only on disturbances. The second one

$$\begin{cases} \dfrac{\partial y_2}{\partial t} = \Delta y_2 \ \text{ in } \ Q = (0,T) \times \Omega, \\ y_2(0,x) = 0, \\ y_2(t,\xi) = u(t,\xi), \ (t,\xi) \in \Sigma \end{cases} \tag{8}$$

is generated by boundary controls and does not involve disturbances. It is easy to see that for any $(u,w) \in U_{ad} \times W_{ad}$ one has

$$y(t,x) = y_1(t,x) + y_2(t,x) \tag{9}$$

for the corresponding trajectories of systems (1), (7), and (8).

Let $(\bar{u}, \bar{w}, \bar{y})$ be an optimal triple in problem (P). Then it follows from (3) and (9) that $\bar{y}_1$ and $\bar{y}_2$ are trajectories for systems (7) and (8), respectively. Now we define the cost functionals

$$J_1(w) := \iint_Q [g(t,x,y_1(t,x) + \bar{y}_2(t,x)) + \varphi(t,x,w(t,x))]dtdx \tag{10}$$

for disturbances w and

$$J_2(u) := \iint_Q g(t,x,\bar{y}_1(t,x) + y_2(t,x))dtdx + \iint_\Sigma h(t,\xi,u(t,\xi))dtd\xi \tag{11}$$

for boundary controls u.

Let us consider two optimization problems corresponding to the cost functionals introduced. The first one is:

(P_1) maximize $J_1(w)$ in (10) over $w \in W_{ad}$ subject to system (7) and the state constraints

$$a - \bar{y}_2(t,x) \le y_1(t,x) \le b - \bar{y}_2(t,x) \text{ a.e. } (t,x) \in Q. \tag{12}$$

The second problem is:

(P_2) minimize $J_2(u)$ in (11) over $u \in U_{ad}$ subject to system (8) and the state constraints

$$a - \bar{y}_1(t,x) \le y_2(t,x) \le b - \bar{y}_1(t,x) \text{ a.e. } (t,x) \in Q. \tag{13}$$

The following assertion shows that, from the viewpoint of necessary optimality conditions, the original minimax problem (P) can be splitted into two optimization problems (P_1) and (P_2) separated on disturbances and controls.

Proposition 3. *Let $(\bar{u}, \bar{w}, \bar{y})$ be an optimal solution to problem (P), and let $\bar{y}_1$ and $\bar{y}_2$ be the corresponding trajectories for systems* (7) *and* (8). *Then $(\bar{w}, \bar{y}_1)$ solves problem (P_1) and $(\bar{u}, \bar{y}_2)$ solves problem (P_2).*

Proof. This follows directly from definition (3) of solutions to (P) and relationship (9) generated by the linearity of the original system (1). □

Now we deal with the splitting subproblems (P_1) and (P_2). To approximate state constraints (12) and (13), we intend to use penalization procedures. Let $\alpha \subset \mathbf{R} \times \mathbf{R}$ be a maximal monotone operator of the form

$$\alpha(r) = \begin{cases} [0,\infty) & \text{if } r = b, \\ (-\infty, 0] & \text{if } r = a, \\ 0 & \text{if } a < r < b, \\ \emptyset & \text{if either } r < a \text{ or } r > b. \end{cases}$$

Suppose that $\alpha_\epsilon(\cdot)$, $\epsilon > 0$, is a family of smooth approximations of the operator $\alpha(\cdot)$ with the following properties:

$$\begin{cases} \alpha_\epsilon(r) \to 0 & \text{if } a < r < b, \epsilon \to 0, \\ \alpha_\epsilon(r) \to -\infty & \text{if } r < a, \epsilon \to 0, \\ \alpha_\epsilon(r) \to \infty & \text{if } r > b, \epsilon \to 0, \\ \alpha'_\epsilon(r) \geq 0 & \text{if } r \in \mathbf{R}. \end{cases} \tag{14}$$

By Barbu [1, 2], we can always build approximations (14) with bounded derivatives $\alpha'_\epsilon(r)$ for any $r \in \mathbf{R}$ and $\epsilon \to 0$. Let ρ be a function satisfying the requirememts:

(i) $\rho(t,x) \geq 0$, $\rho(t,x) \in C^\infty(\mathbf{R}^{N+1})$ with the support containing in the unit closed ball of $\mathbf{R}^{N+1}$;

(ii) $\iint \rho(t,x)dtdx = 1$.

Let $\rho_\epsilon = (1/\epsilon)^{N+1}\rho(t/\epsilon, x/\epsilon)$. It is well known that

$$\bar{y}_{2\epsilon}(t,x) := (\rho_\epsilon * \bar{y}_2)(t,x) = \iint \rho_\epsilon(t-\tau, x-s)\bar{y}_2(\tau,s)d\tau ds$$

is a C^∞ function and

$$\bar{y}_{2\epsilon} \to \bar{y}_2 \text{ in } C(0,T;X) \text{ as } \epsilon \to 0.$$

Now we consider a parametric family of penalized problems as follows:

$(P_{1\epsilon})$ maximize $J_{1\epsilon}(w,y_1) := \iint_Q [g(t,x,y_1(t,x) + \bar{y}_2(t,x)) + \varphi(t,x,w)]dtdx - \|w - \bar{w}\|^p_{L^p(0,T;W)}$ subject to

$$\begin{cases} \dfrac{\partial y_1}{\partial t} = \Delta y_1 + Bw + f + \alpha_\epsilon(y_1 + \bar{y}_{2\epsilon}) \text{ in } Q, \\ y_1(0,x) = y_0(x) \in H_0^1(\Omega) \cap H^2(\Omega), \\ y_1(t,\xi) = 0, \ (t,\xi) \in \Sigma, \\ w \in W_{ad}. \end{cases} \tag{15}$$

Similarly to corresponding arguments in [1, 5, 12], one can establish the following results.

Lemma 4. *$(P_{1\epsilon})$ has at least one optimal pair $(w_\epsilon, y_{1\epsilon})$.*

Lemma 5. *Suppose that $(\bar{w}, \bar{y}_1)$ is an optimal pair to (P_1). Then one has*

$$w_\epsilon \to \bar{w} \ \textit{strongly in} \ L^p(0,T;W),$$

$$y_{1\epsilon} \to \bar{y}_1 \ \textit{uniformly in} \ C(0,T;X) \ \&$$

$$J_{1\epsilon}(w_\epsilon, y_{1\epsilon}) \to J_1(\bar{w}, \bar{y}_1) \ \textit{as} \ \epsilon \to 0.$$

So, optimal solutions of approximating penalized problems $(P_{1\epsilon})$ always exist and appear to be suboptimal solutions to problem (P_1) with state constraints. Let us obtain necessary optimality conditions for w_ϵ which provide a useful information about the worst disturbances in the original minimax problem.

Consider the following linear parabolic system adjoint to the perturbed system (15):

$$\begin{cases} \dfrac{\partial \psi}{\partial t} + \Delta\psi + \alpha_\epsilon'(y_{1\epsilon} + \bar{y}_{2\epsilon})\psi = -\dfrac{\partial g}{\partial y}(t, x, y_{1\epsilon} + \bar{y}_2), \\ \psi(T, x) = 0, \ x \in \Omega, \\ \psi(t, \xi) = 0, \ (t, \xi) \in \Sigma. \end{cases} \tag{16}$$

Due to the boundedness of $\alpha_\epsilon'(y_{1\epsilon} + \bar{y}_{2\epsilon})$ in Q, the operator $\Delta + \alpha'(y_{1\epsilon} + \bar{y}_{2\epsilon})$ generates an analytic semigroup; see, e.g., [13, p. 80]. This implies that (16) admits a unique mild solution for any $y_{1\epsilon} \in C(0,T;X)$; cf. [2, pp. 31–33].

Theorem 6. *Let $(w_\epsilon, y_{1\epsilon})$ be an optimal pair to problem $(P_{1\epsilon})$ for any $\epsilon > 0$, and let $\psi_\epsilon \in C(0,T;X)$ be the corresponding solution to the adjoint system (16) under assumptions (H1)–(H5). Then one has*

$$\iint_Q (\psi_\epsilon B w - \frac{\partial \psi}{\partial w}(t, x, w_\epsilon) w) dt dx - p \int_0^T \|w_\epsilon - \bar{w}\|^{p-2} (\int_\Omega (w_\epsilon - \bar{w}) w dx) dt \leq 0 \tag{17}$$

for any $w \in L^p(0,T;W)$ *such that* $w_\epsilon + \theta w \in W_{ad}$ *for all* $\theta \in (0,\delta_0)$ *with some* $\delta_0 > 0$. *Moreover,* w_ϵ *satisfies the following bang-bang relations:*

$$w_\epsilon(t,x) = c \ \text{ a.e. } \{(t,x) \in Q | (B^\star \psi_\epsilon)(t,x) - \frac{\partial \varphi}{\partial w}(t,x,w_\epsilon) - p\|w_\epsilon - \bar{w}\|_W^{p-2}(w_\epsilon(t,x) - \bar{w}(t,x)) < 0\},$$

$$w_\epsilon(t,x) = d \ \text{ a.e. } \{(t,x) \in Q | (B^\star \psi_\epsilon)(t,x) - \frac{\partial \varphi}{\partial w}(t,x,w_\epsilon) - p\|w_\epsilon - \bar{w}\|_W^{p-2}(w_\epsilon(t,x) - \bar{w}(t,x)) > 0\}$$

where $B^\star$ *is the adjoint operator to* B.

Proof. Due to optimality of $(w_\epsilon, y_{1\epsilon})$ in $(P_{1\epsilon})$, we have

$$\limsup_{\theta \downarrow 0} \frac{J_{1\epsilon}(w_\epsilon + \theta w)) - J_{1\epsilon}(w_\epsilon)}{\theta} \leq 0$$

for any w in the conditions of the theorem. For any such w, let us consider the auxiliary linear system

$$\begin{cases} \dfrac{\partial z_{1\epsilon}}{\partial t} = \Delta z_{1\epsilon} + Bw + \alpha'_\epsilon(y_{1\epsilon} + \bar{y}_{2\epsilon}) z_{1\epsilon}, \\ z_{1\epsilon}(0,x) = 0, \ x \in \Omega, \\ z_{1\epsilon}(t,\xi) = 0, \ (t,\xi) \in \Sigma. \end{cases}$$

which has a unique mild solution; see above. Moreover, we can show that

$$\frac{y_{1\epsilon}(w_\epsilon + \theta w) - y_{1\epsilon}(w_\epsilon)}{\theta} \to z_{1\epsilon} \ \text{ in } \ C(0,T;X) \ \text{ as } \ \theta \to 0.$$

Involving these facts in a variational analysis similar to Friedman [5], we get inequality (17). The bang-bang conditions follow from (17) by taking into account that for any $v \in W_{ad}$ the disturbance $w = v - w_\epsilon$ is available in (17) with $\delta_0 = 1$. □

Now let us develop an approximation procedure for problem (P_2). We consider a parametric family of penalized problems as follows:

$(P_{2\epsilon})$ Minimize $J_{2\epsilon}(u, y_2) := \iint_Q g(t, x, \bar{y}_1(t, x) + y_2(t, x))dtdx + \iint_\Sigma h(t, \xi, u(t, \xi))dtd\xi +$

$\|u - \bar{u}\|^p_{L^p(0,T;U)}$ over $u \in U_{ad}$ subject to

$$\begin{cases} \dfrac{\partial y_2}{\partial t} = \Delta y_2 + \alpha_\epsilon(\bar{y}_1 + y_2) \text{ in } Q, \\ y_2(0, x) = 0, \ x \in \Omega, \\ y_2(t, \xi) = u(t, \xi), \ (t, \xi) \in \Sigma. \end{cases} \tag{18}$$

Following the above line, we consider solutions of the nonlinear system (18) in the mild sense, i.e., $y_2 \in C(0, T; X)$ such that

$$y_2(t) = \int_0^t S(t - \tau)\alpha_\epsilon(\bar{y}_1 + y_2)d\tau + \int_0^t \Delta S(t - \tau)(Du)(\tau)d\tau.$$

According to Proposition 3, the pair $(\bar{u}, \bar{y}_2)$ is an optimal solution to problem (P_2). Moreover, it is a feasible solution to the approximation problem $(P_{2\epsilon})$ for any ϵ. Using constructions of Theorem 2 proved above, we can establish the existence of optimal solutions to perturbed problems $(P_{2\epsilon})$.

Lemma 7. *Let assumptions* (H1)–(H3), (H4a), *and* (H6a) *be fulfilled and let* $p > 8$. *Then the approximation problem* $(P_{2\epsilon})$ *admits an optimal solution for any* $\epsilon > 0$.

Hereafter we always suppose that the assumptions of Lemma 7 hold and consider a given sequences of optimal solutions $(u_\epsilon, y_{2\epsilon})$ to $(P_{2\epsilon})$. Involving compactness and penalization arguments as well as properties of the Dirichlet operator D (see [12]), we can prove the following approximation result ensuring the suboptimality of controls u_ϵ in the original minimax problem.

Lemma 8. *One has*

$$u_\epsilon \to \bar{u} \ \textit{strongly in } L^p(0, T; U),$$

$$y_{2\epsilon} \to \bar{y}_2 \; uniformly \; in \; C(0,T;X) \;\; \&$$

$$J_{2\epsilon}(u_\epsilon, y_{2\epsilon}) \to J_2(\bar{u}, \bar{y}_2) \;\; as \;\; \epsilon \to 0.$$

Let us obtain necessary conditions for optimal controls u_ϵ in approximation problems $(P_{2\epsilon})$ as $\epsilon > 0$ which provide suboptimality conditions for the original state-constrained problem. For any ϵ, let us consider the auxiliary linear system with Dirichlet boundary controls:

$$\begin{cases} \dfrac{\partial z_{2\epsilon}}{\partial t} - \Delta z_{2\epsilon} - \alpha'_\epsilon(\bar{y}_1 + y_{2\epsilon}) z_{2\epsilon} = 0 \text{ in } Q, \\ z_{2\epsilon}(0,x) = 0, \;\; x \in \Omega, \\ z_{2\epsilon}(t,\xi) = u(t,\xi), \;\; (t,\xi) \in \Sigma. \end{cases} \tag{19}$$

Using the Dirichlet map D_ϵ and analytic semigroup $S_{2\epsilon}$ generated by the operator $\Delta + \alpha'_\epsilon(\bar{y}_1 + y_{2\epsilon})$ and boundary controls $u \in L^p(0,T;U)$, we get a unique mild solution $z_{2\epsilon} \in C(0,T;X)$ to (19) by the representation

$$z_{2\epsilon}(t,x) = \int_0^t (\Delta + \alpha'_\epsilon(\bar{y}_1 + y_{2\epsilon})) S_{2\epsilon}(t-\tau)(D_\epsilon u)(\tau) d\tau;$$

cf. [7, 15]. Moreover, Proposition 1 ensures the continutity of the operator

$$(L_\epsilon u)(t,x) := z_{2\epsilon}(t,x) \tag{20}$$

from $L^p(0,T;U)$ into $C(0,T;X)$ as $p > 8$.

Now let us consider the adjoint operator $L^*_\epsilon : C^*(0,T;X) \to L^q(0,T;U)$ to L_ϵ in (20) whose continuity follows from Proposition 1 when $q < 7/8$. One can check that L^*_ϵ admits the representation

$$(L^*_\epsilon v)(t,\xi) = -(\int_t^T D^*_\epsilon[(\Delta + \alpha'_\epsilon(\bar{y}_1 + y_{2\epsilon}))S_{2\epsilon}]^*(\tau - t) y(\tau) d\tau)(\xi). \tag{21}$$

In the results stated below, we use this representation for the functions

$$v_\epsilon = v_\epsilon(t,x) := \frac{\partial g}{\partial y}(t,x,\bar{y}_1(t,x) + y_{2\epsilon}(t,x)), \quad \epsilon > 0, \tag{22}$$

belonging to the space $L^q(0,T;X) \subset C^\star(0,T;X)$ by virtue of (H4b).

Now we are able to provide necessary optimality conditions for approximation problems $(P_{2\epsilon})$.

Theorem 9. *Given $\epsilon > 0$, let $(u_\epsilon, y_{2\epsilon})$ be an optimal solution to problem $(P_{2\epsilon})$ under assumptions* (H1)–(H4), (H6), $p > 8$, *and let $L_\epsilon^\star$ and v_ϵ be defined in* (21), (22). *Then one has*

$$\iint_\Sigma [(L_\epsilon^\star v_\epsilon)(t,\xi)u(t,\xi) + \frac{\partial h}{\partial u}(t,\xi,u_\epsilon(t,\xi))u(t,\xi)]dtd\xi + \tag{23}$$

$$p\int_0^T \|u_\epsilon - \bar{u}\|_U^{p-2}(\int_\Gamma (u_\epsilon - \bar{u})u(t,\xi)d\xi)dt \le 0$$

for any $u \in L^p(0,T;U)$ such that $u_\epsilon + \theta u \in U_{ad}$ for all $\delta \in (0,\delta_0)$ with some $\delta_0 > 0$. Moreover, u_ϵ satisfies the following bang-bang conditions:

$$u_\epsilon(t,\xi) = \nu \text{ for a.e. } (t,\xi) \in \Sigma \text{ with } \tfrac{\partial h}{\partial u}(t,\xi,u_\epsilon) - (L_\epsilon^\star v_\epsilon)(t,\xi) + p\|u_\epsilon - \bar{u}\|_U^{p-2}(u_\epsilon - \bar{u}) < 0,$$

$$u_\epsilon(t,\xi) = \mu \text{ for a.e. } (t,\xi) \in \Sigma \text{ with } \tfrac{\partial h}{\partial u}(t,\xi,u_\epsilon) - (L_\epsilon^\star v_\epsilon)(t,\xi) + p\|u_\epsilon - \bar{u}\|_U^{p-2}(u_\epsilon - \bar{u}) > 0.$$

Proof. Due to the optimality of $(u_\epsilon, y_{2\epsilon})$ in problem $(P_{2\epsilon})$, one has

$$\frac{J_{2\epsilon}(u_\epsilon + \theta u) - J_{2\epsilon}(u_\epsilon)}{\theta} \ge 0 \text{ as } \theta > 0$$

for any u satisfying the assumptions in (23). Now using a basic variational analysis and properties of mild solutions, we get

$$\iint_Q \frac{\partial g}{\partial y}(t,x,\bar{y}_1(t,x) + y_{2\epsilon}(t,x))z_{2\epsilon}(t,x)dtdx + \iint_\Sigma \frac{\partial h}{\partial u}(t,\xi,u_\epsilon(t,\xi))u(t,\xi)dtd\xi +$$

$$p\int_0^T \|u_\epsilon - \bar{u}\|_U^{p-2}(\int_\Gamma (u_\epsilon - \bar{u})u(t,\xi)d\xi)dt \geq 0$$

where $z_{2\epsilon}$ is a solution to the Dirichlet boundary problem (19) with given u; see [12] for more details. The latter implies the necessary condition (23) by virtue of constructions (20)–(22). The bang-bang conditions of the theorem follow directly from (23); cf. the proof of Theorem 6 above. □

In conclusion of the paper, let us formulate the result which summarizes approximation and necessary suboptimality conditions for the original minimax problem.

Theorem 10. *Let* $(\bar{u}, \bar{w})$ *be an optimal solution to the minimax control problem (P) under hypothethes* (H1)–(H6) *and* $p > \max\{8, (N+2)/2\}$. *Then there exists a sequence of suboptimal solutions* (u_ϵ, w_ϵ) *to (P) for which*

$$(u_\epsilon, w_\epsilon) \to (\bar{u}, \bar{w}) \text{ strongly in } L^p(0,T;U) \times L^p(0,T;W) \text{ as } \epsilon \to 0$$

and the necessary conditions in Theorems 6 *and* 9 *hold.*

References

1. Barbu V (1981) Necessary conditions for distributed control problems governed by parabolic variational inequalities. SIAM J Control Optim 19:64–86

2. Barbu V (1993) Analysis and Control of Nonlinear Infinite Dimensional Systems. Academic Press, Boston

3. Barbu V, Precupanu T (1986) Convexity and Optimization in Banach Spaces. Reidel Publishing, Dordrecht-Boston-Lancaster

4. Bonnans JF, Tiba D (1987) Equivalent control problems and applications. Lecture Notes in Control and Information Sciences 97:154–161, Springer-Verlag, Berlin

5. Friedman A (1987) Optimal control for parabolic variational inequalities. SIAM J Control Optim 25:482–497

6. He ZX (1987) State constrained control problems governed by variational inequalities. SIAM J Control Optim 25:119–144

7. Lasiecka I, Triggiani R (1983) Dirichlet boundary control problem for parabolic equations with quadratic cost: analicity and Riccati's feedback synthesis. SIAM J Control Optim 21:41–67

8. Lasiecka I, Triggiani R (1987) The regulator problem for parabolic equations with Dirichlet boundary control. Part I: Riccati's feedback synthesis and regularity of optimal solutions. Appl Math Optim 16:147–168

9. Lions JL (1971) Optimal Control of Systems Governed by Partial Differential Equations. Springer-Verlag, Berlin

10. Mordukhovich BS (1989) Minimax design of a class of distributed control systems. Autom Remote Control 50:1333–1340

11. Mordukhovich BS, Zhang K (1993) Robust optimal control for a class of distributed parameter systems. Proceedings of 1993 ACC 1:466–467

12. Mordukhovich BS, Zhang K (1993) Minimax Control of Parabolic Systems with Dirichlet Boundary Conditions and State Constraints. Dept. of Math., Wayne State Univer-

sity, preprint

13. Pazy N (1983) Semigroups of Linear Operators and Applications to Partial Differential Equations. Springer-Verlag, New York

14. Tiba D (1990) Optimal Control of Nonsmooth Ditributed Parameter Systems. Lecture Notes in Mathematics 1459, Springer-Verlag, Berlin

15. Washburn D (1979) A bound on the boundary input map for parabolic equations with applications to time optimal control. SIAM J Control Optim 17: 652–671.

17 Optimal Control Problems for Some First and Second Order Differential Equations

Nicolae H. Pavel* Ohio University, Athens, Ohio

G. S. Wang Ohio University, Athens, Ohio

Yong Kang Huang The Ohio State University at Marion, Marion, Ohio

1 INTRODUCTION

This paper is concerned with the following optimal control problems.

The first order case

$$\operatorname*{Min}_{(y,u)} \left\{ \int_0^T L(y,u)\,dt \right\} \tag{P$_1$}$$

subject to anti-periodic boundary problems for nonlinear ordinary differential equations of the form

$$\begin{cases} y'(t) + \partial\varphi(y(t)) = Bu(t) + f(t) \\ y(0) + y(T) = 0, \end{cases} \tag{1.1}$$

where $u \in L^2(0,T;U)$, $y \in C(0,T;H)$, H and U are Hilbert spaces of inner product $\langle , \rangle$, $B : U \to H$ is a linear bounded operator and $\partial\varphi$ is the subdifferential of φ.

* Research supported by NSF Grant DMS-91-11794.

The following assumption will guarantee the existence of optimal pairs for (P_1):

(J1) $\varphi : H \to (-\infty, +\infty]$ is even, continuous and convex with $\varphi(0) = 0$ and satisfies: for any $\ell > 0$, the level set $\{x \in D(\varphi) : \varphi(x) \leq \ell, \|x\| \leq \ell\}$ is compact in H.

The second order case

$$\underset{(x,u)}{\text{Min}} \left\{ \int_0^1 (\varphi_1(x(t)) + h(u(t)))\, dt \right\} \tag{P_2}$$

subject to "elliptic type" linear equations

$$\begin{cases} x''(t) = Ax(t) + Bu(t) + f(t) \\ x^{(i)}(0) + x^{(i)}(1) = 0, \quad j = 0, 1 \end{cases} \tag{1.2}$$

under the following assumptions:

(H_1) $A : D(A) \subset H \to H$ is a linear maximal monotone operator. B is a linear bounded operator from the Hilbert space V into the Hilbert space H and $f \in L^2(0, 1; H)$.

(H_2) $\varphi_1 : H \to \mathbb{R}$ and $h : H \to \mathbb{R}$ are lower semicontinuous and (at least one) strictly convex. In addition there exist $a_1 > 0$, $b_1 \in \mathbb{R}$ such that

$$b_1 + a_1 |u|_V^2 \leq h(u), \quad u \in U.$$

For (P_1), we discuss the existence of the optimal pair and the necessary conditions for the pair (y^*, u^*) to be an optimal pair (Theorem 2.1).

For (P_2), we first obtain the existence, uniqueness and continuous dependence (via monotonicity method) of solution of (1.2), then we discuss the necessary and sufficient condition for the pair (y^*, u^*) to be the optimal pair (Theorem 2.2 and Theorem 2.3).

The extension of Theorem 2.1 to infinite dimensional Hilbert space H under Hypotheses (J1) remains an important open problem.

2 MAIN RESULTS

First we will mention the results obtained in [1].

LEMMA 2.1. *Under assumption* (J1), *there exists a unique solution for* (2.1) *and* $y_n \to y$ *strongly in* $C(0,T;H)$ *if* $u_n \to u$ *weakly in* $L^2(0,T;U)$, *for each* $f \in L^2(0,T;H)$.

The existence of optimal pairs (y^*, u^*) for (P_1) under the hypotheses (J1) and some standard conditions on L (like convexity, coercivity and lower semicontinuity) follows from Lemma 2.1.

In order to get necessary conditions for (y^*, u^*) to be an optimal pair for Problem (P_1) we need additional restrictions on φ (which will exclude applications to PDE), namely:

(J2) φ is even, convex and twice Frechet continuously differentiable on $H = \mathbb{R}^n$ and its Hessian matrix $\ddot{\varphi}(y)$ is positive definite, i.e.,

$$\langle \ddot{\varphi}(y)z, z\rangle \geq 0 \quad \text{for } y, z \in \mathbb{R}^n.$$

Obviously, in the case $H = \mathbb{R}^n$ the compactness of the level sets in (J1) is automatically satisfied.

In what follows, $(\ddot{\varphi}(y))^*$ denotes the adjoint matrix of $\ddot{\varphi}(y)$.

To simplify the proof, we only consider the following particular cost functional

$$\operatorname*{Min}_{(y,u)} \int_0^T \frac{1}{2}(|y(t)|^2 + |u(t)|^2)\, dt \tag{P_1}'$$

subject to (1.1). The uniqueness of (y^*, u^*) is a consequence of strict convexity of L.

THEOREM 2.2. *Let* φ *satisfy* (J2). *If* (y^*, u^*) *is the optimal pair of* $(\mathrm{P}_1)'$, *then there exists* $p \in L^2(0,T;\mathbb{R}^n)$ *such that*

$$\begin{aligned} &p'(t) - (\ddot{\varphi}(y^*(t)))^*(p(t)) = y^* \\ &p(0) + p(T) = 0 \\ &B^*p(t) = u^*(t) \quad \text{a.e. } t \in [0,T]. \end{aligned} \tag{2.1}$$

Proof. Let $M = \{(y,u) : (y,u) \text{ satisfies } (1.1)\}$ and

$$L(y,u) = \int_0^T \frac{1}{2}(|y(t)|^2 + |u(t)|^2)\, dt.$$

Then we have

$$\dot{L}(y^*, u^*)(r, w) \geq 0 \quad \text{for all } (r,w) \in M_T, \tag{2.2}$$

where $\dot{L}$ is the Frechet derivative of L and M_T is the tangent cone of M at (y^*, u^*). This implies: $\int_0^T (\langle y^*, r\rangle + \langle u^*, w\rangle)\, dt \geq 0$.

$$(y^*, u^*) + \lambda((r,w) + (r_1(\lambda), r_2(\lambda)) \in M,$$

where $r_i(\lambda) \to 0$ as $\lambda \to 0$, $i = 1, 2$.

One can get by using the Frechet derivative of $\partial\varphi = \dot{\varphi}$:

$$r(0) = -r(T) \tag{2.3}$$

$$r'(t) + \ddot{\varphi}(y^*(t))(r(t)) = Bw(t). \tag{2.4}$$

Now, let p be the solution of

$$\begin{cases} p'(t) - (\ddot{\varphi}(y^*(t)))^*(p(t)) = y^*(t) \\ p(0) + p(T) = 0. \end{cases} \tag{2.5}$$

The existence and uniqueness of solutions of the problems (2.3)+(2.4) and (2.5) can be proved as in [1]. Namely, define the linear operators in $L^2(0, T; \mathbb{R}^n)$: $A : L^2(0, T; \mathbb{R}^n) \to L^2(0, T; \mathbb{R}^n)$ by

$$Az = z', \quad D(A) = \{z \in H^1(0, T; \mathbb{R}^n); \quad z(0) + z(T) = 0\}$$

$$(Bv)(t) = \ddot{\varphi}(y^*(t))v(t), \quad \text{a.e. on } [0, T], \quad v \in L^2(0, T; \mathbb{R}^n).$$

It is easy to check that A is maximal monotone (and dissipative) and B is continuous and monotone. In addition A is coercive so $A + B$ is surjective and so, for every $f \in L^2(0, T; \mathbb{R})$ there is a unique $g \in D(A)$ such that $(A+B)g = f$.

Now, let us continue the proof.

Multiplying (2.5) by r and using (2.3) and (2.4), one obtains successively

$$\langle y^*, r\rangle = -\int_0^T [\langle p, r'\rangle + \langle p, \ddot{\varphi}(y^*(t))r\rangle]\, dt = \int_0^T \langle -Bw, p\rangle.$$

So,

$$\int_0^T \langle -B^*p + u^*, w\rangle\, dt \geq 0, \quad \forall w \in L^2(0, T; U),$$

which yields

$$u^*(t) = B^*p(t) \quad \text{a.e. } t \in [0, T].$$

This completes the proof. □

Now, we turn to the problem (P_2).

The discussion of (P_2) will be based on the study of the following anti-periodic elliptic problem. (For the periodic cases, see [4].)

$$\begin{cases} x''(t) = Ax(t) + f(t) \\ x^{(j)}(0) + x^{(j)}(1) = 0, \quad j = 0, 1. \end{cases} \tag{2.6}$$

THEOREM 2.2. *Under assumption* (H_1), *we have the following results.*

(1) *For every* $f \in L^2(0,1;H)$, (2.6) *has a unique solution* $x \in W^{2,2}([0,1];H)$.

(2) *If* $f_n \to f$ *strongly in* $L^2(0,1;H) = L^2$, *then* $x_n \to x$ *strongly in* L^2.

(3) *If* $f_n \to f$ *weakly in* L^2, *then* $x_n \to x$ *weakly in* L^2.

Proof. Part (1) has been proved in [1].

Part (2): Here we need the key estimate

$$\|x''\|_{L^2} \leq \|f\|_{L^2}. \tag{2.7}$$

In order to get (2.7), multiply (2.6) by x'', integrate over $[0,1]$ and use the following result:

$$\int_0^1 \langle Ax(t), x''(t)\rangle_H dt = \langle \tilde{A}x, x''\rangle_{L^2} \leq 0, \quad \forall x \in D(\tilde{A}), \tag{2.8}$$

where $\tilde{A}$ is the realization of A in L^2. To prove (2.8), we integrate $\langle \tilde{A}_\lambda x, x''\rangle$ by parts to get

$$\langle \tilde{A}_\lambda x, x''\rangle = -\int_0^1 \langle (A_\lambda x(t))', x'(t)\rangle\, dt, \quad \forall \lambda > 0, x \in D(\tilde{A}), \tag{2.9}$$

where A_λ is the Yosida approximation of A and also observe that A_λ is odd while $t \to A_\lambda x(t)$ is absolutely continuous on $[0,1]$. It follows that

$$\langle \frac{d}{dt} A_\lambda x(t), x'(t)\rangle_H \geq 0, \quad \text{a.e. } t \in [0,1]. \tag{2.10}$$

So, (2.8) (and consequently (2.7)) follows by letting $\lambda \downarrow 0$ and using the fact that $\tilde{A}_\lambda x \to \tilde{A}x$ as $\lambda \downarrow 0$, $\forall x \in D(\tilde{A})$.

In view of (2.7), (2.6) yields

$$\|x''\|_{L^2} \leq \|f_n\|_{L^2} \leq k, \quad \forall n = 1,2,\ldots \text{ for some } k > 0.$$

So, both x_n and x_n' are bounded in $C([0,1];H)$. Moreover, from (2.7) we get

$$\|x_n' - x_m'\| \leq \frac{1}{2}\|f_n - f_m\|.$$

It follows that $x_n \to x$, $x_n' \to x'$ strongly in $L^2(0,1;H)$ while $x_n'' \to x''$ weakly in L^2 as $n \to \infty$. It follows that actually $x_n \to x$ and $x_n' \to x'$ in $C([0,1];H)$. With standard argument, one can prove that x is the solution of (2.6).

Part (3): Define the operator B via

$$Bx = -x''$$
$$D(B) = \{x \in W^{2,2}([0,1];H), x^{(j)}(0) + x^{(j)}(1) = 0, j = 1,2\}.$$

One can show easily that $B + \tilde{A}$ is a linear maximal monotone operator in $L^2(0,1;H)$. If $f_n \to f$ weakly in L^2, then $\|f_n\| \le k$ for some $k > 0$, which yields (on the basis of (2.10)) $\|x_n''\| \le k$. This implies the boundedness of x_n and x_n' in $C([0,1];H)$. Say $x_n \to x$ weakly in $L^2(0,1;H)$ (relabeling if necessary), and write (2.6) in the form

$$(B + \tilde{A})x_n = -f_n \quad n = 1,2,\dots . \tag{2.11}$$

Pass to limit in (2.11) for $n \to \infty$ to get

$$(B + \tilde{A})x = -f.$$

This completes our proof. □

REMARK. In the proof of Theorem 2.2, we only considered the case where A is single-valued. A similar proof can be carried out in the multivalued A case.

The following theorem is concerned with the necessary and sufficient condition for (x^*, u^*) to be the optimal pair of (P_2). The problem (P_2) can be regarded as the optimal control of the deflection velocity ψ' of a three layer beam [2]. Indeed if $x(t) = \psi'(t)$, then the problem (P_2) can be rewritten as [2]

$$\text{Min}\,\{L(\psi,u), (\psi,u) \in W^{3,2}([0,1];H) \times L^2(0,1;V)\} \tag{P_2)'}$$

subject to

$$\begin{cases} \psi'''(t) = A\psi'(t) + Bu(t) + f(t), \\ \psi^{(j)}(0) + \psi^{(j)}(1) = 0, \quad j = 0,1,2, \end{cases}$$

with the cost functional

$$L(\psi,u) = \int_0^1 (\varphi_1(\psi'(t)) + h(u(t)))\,dt.$$

THEOREM 2.3. *Let A, B, φ_1 and h satisfy hypotheses (H_1) and (H_2). Then (x^*, u^*) is an optimal pair of the problem (P_2) if and only if there is a $p \in W^{2,2}([0,1];H)$ such that*

$$B^*p(t) \in \partial h(u^*(t)), \quad \text{a.e. on } [0,1] \tag{2.12}$$
$$p''(t) \in A^*p(t) - \partial\varphi_1(x^*(t)) \quad \text{a.e. on } [0,1] \tag{2.13}$$
$$p^{(j)}(0) + p^{(j)}(1) = 0, \quad j = 0,1, \tag{2.14}$$

where B^* and A^* are the adjoints of B and A, respectively.

Proof. The existence of optimal pair (x^*, u^*) for (P_2) can be easily proved. Indeed, in view of the inequality (2.7), we conclude that (for some $c_1 > 0$)

$$\|x(t)\| \leq c_1(\|u\|_V + 1), \quad \forall t \in [0,1]. \tag{2.15}$$

Combining (2.15) and (H_2), we conclude that

$$L(x,u) = \int_0^1 (\varphi_1(x(t)) + h(u(t)))\,dt \geq c_1\|u\|_V^2 + c_2\|u\|_V + c_3, \tag{2.16}$$

for some c_1, c_2 and $c_3 \in \mathbb{R}$, which shows that

$$\text{Min}\,\{L(x,u), (x,u) \text{ subject to } (1.2)\} = d > -\infty.$$

Let (x_n, u_n) be such that

$$L(x_n, u_n) \to d \quad \text{as } n \to \infty \tag{2.17}$$

$$x''(t) = Ax_n(t) + Bu_n(t) + f(t), n = 1, 2, \ldots, \text{ a.e. on } [0,1] \tag{2.18}$$

$$x_n^{(j)}(0) + x_n^{(j)}(1) = 0, \quad j = 0, 1, \quad n = 1, 2, \ldots. \tag{2.19}$$

It follows from (2.16) that u_n is bounded in $L^2(0,1;V)$. So we may assume that u_n (relabeling if necessary) is weakly convergent in $L^2(0,1;V)$ to a function $u^* \in L^2(0,1;V)$. Then $\tilde{B}u_n \rightharpoonup \tilde{B}u^*$ in $L^2(0,1;H)$, where $\tilde{B}$ is the realization of B in $L^2(0,1;V)$. This implies on the basis of Theorem 2.2 (Part (3)) that $x_n \rightharpoonup x^*$ weakly in $L^2(0,1;H)$, and (x^*, u^*) satisfies (1.2). Taking into account that L is a lower semicontinuous convex function, one can pass to limit in (2.17) to get $L(x^*, u^*) = d$. In order to prove the theorem, we assume first that φ_1 and h are continuously Frechet differentiable. Then we have

$$(\dot{L}(x^*, u^*))(v, w) = \int_0^1 (\langle \dot{\varphi}_1(t)), v(t)\rangle_H + \langle \dot{h}(u^*(t)), w(t)\rangle_V)\,dt = 0, \tag{2.20}$$

for all $v \in W^{2,2}([0,1];H)$ and $v \in L^2(0,1;V)$ satisfying

$$v''(t) = Av(t) + Bw(t), \quad \text{a.e. on } [0,1] \tag{2.21}$$

$$v^{(j)}(0) + v^{(j)}(1) = 0, \quad j = 0, 1. \tag{2.22}$$

Therefore, in this case

$$\partial\varphi_1(x^*(t)) = \dot{\varphi}_1(x^*(t)); \ \partial h(u^*(t)) = \dot{h}(u^*(t)) \text{ a.e. on } [0,1] \tag{2.23}$$

and $t \to \dot{\varphi}_1(x^*(t))$ is continuous on $[0,1]$. In view of Theorem 2.2, there is a unique function $p \in W^{2,2}([0,1];H)$ satisfying (2.13) and (2.14). This is because A^* is also maximal monotone. Multiplying (2.13) by $v(t)$ and substituting $\langle \dot{\varphi}_1(x^*(t)), v(t)\rangle_H$ into (2.16), one derives

$$B^*p(t) = \dot{h}(u^*(t)) \quad \text{for all } t \in [0,1].$$

In the general case, we will only sketch the proof. (The details will be given in [2].) One considers the approximating control problem with $(\partial\varphi_1)_\lambda = \partial\varphi_{1\lambda}$, the Yosida approximation of $\partial\varphi_1$. Namely, let us introduce

$$L_\lambda(x,u) = \int_0^1 (\varphi_{1\lambda}(x(t)) + h_\lambda(u(t)))\,dt + \frac{1}{2}\|u - u^*\|_V^2, \quad \lambda > 0,$$

where $\varphi_{1\lambda}$ and h_λ are the regularization of φ_1 and h respectively [3]. Then $\varphi_{1\lambda}$ and h_λ are continuously Frechet differentiable. It can be seen that the optimal pair $(x_\lambda^*, u_\lambda^*)$ of the problem (P_λ) (with L_λ in place of L in (P_2)) is characterized by the analogous of (2.12)–(2.14), i.e.:

$$B^*p_\lambda = \dot{h}_\lambda(u_\lambda^*(t)) + u_\lambda^*(t) - u^*(t), \quad \forall t \in [0,1],\ \lambda > 0 \tag{2.24}$$

$$p_\lambda''(t) = A^*p_\lambda(t) - \dot{\varphi}_{1\lambda}(x_\lambda(t)), \quad \text{a.e. on } [0,1],\ \lambda > 0 \tag{2.25}$$

$$p_\lambda^{(j)}(0) + p_\lambda^{(j)}(1) = 0, \quad j = 0,1,\ \lambda > 0. \tag{2.26}$$

Therefore, we have

$$x_\lambda^{*''}(t) = Ax_\lambda^*(t) + Bu_\lambda^*(t) + f(t) \quad \text{a.e. on } [0,1] \tag{2.27}$$

$$x_\lambda^{*(j)}(0) + x_\lambda^{*(j)}(1) = 0, \quad j = 0,1,\ \lambda > 0. \tag{2.28}$$

First of all, one derives that $u_\lambda^* \to u^*$ strongly in $L^2(0,1;V)$. Then, by Theorem 2.2, it follows that $x_\lambda^* \to x^*$ also strongly in $L^2(0,1;H)$ as $\lambda \downarrow 0$. One also proves that $t \to \dot{\varphi}_{1\lambda}(x_\lambda^*(t))$ is bounded in $L^2(0,1;H)$ independently of λ. Passing to the limit in (2.27) and (2.28), we get

$$x^{*''}(t) = Ax^*(t) + Bu^*(t) + f(t), \quad \text{a.e. on } [0,1]$$

$$x^{*(j)}(0) + x^{*(j)}(1) = 0, \quad j = 0,1.$$

Passage to the limit for $\lambda \downarrow 0$ in (2.24)+(2.26). We have

$$\dot{\varphi}_{1\lambda}(x_\lambda^*(t)) = (\partial\varphi_1)_\lambda(x_\lambda^*(t)) \in (\partial\varphi_1)(J_\lambda x_\lambda^*(t)). \tag{2.29}$$

The realization of (2.29) in $L^2(0,1;H)$ has the form

$$(\partial\tilde{\varphi}_1)_\lambda(x_\lambda^*) \in (\partial\tilde{\varphi}_1)(\tilde{J}_\lambda x_\lambda^*).$$

Say, $(\partial\tilde{\varphi}_1)_\lambda(x^*_\lambda) \rightharpoonup y$ weakly in $L^2(0,1;H)$. It follows that $\tilde{J}_\lambda x^*_\lambda \to x^*$, strongly in $L^2(0,1;H)$. So, we can pass to the limit in (2.29) for $\lambda \downarrow 0$. Hence, we conclude that $y(t) \in \partial\varphi_1(x^*(t))$, a.e. on $[0,1]$. In view of Theorem 2.2 (Part(3)), (2.24)+(2.26) yields

$$p''(t) = A^*p(t) - y(t), \quad \text{a.e. on } [0,1].$$

So, one obtains (2.12)+(2.14), which completes our proof. □

REMARK. A standard example of A in Theorem 2.3 is $A = -\Delta$ (Laplace operator in $L^2(\Omega)$). An example of φ in Theorem 2.1 is $\varphi(x) = x^4$, $x \in \mathbb{R}$ (i.e., $H = \mathbb{R}$).

REFERENCES

[1] S. Aizicovici and N. H. Pavel, *Anti-periodic solutions to a class of nonlinear differential equations in Hilbert space,* J. Functional Anal., 99 (1991), 387–408.

[2] R. Aftabizadeh, Y. K. Huang and N. H. Pavel, *Nonlinear third-order differential equations with anti-periodic boundary conditions and some optimal control problems,* in preparation.

[3] V. Barbu and N. H. Pavel, *Optimal control problems with two point boundary conditions,* J. Optimiz. Th. Appl., 77 (1993), 51–78.

[4] V. Barbu and N. H. Pavel, *Optimal control of an elliptic governed problem,* Libertas Mathematica, 13 (1993), 131–139.

18 A Variational Approach to Shape Optimization for the Navier-Stokes Equations

Srdjan Stojanovic* University of Cincinnati, Cincinnati, Ohio

Thomas P. Svobodny† Wright State University, Dayton, Ohio

Abstract

A shape optimization problem for viscous incompressible flow is studied. The goal is to find a shape of the part of the boundary of the flow region, so that the given force field is produced there. The variable domain problem is relaxed so that it becomes a nonsmooth optimization problem on the fixed domain for the Navier-Stokes system with somewhat singular right-hand side. This formulation is useful for flows of highly viscous materials or in the case where we are not too far off-design.

1 PROBLEM STATEMENT

In this paper we analyze what may be termed the wing-design question: how to design the shape of a wing, so that a given surface stress is produced on (at least part of) the wing. We will show that when this question is reformulated as a variational problem on a fixed domain

*Supported in part by the NSF Grant DMS-91-11794.

†Supported in part by the ONR Grant N00014-91-1494

that it involves the minimization of a functional that is not differentiable in the usual sense. We consider motionless body $\mathcal{B}$ in a viscous incompressible fluid moving with respect to the fluid that is far from the body at a uniform velocity $\mathbf{h}$. The flow is considered in a bounded region Λ containing $\mathcal{B}$. The boundary of the region Λ will be denoted as $\partial\Lambda$. The boundary of the body $\partial\mathcal{B}$ consists of two disjoint and connected parts Σ and Γ, $\partial\mathcal{B} = \Sigma \cup \Gamma$. We shall suppose that Γ can be described as a Monge patch on the inner surface

$$\Gamma = \Gamma_\phi = \{(x, \phi(x)); x \in D \subset R^2\}, \tag{1.1}$$

given by some function $\phi \in U$ where

$$U = \left\{\phi \in H_0^3(D); 0 \le \phi(x) \le 1, \right\}. \tag{1.2}$$

So, if we want to emphasize the dependence on $\phi \in U$ we shall write also $\mathcal{B} = \mathcal{B}_\phi$. In this paper, we shall refer to ϕ as the *control.* Denote by Ω_ϕ the actual flow region

$$\Omega_\phi \stackrel{\text{def}}{=} \Lambda \setminus \bar{\mathcal{B}}_\phi. \tag{1.3}$$

Also, we assume that Σ is such that $\partial\mathcal{B}$ is sufficiently regular. Finally, denote,

$$\Omega = \{(x_1, x_2, x_3);\ (x_1, x_2) \in D,\ 0 < x_3 < \phi(x_1, x_2)\} \cup \Omega_\phi \cup \Gamma_\phi, \tag{1.4}$$

so that $\Omega \setminus \bar{\Omega_\phi} = \{(x_1, x_2, x_3);\ (x_1, x_2) \in D\ 0 < x_3 < \phi(x_1, x_2)\}$. Thus Ω consists of the original domain plus the 'movable' region underneath the controlled part of the boundary, Γ. Note that $\Omega = \Omega_0$.

Now, the velocity vector field of the fluid $\mathbf{w} = \mathbf{w}^\phi$, and the pressure p, are the solution of the Navier-Stokes system

$$\begin{aligned} -\nu\Delta\mathbf{w} + \mathbf{w}\cdot\nabla\mathbf{w} + \nabla p &= \mathbf{f} \text{ in } \Omega_\phi \\ \nabla \cdot \mathbf{w} &= 0 \text{ in } \Omega_\phi \\ \mathbf{w} &= 0 \text{ in } \Gamma_\phi \cup \Sigma \\ \mathbf{w} &= \mathbf{h} \text{ in } \partial\Lambda \end{aligned} \tag{1.5}$$

We observe that the pressure p in (1.5) is determined only uniquely up to the additive constant.

The problem we propose is the following: we wish to design a surface so that flow across it will impart a prescribed surface stress force field: i.e., given $\mathbf{g}$ such that

$$\begin{aligned} &\mathbf{g} \in \mathbf{W}^{2,q}(\Omega), \quad q > 2 \\ \mathbf{g} = \ &\mathbf{0} \quad \text{outside some normal tubular neighborhood of } \Gamma, \end{aligned} \tag{1.6}$$

find (if possible) $\phi \in U$ such that, if $\mathbf{w}^\phi$ is the corresponding solution of (1.5), then also

$$-pn_j + \nu\left(\frac{\partial w_j}{\partial x_i} + \frac{\partial w_i}{\partial x_j}\right) n_i = g^j \text{ in } \Gamma_\phi, \quad j = 1, 2, 3. \tag{1.7}$$

We observe now that if (1.7) is prescribed in addition to (1.5) (and if (1.7) and (1.5) have a solution) then pressure p is determined *uniquely.* Also, we note that condition (1.7) means that fluid motion exerts force distribution $\mathbf{g}$ on the boundary Γ_ϕ. So, the problem we propose is to find a shape of the immersed body so that the prescribed force field is generated at the part of the boundary.

2 PROBLEM REFORMULATION

In this section we assume that

$$\mathbf{g} \in \mathbf{H}^2(\Omega). \tag{2.1}$$

Suppose that there exist a chart ϕ and a pair $(\mathbf{w}^\phi, p^\phi)$ which constitute a solution of (1.5) and that also satisfy (1.7) . In other words, we suppose existence for the free boundary problem (1.5), (1.7). To refer to such an assumption we shall say that ϕ is supposed to be an *exactly compliant control.* Extend $\mathbf{w}^\phi$ from Ω_ϕ to Ω as $\mathbf{z}^\phi$:

$$\mathbf{z}^\phi = \begin{cases} 0 & \text{on } \Omega \setminus \bar{\Omega}_\phi \\ \mathbf{w}^\phi & \text{on } \Omega_\phi \end{cases}. \tag{2.2}$$

We define the space of test functions

$$V = V(\Omega) = \{\mathbf{u} \in \mathbf{H}_0^1(\Omega) : \nabla \cdot \mathbf{u} = \mathbf{0}\}$$

Lemma 2.1 *If ϕ is an exactly compliant control, then $\mathbf{z}^\phi \in H^1(\Omega)^n$, and is a weak solution of the Navier-Stokes system (with singular right hand side)*

$$\frac{\nu}{2}\int_\Omega D(\mathbf{z}^\phi) : D(\boldsymbol{\tau}) + \int_\Omega \mathbf{z}^\phi \cdot \nabla \mathbf{z}\phi \cdot \boldsymbol{\tau} = \int_\Omega \mathbf{f} \cdot \boldsymbol{\tau} + \xi_\phi(\boldsymbol{\tau}), \ \ \forall \boldsymbol{\tau} \in V, \tag{2.3}$$

where $\xi_\phi \in H^{-1}(\Omega)^3$ is a signed vector measure given by

$$\xi_\phi(\boldsymbol{\tau}) = \int_{\Gamma_\phi} \mathbf{g} \cdot \boldsymbol{\tau} d\sigma. \tag{2.4}$$

Here, $D(\mathbf{z}^\phi) : D(\boldsymbol{\tau}^\phi) = \left(\frac{\partial z_j}{\partial x_i} + \frac{\partial z_i}{\partial x_j}\right)\left(\frac{\partial \tau_j}{\partial x_i} + \frac{\partial \tau_i}{\partial x_j}\right)$, and throughout the paper the summation convention is assumed. Also,

$$V \stackrel{\text{def}}{=} \left\{\mathbf{u} \in H_0^1(\Omega)^2; \nabla \cdot \mathbf{u} = 0\right\}. \tag{2.5}$$

Proof of the Lemma: By *a priori* estimates (see e.g. [2],[4],[8]) $(\mathbf{w}^\phi, p^\phi)$ is regular in Ω_ϕ. That implies $\mathbf{z}^\phi \in C^{0,1}(\Omega)$, and in particular $\mathbf{z}^\phi \in H^1(\Omega)$.

By the Trace Theorem,

$$\begin{aligned} |\xi_\phi(\boldsymbol{\tau})| = \left|\int_{\Gamma_\phi} \mathbf{g} \cdot \boldsymbol{\tau} d\sigma\right| &\leq \|\mathbf{g}\|_{L^2(\Gamma_\phi)^2} \|\boldsymbol{\tau}\|_{L^2(\Gamma_\phi)^2} \\ &\leq c_\phi \|\mathbf{g}\|_{C^0(\Omega)^2} \|\boldsymbol{\tau}\|_{H^1(\Omega)^2}. \end{aligned} \tag{2.6}$$

So, in particular, $\xi_\phi \in H^{-1}(\Omega)^3$.

Proceeding to verify (2.3), we get that

$$\begin{aligned}\frac{\nu}{2}\int_{\Omega} D(\mathbf{z}^{\phi}) : D(\boldsymbol{\tau}) &= \frac{\nu}{2}\int_{\Omega_\phi} D(\mathbf{w}^{\phi}) : D(\boldsymbol{\tau}) \\ &= -\int_{\Omega_\phi}\left(\nu\Delta\mathbf{w}^{\phi} - \nabla p\right)\cdot\boldsymbol{\tau} \\ &\quad -\int_{\partial\Omega_\phi}\left[p\delta_i^j - \nu\left(\frac{\partial w_i}{\partial x_j} + \frac{\partial w_j}{\partial x_i}\right)\right]\tau_i n_j d\sigma. \end{aligned} \tag{2.7}$$

We note that (2.7) holds for any p, and for $\boldsymbol{\tau} \in V$. Proceeding, we choose p from (1.5) and (1.7), to conclude

$$\begin{aligned}\frac{\nu}{2}\int_{\Omega} D(\mathbf{z}^{\phi}) : D(\boldsymbol{\tau}) &+ \int_{\Omega}\mathbf{z}^{\phi}\cdot\nabla\mathbf{z}^{\phi}\cdot\boldsymbol{\tau} \\ &= \int_{\Omega}\mathbf{f}\cdot\boldsymbol{\tau} - \int_{\Gamma_\phi}\left[pn_i - \nu\left(\frac{\partial w_i}{\partial x_j} + \frac{\partial w_j}{\partial x_i}\right)n_j\right]\tau_i d\sigma \\ &= \int_{\Omega}\mathbf{f}\cdot\boldsymbol{\tau} + \int_{\Gamma_\phi}\mathbf{g}\cdot\boldsymbol{\tau} d\sigma \\ &= \int_{\Omega}\mathbf{f}\cdot\boldsymbol{\tau} + \xi_\phi(\boldsymbol{\tau}), \end{aligned} \tag{2.8}$$

which completes the proof of the Lemma. □

The next lemma supplies a partial converse to the above lemma and is the main justification for our method. In the proof we refer to and make use of a uniqueness condition that is developed in the next section.

Lemma 2.2 *Let $\phi \in U$ be given. Then, if $\mathbf{z}^{\phi}$ is a solution of (2.3) with $\mathbf{f} = \mathbf{0}$ and if it happens that $\mathbf{z}^{\phi}|_{\Gamma_\phi} = 0$, then there exists pressure p such that $\left(\mathbf{z}^{\phi}|_{\Omega_\phi}, p|_{\Omega_\phi}\right)$ is a solution of (2.3), (1.7), i.e., ϕ is an exactly compliant control.*

Proof: Observe that any test function $\boldsymbol{\tau} \in V$ must satisfy

$$0 = \int_{\Omega\setminus\bar{\Omega}_\phi}\nabla\cdot\boldsymbol{\tau} = \int_{\partial(\Omega\setminus\bar{\Omega}_\phi)}\boldsymbol{\tau}\cdot\mathbf{n}d\sigma = \int_{\Gamma_\phi}\boldsymbol{\tau}\cdot\mathbf{n}d\sigma. \tag{2.9}$$

From (2.3) we see (here we assume regularity of $\mathbf{z}^{\phi}|_{\Omega\setminus\bar{\Omega}_\phi}$ and $\mathbf{z}^{\phi}|_{\Omega_\phi}$; it can be shown ([6]) that the integrals presented here are well-defined),

$$\int_{\Gamma_\phi}\mathbf{g}\cdot\boldsymbol{\tau} d\sigma = \frac{\nu}{2}\left(\int_{\Omega\setminus\bar{\Omega}_\phi} D(\mathbf{z}^{\phi}) : D(\boldsymbol{\tau}) + \int_{\Omega_\phi} D(\mathbf{z}^{\phi}) : D(\boldsymbol{\tau})\right) + \int_{\Omega}\mathbf{z}^{\phi}\cdot\nabla\mathbf{z}^{\phi}\cdot\boldsymbol{\tau}$$

$$= \int_{\Omega\setminus\bar{\Omega}_\phi} \left(-\nu\Delta\mathbf{z}^\phi\right)\cdot\boldsymbol{\tau} + \int_{\Omega_\phi}\left(-\nu\Delta\mathbf{z}^\phi\right)\cdot\boldsymbol{\tau}$$

$$+\int_{\partial(\Omega\setminus\bar{\Omega}_\phi)} \nu D(\mathbf{z}^\phi)n\cdot\boldsymbol{\tau}d\sigma + \int_{\partial\Omega_\phi}\nu D(\mathbf{z}^\phi)n\cdot\boldsymbol{\tau}d\sigma + \int_\Omega \mathbf{z}^\phi\cdot\nabla\mathbf{z}^\phi\cdot\boldsymbol{\tau} \quad (2.10)$$

Recall that $\mathbf{n}$ is the unit normal exterior to Ω_ϕ. Then (2.10) is equal to

$$\int_{\Omega\setminus\bar{\Omega}_\phi} \left(-\nu\Delta\mathbf{z}^\phi\right)\cdot\boldsymbol{\tau} \;+\; \int_{\Omega_\phi}\left(-\nu\Delta\mathbf{z}^\phi\right)\cdot\boldsymbol{\tau}$$

$$+\;\int_{\Gamma_\phi}\nu\left(D(\mathbf{z}^\phi)^{ext} - D(\mathbf{z}^\phi)^{int}\right)\mathbf{n}\cdot\boldsymbol{\tau}d\sigma + \int_\Omega \mathbf{z}^\phi\cdot\nabla\mathbf{z}^\phi\cdot\boldsymbol{\tau} \quad (2.11)$$

where $f^{int} \overset{\text{def}}{=} f|_{\Omega\setminus\bar{\Omega}_\phi}$ and $f^{ext} \overset{\text{def}}{=} f|_{\Omega_\phi}$.

We conclude that there exists $p \in L^2(\Omega)$ such that

$$-\nu\Delta\mathbf{z}^\phi + \mathbf{z}^\phi\cdot\nabla\mathbf{z}^\phi + \nabla p = \mathbf{0} \text{ in } \Omega\setminus\bar{\Omega}_\phi \quad (2.12)$$

$$-\nu\Delta\mathbf{z}^\phi + \mathbf{z}^\phi\cdot\nabla\mathbf{z}^\phi + \nabla p = \mathbf{0} \text{ in } \Omega_\phi. \quad (2.13)$$

For such p we have

$$\int_{\Gamma_\phi}\mathbf{g}\cdot\boldsymbol{\tau}d\sigma$$

$$= \int_{\Omega\setminus\bar{\Omega}_\phi}(-\nabla p)\cdot\boldsymbol{\tau} + \int_{\Omega_\phi}(-\nabla p)\cdot\boldsymbol{\tau}$$

$$+\int_{\Gamma_\phi}\nu\left(D(\mathbf{z}^\phi)^{ext} - D(\mathbf{z}^\phi)^{int}\right)\mathbf{n}\cdot\boldsymbol{\tau}d\sigma$$

$$= \int_\Omega p\nabla\cdot\boldsymbol{\tau} - \int_{\partial(\Omega\setminus\bar{\Omega}_\phi)} p\boldsymbol{\tau}\cdot\mathbf{n}d\sigma - \int_{\partial\Omega_\phi} p\boldsymbol{\tau}\cdot\mathbf{n}d\sigma$$

$$+\int_{\Gamma_\phi}\nu\left(D(\mathbf{z}^\phi)^{ext} - D(\mathbf{z}^\phi)^{int}\right)\mathbf{n}\cdot\boldsymbol{\tau}d\sigma. \quad (2.14)$$

Let $[f]_{int}^{ext} \overset{\text{def}}{=} f^{ext}|_{\Gamma_\phi} - f^{int}|_{\Gamma_\phi}$ denote the jump across Γ_ϕ. Then (2.14) implies that

$$\int_{\Gamma_\phi} g_j\boldsymbol{\tau}_j d\sigma = \int_{\Gamma_\phi}\left[-pn_j + \nu\left(\frac{\partial z_j}{\partial x_i} + \frac{\partial z_i}{\partial x_j}\right)n_i\right]_{int}^{ext}\boldsymbol{\tau}_j d\sigma \quad (2.15)$$

for all $\boldsymbol{\tau}$ such that $\int_{\Gamma_\phi}\boldsymbol{\tau}_j n_j = 0$. This implies

$$\left[-pn_j + \nu\left(\frac{\partial z_j}{\partial x_i} + \frac{\partial z_i}{\partial x_j}\right)n_i\right]_{int}^{ext} - g^j = (const.)n_j,\; j = 1,2,3. \quad (2.16)$$

We note that (2.16) holds for any p such that (2.12), (2.13) hold.

Now suppose that $\mathbf{z}^{\phi}|_{\Gamma_\phi} = 0$. Then $\mathbf{z}^{\phi}|_{\Omega\setminus\bar{\Omega}_\phi} = 0$ [1] and without loss of generality we can choose $p|_{\Omega\setminus\bar{\Omega}_\phi} = 0$. Then (2.16) implies

$$\left(-pn_j + \nu\left(\frac{\partial z_j}{\partial x_i} + \frac{\partial z_i}{\partial x_j}\right) n_i\right)^{ext}\Bigg|_{\Gamma_\phi} - g^j = (const.)n_j,\ j = 1,2,3 \tag{2.17}$$

for any p satisfying (2.13). So, there exists p such that (2.13) holds and such that

$$\left(-pn_j + \nu\left(\frac{\partial z_j}{\partial x_i} + \frac{\partial z_i}{\partial x_j}\right) n_i\right)^{ext}\Bigg|_{\Gamma_\phi} - g^j = 0,\ j = 1,2,3 \tag{2.18}$$

which is nothing but (1.7). □

Lemma 2.2 motivates the following definition.

Definition 2.1 *$\phi^* \in U$ is said to solve the* relaxed shape optimization problem *if the corresponding $\mathbf{z}^{\phi}$ defined by (2.3), is such that if*

$$\Phi(u) = \frac{1}{2}\int_{\Gamma_\phi} |\mathbf{z}^{\phi}|^2 d\sigma \tag{2.19}$$

then

$$\Phi(\phi^*) = \min_{\phi\in U} \Phi(\phi). \tag{2.20}$$

Of course, an exactly compliant control (if it exists) is a minimizer, that is, a solution of (2.20). On the other hand, a solution of (2.20) is an exactly compliant control provided an exactly compliant control exists. We do not consider the controllability, which would be straightforward if one could pose it in a variational setting, but otherwise seems quite formidable. Rather, we shall study the relaxed problem introduced in Definition 1.

3 The STATE EQUATION

In this section we gather together some information about the Navier-Stokes equations that we will use in the following sections. Most of the estimations to be derived here are based on established techniques, to be found in, say, [4]. In the following, $\mathbf{w}$ will denote the unique solution to the following Stokes problem

$$-\nu\Delta \mathbf{w} + \nabla p = 0 \text{ in } \Omega_\phi$$

[1] From a technical standpoint, we should emphasize that we are in the regime of uniqueness that is discussed in the next section. Practically speaking, to ensure such uniqueness, we want a situation with large viscosity or small applied force

$$\nabla \cdot \mathbf{w} = 0 \text{ in } \Omega_\phi$$
$$\mathbf{w} = \mathbf{h} \text{ in } \partial\Lambda \tag{3.1}$$

It is known ([1]) that

$$||\mathbf{w}||_1 \leq C_S||\mathbf{h}||_{1/2}.$$

We will first establish an *a priori* estimate for the solution of

$$-\nu\Delta\mathbf{u} + \mathbf{u}\cdot\nabla\mathbf{u} + \nabla p = \xi_\mathbf{g} \text{ in } \Omega_\phi$$
$$\nabla\cdot\mathbf{u} = 0 \text{ in } \Omega_\phi$$
$$\mathbf{u} = \mathbf{h} \text{ in } \partial\Lambda \tag{3.2}$$

Consider $\mathbf{v} = \mathbf{u} - \mathbf{w}$; then $\mathbf{v}$ satisfies

$$-\nu\Delta\mathbf{v} + \mathbf{v}\cdot\nabla\mathbf{v} + \nabla p + \mathbf{w}\cdot\nabla\mathbf{v} + \mathbf{v}\cdot\nabla\mathbf{w} = \xi_\mathbf{g} \text{ in } \Omega_\phi$$
$$\nabla\cdot\mathbf{v} = 0 \text{ in } \Omega_\phi$$
$$\mathbf{v} = \mathbf{0} \text{ in } \partial\Lambda, \tag{3.3}$$

so that $\mathbf{v}$ satisfies the variational equality

$$\int_\Omega \nu\nabla\mathbf{v}\cdot\nabla\boldsymbol{\tau} + \mathbf{v}\cdot\nabla\mathbf{v}\cdot\boldsymbol{\tau} + \mathbf{w}\cdot\nabla\mathbf{v}\cdot\boldsymbol{\tau} + \mathbf{v}\cdot\nabla\mathbf{w}\cdot\boldsymbol{\tau}\,dx = \int_\Gamma \mathbf{g}\cdot\boldsymbol{\tau}, \quad \forall\boldsymbol{\tau}\in V$$

Letting $\boldsymbol{\tau} = \mathbf{v}$, we get the estimate

$$\begin{aligned}
||\mathbf{v}||_1^2 &\leq |\int_\Omega \mathbf{v}\cdot\nabla\mathbf{v}\cdot\mathbf{w}\,dx| + |\int_\Gamma \mathbf{g}\cdot\boldsymbol{\tau}| && (3.4)\\
&\leq \sqrt{3}(\sqrt{2}||\mathbf{v}||_{L^2}^{1/4}||\mathbf{v}||_1^{3/4})||\mathbf{w}||_{L^4}||\mathbf{v}||_1 + C(\Gamma)||\mathbf{g}||_{L^2}||\mathbf{v}||_1 && (3.5)\\
&\leq \sqrt{6}\mu^{-1/8}||\mathbf{v}||_1||\mathbf{w}||_{L^4}||\mathbf{v}||_1 + C(\Gamma)||\mathbf{g}||_{L^2}||\mathbf{v}||_1 && (3.6)\\
&\leq \sqrt{12}\mu^{-1/4}||\mathbf{w}||_1||\mathbf{v}||_1^2 + C(\Gamma)||\mathbf{g}||_{L^2}||\mathbf{v}||_1, && (3.7)
\end{aligned}$$

so that

$$||\mathbf{v}||_1 \leq \frac{C(\Gamma)||\mathbf{g}||_{L^2}}{\nu - \sqrt{12}\mu^{-1/4}||\mathbf{w}||_1},$$

(where μ is the smallest eigenvalue for the Dirichlet Laplace operator on Ω [4]) as long as

$$\sqrt{12}\mu^{-1/4}||\mathbf{w}||_1 < \nu.$$

In the next section, we will need a uniqueness result; so suppose that $\mathbf{u}$ and $\mathbf{u}'$ both satisfy the Navier-Stokes equations; then $\mathbf{v} = \mathbf{u} - \mathbf{w}$ and $\mathbf{v}' = \mathbf{u}' - \mathbf{w}$ both satisfy the variational equality; set $\mathbf{z} = \mathbf{v} - \mathbf{v}'$, then $\mathbf{z}$ satisfies homogeneous boundary conditions, and so the following variational equality

$$\int_\Omega \nu\nabla\mathbf{u}\cdot\nabla\boldsymbol{\tau} + \mathbf{u}\cdot\nabla\mathbf{v}\cdot\boldsymbol{\tau} + \mathbf{v}'\cdot\nabla\mathbf{u}\cdot\boldsymbol{\tau} + \mathbf{w}\cdot\nabla\mathbf{u}\cdot\boldsymbol{\tau} + \mathbf{u}\cdot\nabla\mathbf{w}\cdot\boldsymbol{\tau}\,dx = 0, \quad \forall\boldsymbol{\tau}\in V$$

Choose $\boldsymbol{\tau} = \mathbf{u}$ in this equation, then

$$0 = \nu||\mathbf{u}||_V^2 - \int_\Omega \mathbf{u} \cdot (\mathbf{v} \cdot \nabla \mathbf{u})\, dx - \int_\Omega \mathbf{u} \cdot (\mathbf{w} \cdot \nabla \mathbf{u})\, dx,$$

and so,

$$\nu||\mathbf{u}||_V^2 \leq \sqrt{12}\mu^{-1/4}||\mathbf{u}||_V^2(||\mathbf{v}||_1 + ||\mathbf{w}||_1).$$

Using previous estimates, we get that

$$||\mathbf{u}||_1^2 \leq \sqrt{12}\mu^{-1/4}\nu^{-1}\{\frac{C(\Gamma)||\mathbf{g}||_{L^2}}{\nu - \sqrt{12}\mu^{-1/4}C_S||\mathbf{h}||_{1/2}} + C_S||\mathbf{h}||_{1/2}\}||\mathbf{u}||_1^2,$$

and so that $\mathbf{u}$ is identically the zero vector if $\mathcal{R} < 1$ where

$$\mathcal{R} = \nu^{-1}(\sqrt{12}\mu^{-1/4}C_S||\mathbf{h}||_{1/2}) + \nu^{-2}(\sqrt{12}\mu^{-1/4}C(G)||\mathbf{g}||_{L^2} - 12\mu^{-1/2}C_S^2||\mathbf{h}||_{1/2}),$$

which is to be considered as a generalized Reynold's number, being small with small boundary data, small dimensions and high viscosity.

4 EXISTENCE of a MINIMUM

We introduce the set of admissible bounding surfaces

$$U_b = \{\phi \in U : ||\phi||_{H^4(D)} \leq b\}$$

where b is some positive constant.

Proposition 4.1 *Let* $\mathbf{g} \in W^{1,q}(\Omega)$, *where* $q > 3$; *suppose that* $\mathcal{R} < 1$. *Then, there exists a* $\phi^* \in U_b$ *such that*

$$\Phi(\phi^*) = \min_{\phi \in U_b} \Phi(\phi).$$

Proof: Let $\{\phi_n\}$ be a minimizing sequence. By regularity, we know that

$$||z^{\phi_n}||_{H^1(\Omega)} + ||z^{\phi_n}||_{C^{0,\lambda}(\Omega)} \leq c$$

And so, by passing to subsequences,

$$\phi_n \to \phi^* \text{ in } H^3(D),$$

$$\mathbf{z}^{\phi_n} \rightharpoonup \mathbf{z}^* \text{ in } H^1(\Omega),$$

$$\mathbf{z}^{\phi_n} \to \mathbf{z}^* \text{ in } C(\bar{\Omega})$$

If $\boldsymbol{\tau} \in C^1(\bar{\Omega})$ then clearly

$$\int_{\Gamma_{\phi_n}} \mathbf{g} \cdot \boldsymbol{\tau}\, d\sigma \to \int_{\Gamma_{\phi^*}} \mathbf{g} \cdot \boldsymbol{\tau}\, d\sigma.$$

Because of the identity

$$\int_{\Omega} \mathbf{z} \cdot \nabla \mathbf{z} \cdot \boldsymbol{\tau} = -\int_{\Omega} \mathbf{z} \cdot \nabla \boldsymbol{\tau} \cdot \mathbf{z} + \int_{\partial\Omega} (z \cdot \boldsymbol{\tau})(z \cdot \mathbf{n}),$$

we can pass to the limit in the equation for $\mathbf{z}^{\phi_n}$, to get

$$\frac{\nu}{2} \int_{\Omega} D(z^*) : D(\boldsymbol{\tau}) + \int_{\Omega} \mathbf{z}^* \cdot \nabla \mathbf{z}^* \cdot \boldsymbol{\tau} = \int_{\Gamma_{\phi^*}} \mathbf{g} \cdot \boldsymbol{\tau} \, d\sigma$$

which holds for all $\boldsymbol{\tau} \in V \cap C^1(\bar{\Omega})$; by density it holds for all $\boldsymbol{\tau} \in V$ also. Now if $\mathcal{R} < 1$ then the solutions to this equation are unique and we conclude that $\mathbf{z}^* = \mathbf{z}^{\phi^*}$. And so

$$\lim_{n \to \infty} \Phi(\phi_n) = \Phi(\phi^*).$$

5 DIRECTIONAL DIFFERENTIABILITY of the VARIATIONAL FUNCTIONAL Φ

It turns out that the functional is non-differentiable in general (see [6]). That this is so makes intuitive sense if we realize that the solution to the relaxed problem is possibly tent-like with a jump in its derivative across the surface Γ. (It is however (Gateau) differentiable at the exactly compliant control). The non-differentiability is exhibited most clearly by the form of the directional derivative; moreover one can glean from that expression the conditions that are necessary for differentiablity. Let us compute the directional derivative of the variational functional Φ.

$$\begin{aligned}
\Phi'(\phi;\psi) &= \lim_{\lambda \downarrow 0} \frac{1}{2\lambda} \left(\int_{\Gamma_{\phi+\lambda\psi}} |\mathbf{z}^{\phi+\lambda\psi}|^2 d\sigma - \int_{\Gamma_\phi} |\mathbf{z}^{\phi}|^2 d\sigma \right) \\
&= \int_D \left(\mathbf{z}^{\phi} \cdot \left(\mathbf{z}_z^{\phi,ext} \psi^+ - \mathbf{z}_z^{\phi,int} \psi^- \right) \sqrt{1 + \nabla\phi^2} + |\mathbf{z}^{\phi}|^2 \frac{\nabla\phi \cdot \nabla\psi}{\sqrt{1+\nabla\phi^2}} \right) d\mathbf{x} \\
&\quad + \lim_{\lambda \downarrow 0} \frac{1}{2\lambda} \int_{\Gamma_\phi} \left(|\mathbf{z}^{\phi+\lambda\psi}|^2 - |\mathbf{z}^{\phi}|^2 \right) d\sigma. \qquad (5.1)
\end{aligned}$$

Now the solution $\mathbf{z}$ is differentiable with respect to ϕ in the sense that

$$\lim_{\lambda \downarrow 0} \frac{1}{\lambda} \int_{\Gamma_\phi} |\mathbf{z}^{\phi+\lambda\psi} - \mathbf{z}^{\phi}|^2 d\sigma = 0. \qquad (5.2)$$

To see this, note that

$$\begin{aligned}
||\mathbf{z}^{\phi+\lambda\psi} - \mathbf{z}^{\phi}||_{L^2}^2 &\leq \mu^{-1} ||\mathbf{z}^{\phi+\lambda\psi} - \mathbf{z}^{\phi}||_1^2 \\
&\leq C ||\xi_{\phi+\lambda\psi} - \xi_\phi||_{-1} \\
&\leq C\lambda^{2\alpha+1}, \quad \alpha > \frac{1}{2},
\end{aligned}$$

by the Holder smoothness of $\mathbf{g}$ ([6]). Then the second term in (5.1) is expanded by use of the adjoint equation:

$$\begin{aligned}
&\lim_{\lambda\downarrow 0}\frac{1}{2\lambda}\int_{\Gamma_\phi}\left(|\mathbf{z}^{\phi+\lambda\psi}|^2-|\mathbf{z}^\phi|^2\right)d\sigma \\
&= \lim_{\lambda\downarrow 0}\frac{1}{\lambda}\int_{\Gamma_\phi}\left(\mathbf{z}^{\phi+\lambda\psi}-\mathbf{z}^\phi\right)\cdot\mathbf{z}^\phi d\sigma \\
&= \lim_{\lambda\downarrow 0}\frac{1}{\lambda}\int_\Omega\{D(\boldsymbol{\zeta}):D\left(\mathbf{z}^{\phi+\lambda\psi}-\mathbf{z}^\phi\right)+\mathbf{z}^\phi\cdot\nabla(\mathbf{z}^{\phi+\lambda\psi}-\mathbf{z}^\phi)\cdot\boldsymbol{\zeta} \\
&\qquad +(\mathbf{z}^{\phi+\lambda\psi}-\mathbf{z}^\phi)\cdot\nabla\mathbf{z}^\phi\cdot\boldsymbol{\zeta}\}d\mathbf{x},
\end{aligned}$$

where $\boldsymbol{\zeta}$ is the adjoint variable defined by the equation

$$\frac{\nu}{2}\int_\Omega D(\boldsymbol{\zeta}):D(\boldsymbol{\tau})+\int_\Omega(\mathbf{z}^\phi\cdot\nabla\boldsymbol{\zeta}\cdot\boldsymbol{\tau}+\boldsymbol{\tau}\cdot\nabla\mathbf{z}^\phi\cdot\boldsymbol{\zeta})=\int_{\Gamma_\phi}\mathbf{z}^\phi\cdot\boldsymbol{\tau}\,d\sigma,\quad\forall\boldsymbol{\tau}\in X$$

Now, using the weak form of the state equation (2.3),

$$\begin{aligned}
&\lim_{\lambda\downarrow 0}\frac{1}{2\lambda}\int_{\Gamma_\phi}\left(|\mathbf{z}^{\phi+\lambda\psi}|^2-|\mathbf{z}^\phi|^2\right)d\sigma \\
&= \lim_{\lambda\downarrow 0}\frac{1}{\lambda}\left(\int_{\Gamma_{\phi+\lambda\psi}}\mathbf{g}\cdot\boldsymbol{\zeta}d\sigma-\int_{\Gamma_\phi}\mathbf{g}\cdot\boldsymbol{\zeta}d\sigma\right) \\
&= \int_D\left\{\left(\left(\mathbf{g}\cdot\boldsymbol{\zeta}^{\phi,ext}\right)_z\psi^+-\left(\mathbf{g}\cdot\boldsymbol{\zeta}^{\phi,int}\right)_z\psi^-\right)\sqrt{1+|\nabla\phi|^2}\right. \\
&\qquad\left.+(\mathbf{g}\cdot\boldsymbol{\zeta})\frac{\nabla\phi\cdot\nabla\psi}{\sqrt{1+|\nabla\phi|^2}}\right\}d\mathbf{x}.
\end{aligned}$$

Thus, the directional derivative is

$$\begin{aligned}
\int_D\{\{[\mathbf{z}^\phi\cdot\mathbf{z}_z^{\phi,ext} &+ (\mathbf{g}\cdot\boldsymbol{\zeta}^{\phi,ext})_z]\psi^+-[\mathbf{z}^\phi\cdot\mathbf{z}_z^{\phi,int}-\left(\mathbf{g}\cdot\boldsymbol{\zeta}^{\phi,int}\right)_z]\psi^-\} \\
&+[|\mathbf{z}|^2-(\mathbf{g}\cdot\boldsymbol{\zeta})]\frac{\nabla\phi\cdot\nabla\psi}{\sqrt{1+|\nabla\phi|^2}}\}d\mathbf{x}.
\end{aligned}\tag{5.3}$$

From this expression, one can deduce convexity-like conditions on the solution so that a sub-differential exists for this functional and one can thus use methods of non-smooth analysis to investigate the minimization of Φ ([5],[6]). On the other hand, it may be possible to use the directional derivatives to generate an algorithm for minimizing Φ ([7]).

References

[1] L. Cattabriga, Su un problema al contorno relativo al sistema di equazione di Stokes, Rend. Mat. Sem. U. Padova, 31, (1961) 308-340

[2] P. Constantin and C. Foias, *Navier-Stokes Equations*, The University of Chicago Press, Chicago, 1990.

[3] L. C. Evans and R. F. Gariepy, *Measure Theory and Fine Properties of Functions*, Studies in Advanced Mathematics, CRC Press, Boca Raton, 1992.

[4] O. Ladyzhenskaya, *The Mathematical Theory of Viscous Incompressible Flow*, Gordon and Breach, New York, 1963

[5] S. Stojanovic, Nonsmooth analysis and shape optimization in a flow problem, IMA Preprint Series # 1046

[6] S. Stojanovic and T.P. Svobodny, A free boundary problem for the Stokes equation via nonsmooth analysis, IMA Preprint Series # 1113, to appear in SIAM J. Math. Anal.

[7] T. Svobodny, Computational hybrid-control techniques in hydrodynamics, Proceedings of world congress of nonlinear analysts, 1992

[8] R. Temam, *Navier-Stokes Equations*, North-Holland, New York, 1979

19 A Strong Version of the Lojasiewicz Maximum Principle

H. J. Sussmann Rutgers University, New Brunswick, New Jersey

1 INTRODUCTION

The purpose of this note is to announce and outline the proof of a version of the Maximum Principle of finite-dimensional Control Theory that requires weaker technical assumptions than previous "classical" and "nonsmooth" formulations —such as those of the book [7] by Pontryagin et al. or the "maximum Principle under minimal hypotheses" due to F.H. Clarke (cf. [2])— and at the same time yields a much stronger conclusion. (Full details of the proof will be given in [10].) The result presented here follows and generalizes a remarkable idea of S. Lojasiewicz Jr. [6], who recently proved a new version of the Maximum Principle, in the form of a necessary condition for the terminal point $q_* = \xi_*(b)$ of a reference trajectory $\xi_* : [a,b] \to \Omega$ of a system Σ to belong to the boundary of $\mathcal{R}^{\Sigma}_{[a,b]}(p_*)$ (where Ω is the state space of Σ, $p_* \stackrel{\text{def}}{=} \xi_*(a)$, and $\mathcal{R}^{\Sigma}_{[a,b]}(p_*)$ is the Σ-reachable set from p_* over $[a,b]$) *without requiring the right-hand side of the system equations to be locally Lipschitz with respect to the state variable x.* In Lojasiewicz's result, only the reference vector field (i.e. the vector field $(x,t) \to f_{\eta_*}(x,t)$ giving rise to the reference trajectory) has to be locally Lipschitz with respect to the state variable x —with the usual integral bounds of Carathédory type for the Lipschitz constants— while the other vector fields f_η are only assumed to be continuous with respect to x. We use the same weak hypotheses as Lojasiewicz and, by generalizing his method of proof, are able to incorporate a number of extra features into the result. Specifically, we prove a general "separation theorem," giving a necessary condition for $\mathcal{R}^{\Sigma}_{[a,b]}(p_*)$ to be separated at q_* from some other set S. (We say that two sets S_1, S_2 are *separated at q* if $S_1 \cap S_2 = \{q\}$.) This includes the Lojasiewicz result on controllability along a reference trajectory, and also easily implies the necessary conditions for all the other properties usually considered in the various versions of the Maximum Principle (such as

Partially supported by NSF grant DMS92-02554.

optimal control problems of the Lagrange, Meyer and Bolza forms, with various kinds of constraints on the endpoints, and with variable as well as fixed time), all of which follow by simple transformations (such as adding the cost as an extra state variable) that recasts all these as separation conditions.

Most importantly, we incorporate *high-order point variations*, thereby making our necessary conditions much stronger than those of the usual "smooth" and "nonsmooth" versions. We do this by abstractly defining a class of "variations" much larger than that of the usual needle variations of the classical Maximum Principle. High-order variations make a significant difference whenever a candidate trajectory ξ_* for a property $\mathcal{P}$ —such as noncontrollability or optimality— passes the test of the usual versions of Maximum Principle but can be proved to violate some other "high-order condition." Two simple examples of this phenomenon are given in §6. As these examples show, the situations where high-order conditions matter are by no means exceptional or pathological.

High-order conditions gave rise to a vast classical body of literature in the 1960's and 70's (cf., e.g., the book [5]). More recently, they have attracted renewed interest due to the development of methods for generating high-order variations expressed in terms of Lie brackets (cf., e.g., Sussmann [8] and Kawski [4] for high-order conditions for local controllability from a point, and Bianchini-Stefani [1] for controllability about a reference trajectory). All this work usually assumes a considerable amount of smoothness, and this is undoubtedly one of the reasons why it has by and large failed to establish contact with the parallel developments based on nonsmooth methods, whose primary concern was precisely the elimination of unnecessary requirements such as continuous differentiability. However, it is our belief that the main reason for this failure is not so much the difference in technical assumptions but the difference in *methods of proof*. Whereas most of the work on high-order conditions was based on the classical approach —needle variations and approximating cones— of [7], and sought to extend it by bringing in new classes of variations, the nonsmooth proofs —especially those based on some application of Ekeland's variational principle— follow a different route which does not lend itself so easily to the incorporation of high-order variations.

The proof presented here follows the classical approach of [7]. As explained in §5, the main ingredient of any proof based on this method is the choice of an appropriate "generalized differential" **GD**, i.e., roughly speaking, a way of defining, for certain pairs (M, p) consisting of a (possibly multivalued) map M and a point in the domain of M, the "differential of M at p." If **GD** satisfies a good chain rule and gives rise to a reasonable separation theorem, then we can consider control systems Σ and reference trajectories ξ_* generated by reference controls η_* such that (I) the flow maps $\Phi^{t,s}$ corresponding to η_* are **GD**-differentiable at every point, and (II) the variations in some class **V** are also **GD**-differentiable. For any such Σ, ξ_*, η_*, the chain rule for **GD** together with the open mapping and separation theorems imply a version of the Maximum Principle. The classical result of [7] corresponds to the ordinary differentiability theory, in which (M, p) is differentiable iff M is continuous near p and differentiable at p in the ordinary sense, and the differential of (M, p) is the ordinary differential of M at p. If the reference control η_* gives rise to a vector field $(x, t) \to f_{\eta_*}(x, t)$ which is of class C^1 with respect to x and satisfies the usual Carathéodory measurability conditions and integral bounds, then the corresponding flow maps are of class C^1, so all the pairs $(\Phi^{t,s}, x)$ are differentiable. If the vector fields f_η corresponding to the other controls are continuous in x, bounded by integrable functions of t, and have uniqueness of trajectories, then the needle variations are also differentiable at almost all times t, by standard arguments involving the Lusin and Scorza-Dragoni theorems, and these facts together imply the classical version of the Maximum Principle.

Our argument is essentially that of the classical proof, except that we use, instead of the ordinary differential, a generalized differential, that we define in §5 and call "semidifferential." (Our definition of is an extension of J. Warga's theory of "derivate containers." An alternative candidate, at least for Lipschitz maps, would be F. Clarke's generalized Jacobian, but this object does not satisfy the correct chain rule and is therefore inadequate for our purposes.)

The class of semidifferentiable pairs is much larger than that of pairs that are differentiable in the ordinary sense. For example, all (M, p) such that M is Lipschitz near p are semidifferentiable, as are the needle variations and high-order variations that occur in our control problems. (Since our needle variations are set-valued, because they are built from continuous vector fields whose trajectories need not be unique, it is crucial for our proof that we are able to differentiate set-valued maps.) Semidifferentials turn out to satisfy the chain rule as well as a separation theorem, and this suffices to imply our results.

The geometric formulation of the Maximum Principle as a separation theorem includes a transversality condition, and this requires the choice of an appropriate concept of "tangent cone" to a set at a point. We choose the type of tangent cone that fits most naturally within our general framework, namely, "generalized approximating cones" (GAC's), defined in §3. For convex sets or submanifolds, where there is a natural object that deserves to be called "tangent cone," this object is a GAC in our sense. In other cases, GAC's fare rather well compared with other cones such as the Clarke tangent cone. (It is easy to give examples of sets that have very large GAC's but very small Clarke cones, so that for such sets a transversality condition formulated in terms of GAC's is stronger than one involving a Clarke cone. Also, GAC's have the right behavior with respect to inclusion —in the sense that any GAC of a set S at a point q is also a GAC at q of any set S' containing q— while this property fails for Clarke cones. It should be pointed out, however, that there are also examples of sets with a large Clarke cone and a small GAC, so neither concept of tangent cone is better that the other one in all cases.) Moreover, it can be shown by simple examples that GAC's are more general than the approximating cones considered in [7].

Our results can be extended to incorporate state space constraints as well, but this requires a more complicated argument, so this extension will not be discussed here. Several other extensions are briefly outlined in §2. In particular, we remark that the extension to systems with jumps is completely straightforward, and yields as a byproduct a simple proof of the Maximum Principle on manifolds.

2 STATEMENT OF THE MAIN RESULTS

We use $\mathbb{R}$, $\mathbb{N}$ to denote, respectively, the real line and the set of strictly positive integers. We write $\mathbb{R}_+ = [0, +\infty)$, $\bar{\mathbb{R}} = \mathbb{R} \cup \{-\infty, +\infty\}$, $\bar{\mathbb{R}}_+ = \mathbb{R}_+ \cup \{+\infty\}$. An *interval* is an arbitrary nonempty connected subset of $\mathbb{R}$.

The abbreviation FDRLS stands for "finite-dimensional real linear space." When working with FDRLS's, we will freely endow them with norms. All the concepts defined using norms will in fact be independent of the choice of norm. If X, Y are FDRLS's, then $L(X,Y)$ is the set of all linear maps from X to Y. If X is a FDRLS, then X^* denotes the dual $L(X,\mathbb{R})$ of X. If $x \in X$, $A \in L(X,Y)$, we write Ax rather than $A(x)$. In particular, if $z \in X^*$, $x \in X$, then zx is an alternative notation for $z(x)$, the value of z at x. If $z \in Y^*$, $A \in L(X,Y)$, we write zA rather than $z \circ A$ for the composite map $x \to z(Ax)$ from X to $\mathbb{R}$, so that $zA = A^*(z)$, where $A^* : Y^* \to X^*$ is the adjoint of A. (Notice that with these conventions the identity $z(Ax) = (zA)x$ holds, so the expression zAx is well defined.)

A *set-valued map* (abbreviated SVM) is a set M all whose members are ordered pairs. If M is a SVM, we write $M(x) = \{y : (x,y) \in M\}$, $\mathrm{Dom}(M) = \{x : M(x) \neq \emptyset\}$, $\mathrm{Im}(M) = \cup_x M(x)$. We call M *single-valued at* x if $M(x)$ consists of a single point. (In that case, if $M(x) = \{y\}$, we also write $M(x) = y$.) We call M *single-valued* if it is single-valued at every $x \in \mathrm{Dom}(M)$. If M is a SVM and A is a set, the *restriction* $M\lceil A$ of M to A is the SVM $M \cap (A \times \mathrm{Im}(M))$, and we define $M(A) \stackrel{\mathrm{def}}{=} \mathrm{Im}(M\lceil A)$. If A, B are sets, a *set-valued map from A to B* is a SVM M such that $\mathrm{Dom}(M) \subseteq A$ and $\mathrm{Im}(M) \subseteq B$. We use $SVM(A,B)$ to denote the set of all SVM's from A to B. An *ordinary map* (or just *map*, or *mapping*) from A to B is a single-valued $M \in SVM(A,B)$. The notation $M : A \to B$ will always mean "M is an ordinary map from A to B which is everyhwhere defined, i.e. such that $\mathrm{Dom}(M) = A$." If A is any set, then id_A denotes the identity map of A, i.e. $\mathrm{id}_A = \{(x,x) : x \in A\}$.

If A is a topological space, I is an interval, L is a FDRLS, $\varphi : A \times I \to L$ is a mapping, and $(x,t) \in A \times I$, we use φ_t, φ^x to denote, respectively, the partial maps $y \to \varphi(y,t)$ and $s \to \varphi(x,s)$. An *L-valued Carathéodory function* (or *map*) *on A with time domain I* is a map $\varphi : A \times I \to L$ such that: (i) the function $\varphi_t : A \to L$ is continuous for every $t \in \mathbb{R}$, and (ii) for every $x \in A$, the function $\varphi^x : I \to L$ is Lebesgue measurable. (If A is separable metric, it then follows easily that φ must be jointly Borel$\times$Lebesgue-measurable.) A Carathéodory function $\varphi : A \times I \to L$ will be called *locally integrably bounded* (LIB) if for every compact subset K of A there exists a locally Lebesgue integrable function $\psi_K^B : I \to \mathbb{R}_+$ such that $||\varphi(x,t)|| \leq \psi_K^B(t)$ for all $(x,t) \in K \times I$. If A is metric, then φ is *locally integrably Lipschitz* (LIL) if it is LIB and for every compact subset K of A there exists a locally Lebesgue integrable function $\psi_K^L : I \to \bar{\mathbb{R}}_+$ such that $||\varphi(x,t) - \varphi(y,t)|| \leq \psi_K^L(t)\mathrm{dist}(x,y)$ for all $(x,y,t) \in K \times K \times I$.

If Ω is an open subset of a FDRLS X, and I is an interval, then a *control vector field* (CVF) on Ω with *time domain* I is a Carathéodory map $f : \Omega \times I \to X$. We use $\mathrm{TDom}(f)$ to denote the time domain of the CVF f, and $CVF(\Omega, I)$ to denote the set of all CVF's on Ω with time domain I. (We reserve the words "vector field," or "ordinary vector field" to mean "continuous map from Ω to X," so vector fields are vector-valued maps $x \to f(x)$ that do not depend on t, whereas CVF's are allowed to depend on t.) The concepts of LIB and LIL control vector field are therefore well defined. We call a CVF *locally integrably of class* C^1 (LIC[1]) if f is LIL and each partial map f_t is of class C^1. We use $CVF_{LIB}(\Omega, I)$ (resp. $CVF_{LIL}(\Omega, I)$, $CVF_{LIC^1}(\Omega, I)$) to denote the set of all $f \in CVF(\Omega, I)$ that are LIB (resp. LIL, LIC[1]).

A *curve* in Ω is a continuous map $\xi : I \to \Omega$, where I is an interval. The *graph* of a curve ξ is the subset $\text{Graph}(\xi) \stackrel{\text{def}}{=} \{(\xi(t), t) : t \in \text{Dom}(\xi)\}$ of $\Omega \times \mathbb{R}$. An *arc* in Ω is an absolutely continuous curve in Ω whose domain is a compact interval. A *trajectory* of an $f \in CVF(\Omega, I)$ is an arc ξ such that $\text{Graph}(\xi) \subseteq \Omega \times I$ and $\dot{\xi}(t) = f(\xi(t), t)$ for almost all $t \in \text{Dom}(\xi)$. If $f \in CVF(\Omega, I)$, we use $\text{Traj}(f)$ to denote the set of all trajectories of f.

A *control system* on a FDRLS X is a triple $\Sigma = (\Omega, \mathcal{U}, f)$, where Ω is an open subset of X, $\mathcal{U}$ is a set, and $f = \{f_\eta : \eta \in \mathcal{U}\}$ is a parametrized family of CVF's on Ω. The set Ω is the *state space* of the system, $\mathcal{U}$ is the *open-loop control space*, the elements of $\mathcal{U}$ are the *open-loop controls* or, simply, *controls*, and f is the *dynamics*.

A *controlled trajectory*, or *controlled arc* of a control system $\Sigma = (\Omega, \mathcal{U}, f)$ is a pair $\gamma = (\xi, \eta)$ such that $\eta \in \mathcal{U}$ and $\xi \in \text{Traj}(f_\eta)$. A *trajectory* of Σ is an arc ξ such that (ξ, η) is a controlled arc of Σ for some $\eta \in \mathcal{U}$. We use $\text{Traj}(\Sigma)$ (resp. $\text{CTraj}(\Sigma)$) to denote the set of all trajectories (resp. controlled trajectories) of Σ. If $\gamma = (\xi, \eta) \in \text{CTraj}(\Sigma)$ then we call ξ (resp. γ) a trajectory (resp. controlled trajectory) *Σ-generated by η* or, simply *generated by η* if the context makes it clear which system Σ is being considered. We call a $\gamma = (\xi, \eta) \in \text{CTraj}(\Sigma)$ *LIB-controlled* (resp. *LIL-controlled*, *IC^1-controlled*) if the CVF f_η is LIB (resp. LIL, IC[1]). The *domain* $\text{Dom}(\gamma)$ of a controlled trajectory $\gamma = (\xi, \eta)$ is, by definition, $\text{Dom}(\xi)$. The *starting time* (resp. *terminal time, starting point, terminal point*) of an arc ξ, or of a controlled trajectory $\gamma = (\xi, \eta)$, with domain $[a, b]$ is a (resp. b, $\xi(a)$, $\xi(b)$). If p, q are the starting and terminal points of ξ, or of $\gamma = (\xi, \eta)$, we will say that ξ (or γ) *goes from p to q*, and that η *steers p to q*.

Let $\Sigma = (\Omega, \mathcal{U}, f)$ be a control system on X. If $p, x \in \Omega$, we say that x is *Σ-reachable from p* if there exists a $\gamma = (\xi, \eta) \in \text{CTraj}(\Sigma)$ that goes from p to x. If, in addition, γ can be chosen so that $\text{Dom}(\gamma)$ is a given interval $[a, b]$, then we say that x is *Σ-reachable from p over $[a, b]$*. We define $\mathcal{R}^\Sigma(p)$ (resp. $\mathcal{R}^\Sigma_{[a,b]}(p)$), the *$\Sigma$-reachable set from p* (resp. the *Σ-reachable set from p over $[a, b]$*), to be the set of all points $x \in M$ that are Σ-reachable from p (resp. Σ-reachable from p over $[a, b]$). More generally, we define reachable sets from subsets of Ω, by letting $\mathcal{R}^\Sigma(S) = \cup_{p \in S} \mathcal{R}^\Sigma(p)$ and $\mathcal{R}^\Sigma_{[a,b]}(S) = \cup_{p \in S} \mathcal{R}^\Sigma_{[a,b]}(p)$.

Let $\mathcal{S}$ be a class of subsets of the real line. The system $\Sigma = (\Omega, \mathcal{U}, f)$ will be said to have the *exchange property with respect to $\mathcal{S}$* if, whenever η_1, η_2 belong to $\mathcal{U}$, $E \in \mathcal{S}$, $E \subseteq \text{TDom}(f_{\eta_1}) \cap \text{TDom}(f_{\eta_2})$, it follows that there exist an $\eta_3 \in \mathcal{U}$ such that (i) $\text{TDom}(f_{\eta_1}) \subseteq \text{TDom}(f_{\eta_3})$ and (ii) the identity

$$f_{\eta_3}(x, t) = \chi_E(t) f_{\eta_2}(x, t) + \Big(1 - \chi_E(t)\Big) f_{\eta_1}(x, t) \tag{2.1}$$

holds for $(x, t) \in \Omega \times \text{TDom}(f_{\eta_1})$. (Here χ_E is, for a set E, the indicator function of E.) If Σ has the exchange property with respect to the class of all compact subsets (resp.

compact subintervals) of $\mathbb{R}$, then Σ will be said to have the *compact* (resp. *interval*) *exchange property*.

The following result is the Lojasiewicz Maximum Principle. In the statement, the standard notation $\partial F(x)$ is used for the Clarke generalized Jacobian of a locally Lipschitz map F at a point x. According to our general conventions, $(f_{\eta_*})_t$ is the map $x \to f_{\eta_*}(x,t)$, which is locally Lipschitz for a.e. t if $f_{\eta_*} \in CVF_{LIL}(\Omega, I)$, so $\partial(f_{\eta_*})_t(x)$ is well defined for all $x \in \Omega$, for a.e. $t \in I$. The basic assumptions are

(A1) $\Sigma = (\Omega, \mathcal{U}, f)$ is a control system on a FDRLS X, such that either: (i) Σ has the compact exchange property, or (ii) Σ has the interval exchange property and every f_η, $\eta \in \mathcal{U}$ is locally integrably bounded;

(A2) $\xi_* : [a,b] \to \Omega$ is a trajectory of Σ, $p_* = \xi_*(a)$, and $q_* = \xi_*(b)$.

Theorem 2.1. (The Lojasiewicz Maximum Principle.) *Assume that (A1) and (A2) hold. Then, if q_* is not an interior point of the reachable set $\mathcal{R}^{\Sigma}_{[a,b]}(p_*)$, it follows that for every $\eta_* \in \mathcal{U}$ such that $\gamma_* = (\xi_*, \eta_*)$ is a LIL-controlled trajectory of Σ there exists a pair (A, ζ) such that A is a measurable function from $[a,b]$ to $L(X,X)$ that satisfies*

$$A(t) \in \partial(f_{\eta_*})_t(\xi_*(t)) \text{ for almost all } t \in [a,b], \tag{2.2}$$

and ζ is an absolutely continuous map from $[a,b]$ to X^* that satisfies the three conditions

(AE) $\dot{\zeta}(t) = -\zeta(t)A(t)$ *for a.e.* $t \in [a,b]$,

(NT) $\zeta(t) \neq 0$ *for all* $t \in [a,b]$,

(WHM) $\Big(\forall \eta \in \mathcal{U}\Big)\Big(\zeta(t)f_{\eta_*}(\xi_*(t),t) \leq \zeta(t)f_\eta(\xi_*(t),t)$ *for* a.e. $t \in [a,b]\Big)$.

We will refer to (AE) as the *adjoint equation* corresponding to $A(\cdot)$, and to (NT) and (WHM) as the *nontriviality* and *weak Hamiltonian minimization* conditions. The word "weak" refers to the fact that the exceptional set of measure zero where the inequality is allowed to fail is allowed to depend on η. Naturally, it is much more desirable to have instead the stronger condition

$$\text{(SHM)} \quad \zeta(t)f_{\eta_*}(\xi_*(t),t) = \min\Big\{\zeta(t)f_\eta(\xi_*(t),t) \,:\, \eta \in \mathcal{U}\Big\} \text{ for a.e. } t \in [a,b],$$

which is the version of (WHM) where the exceptional set can be taken not to depend on η. The question whether one can substitute (SHM) for (WHM) is a separate issue, related to the "separability" of Σ, that will be taken up in [10].

We note the following features of Theorem 2.1:

(L1) Only the "reference CVF" f_{η_*} is assumed to be LIL; the other CVF's f_η are only required to be continuous in x and measurable in t.

(L2) A pair (A, ζ) that satisfies all the conclusions has to exist for every choice of $\eta_* \in \mathcal{U}$ such that ξ_* is a trajectory of Σ generated by η_*. This means, in particular, as pointed out by Kaskosz and Lojasiewicz in [3], that one can often get stronger information by adding new CVF's f_η parametrized by η in some set $\mathcal{U}_{new}$, in such a way that no new trajectories are introduced, and then applying the necessary condition to some η_* in the new class.

We now state the simplest version of our Strong Maximum Principle, and do it directly in the form of a separation theorem. Recall from §1 that two sets S_1, S_2 are *separated at* q if $S_1 \cap S_2 = \{q\}$.

Our statement involves two new kinds of objects, namely, "generalized approximating cones" and "variational vectors," and one new property, namely, "compatibility" of a set of variational vectors. These three concepts are defined in §3 below. In addition to (A1) and (A2), we now assume

(A3) Y is a FDRLS, U is an open subset of X, $g : U \to Y$ is a Lipschitz map, $q_* \in U$, S_0, S are subsets of X, Y such that $p_* \in S_0$ and $g(q_*) \in S$;

(A4) S_0 has a generalized approximating cone C_0 at p_*, and S has a generalized approximating cone C at $g(q_*)$;

(A5) either (i) C is not a linear subspace of Y or (ii) $C = \{0\}$ and $g(q_*)$ belongs to $\mathrm{Clos}(S - \{g(q_*)\})$.

Theorem 2.2. (The Strong Maximum Principle.) *Assume that (A1), (A2), (A3), (A4), (A5) hold. Let $\eta_* \in \mathcal{U}$ be such that $\gamma_* = (\xi_*, \eta_*)$ is a LIL-controlled trajectory of Σ. Let $\mathbf{v}$ be a compatible set of variational vectors for γ_*. Then, if $g(\mathcal{R}^{\Sigma}_{[a,b]}(S_0))$ is separated from S at $g(q_*)$, it follows that there exist: (I) a measurable function $A : [a,b] \to L(X,X)$ that satisfies (2.2), (II) a $B \in L(X,Y)$ such that $B \in \partial g(q_*)$, (III) a $w \in Y^* - \{0\}$, and (IV) an absolutely continuous map $\zeta : [a,b] \to X^*$, such that (AE) and (WHM) hold, and in addition (i) $\zeta(b) = wB$, (ii) $wv \leq 0$ for all $v \in C$, (iii) $\zeta(a)v \geq 0$ for all $v \in C_0$, and (iv) $\zeta(t)v \geq 0$ for all $(v,t) \in \mathbf{v}$.*

Theorem 2.2 implies Theorem 2.1 as can be seen by taking $Y = X$, $g = \mathrm{id}_X$, $C = \{0\}$, $S_0 = \{p_*\}$, and S=any subset of X such that $S \cap \mathcal{R}^{\Sigma}_{[a,b]}(p_*) = \{q_*\}$, and $q_* \in \mathrm{Clos}(S \backslash \{q_*\})$.

Theorem 2.2 can be strengthened in a number of ways. For example:

1. The set $\partial g(q_*)$ can be replaced by an arbitrary derivate container of g at q_*. Actually, g can be assumed to be set-valued and semidifferentiable at q_*, provided that $\mathrm{Dom}(g)$ is a neighborhood of q_*.
2. We can consider *systems with jumps.* For instance, one could consider a finite sequence $(\Sigma_0, \ldots, \Sigma_m)$ of systems $\Sigma_i = (\Omega_i, \mathcal{U}_i, f^i)$ on FDRLS's X_i, together with jump times $t_1 < t_2 < \ldots < t_m$ in $[a,b]$ and jump maps $J_i \in SVM(\Omega_{i-1}, \Omega_i)$. A control would now be an $m+1$-tuple $\eta = (\eta_0, \ldots, \eta_m) \in \mathcal{U}_0 \times \ldots \mathcal{U}_m$, and a trajectory for such an η would be an $m+1$-tuple $\xi = (\xi_0, \ldots, \xi_m)$ such that $\xi_i : [t_{i-1}, t_i] \to \Omega_i$ (where we let $t_0 = a$, $t_{m+1} = b$), $(\xi_i, \eta_i) \in \mathrm{CTraj}(\Sigma_i)$, and $\xi_i(t_i) = J_i(\xi_{i-1}(t_i))$. The maps J_i should be assumed to be semidifferentiable at the jump points $x_i = \xi^*_{i-1}(t_i)$ of the reference trajectory ξ^*, with semidifferentials $\mathbf{D}_i$, and their domains should be neighborhoods of the x_i. (In particular, the J_i could be Lipschitz, in which case we can take $\mathbf{D}_i = \partial J_i(x_i)$.) The definition of the reachable set is obvious, and the conclusion is the same, except that now we get finite sequences $A = (A_0, \ldots, A_m)$, $\zeta = (\zeta_0, \ldots, \zeta_m)$ such that $A_i : [t_{i-1}, t_i] \to L(X_i, X_i)$, $\zeta_i : [t_{i-1}, t_i] \to X_i^*$ with the obvious properties, plus the jump condition $(\exists D_i \in \mathbf{D}_i)(\zeta_{i-1}(t_i) = \zeta_i D_i)$. It is clear that the proof sketched in §4 and §5 applies with no change to this situation.

3. The *Maximum Principle on manifolds* follows easily from the result on linear spaces for systems with jumps. (It suffices to partition the domain $[a,b]$ of the reference trajectory ξ_* into intervals $[t_{i-1},t_i]$ such that there are coordinate patches Ω_i for which $\xi_*([t_{i-1},t_i]) \subseteq \Omega_i$. The J_i can then be taken to be the coordinate changes.)
4. The systems with fixed jumps at fixed times considered above can be replaced by systems with optional jumps at optional times, in which case one obtains extra conditions on ζ.
5. State space constraints can be incorporated as well.

3 GENERALIZED APPROXIMATING CONES AND VARIATIONS

If X and Y are FDRLS's, a set-valued map $F \in SVM(X,Y)$ is said to be *compactly valued* if $F(x)$ is compact for every $x \in \mathrm{Dom}(F)$, and *upper semicontinuous* (USC) if, whenever $x \in X$, $U \subseteq Y$ is open, and $F(x) \subseteq U$, there exists a neighborhood V of x such that $F(y) \subseteq U$ for all $y \in V$. (In particular, if F is USC then $\mathrm{Dom}(F)$ must be closed.) We will call F *compact* if Graph(F) is compact, i.e. if F has compact domain and is compactly-valued and USC. If F is compact, then a *single-valued approximating sequence* for F is a sequence $\{F_j\}_{j=1}^{\infty}$ that satisfies the following three conditions:

(SV1) each F_j is a continuous Y-valued map with domain $\mathrm{Dom}(F)$,
(SV2) the maps F_j, $j = 1,2,\ldots$ are uniformly bounded,
(SV3) if $\{x_\ell\}_{\ell=1}^{\infty}$ is any sequence of points in $\mathrm{Dom}(F)$, $j(\ell)$ are integers going to ∞ as $\ell \to \infty$, and the limit $(x,y) = \lim_{\ell\to\infty}(x_{j(\ell)}, F_{j(\ell)}(x_{j(\ell)}))$ exists, then $y \in F(x)$.

We will say that F *has single-valued approximations* if there exists a single-valued approximating sequence for F.

Definition 3.1. *(Regular maps)* If X and Y are FDRLS's, a set-valued map $F \in SVM(X,Y)$ is *regular* if it is compact and has single-valued approximations. ▮

The next definition involves the condition "$F_i(x) = q + x + o(\|x\|)$ as $x \to 0$ through values in N_i," which should be interpreted at stating that

$$\lim_{x\to 0\,,\, x\in N_i\,,\, x\neq 0}\left(\sup\left\{\frac{\|y-x-q\|}{\|x\|} : y \in F_i(x)\right\}\right) = 0\ . \tag{3.1}$$

Definition 3.2. *(Generalized approximating cones)* Let X be a FDRLS, and let $S \subseteq X$, $q \in S$. A *generalized approximating cone* (GAC) for S at q is a closed convex cone C in X which is the closure of the union of an increasing sequence $C_1 \subseteq C_2 \subseteq C_3 \subseteq \ldots$ of closed convex cones such that for each i there exists a regular map $F_i \in SVM(N_i, X)$, defined on a relative neighborhood N_i of 0 in C_i, such that $F_i(N_i) \subseteq S$, $F_i(0) = q$, and $F_i(x) = q + x + o(\|x\|)$ as $x \to 0$ through values in N_i. ▮

If $k \in \mathbb{N}$ and $r > 0$, we use $\sigma_k(r)$ to denote the closed simplex $\{(\alpha_1, \ldots, \alpha_k) \in \mathbb{R}^k : \alpha_1 \geq 0, \ldots, \alpha_k \geq 0, \alpha_1 + \ldots + \alpha_k \leq r\}$. If X is a FDRLS, a *k-parameter variation* in X is a pair $\mathcal{V} = (x_*, V)$ such that $x_* \in X$ and $V \in SVM(X \times \mathbb{R}^k_+, X)$ with the property that there exist $r > 0$ and a neighborhood K of x_* for which (i) $K \times \sigma_k(r) \subseteq \mathrm{Dom}(V)$, (ii) the restriction $V\lceil K \times \sigma_k(r)$ is regular, and (iii) $V(x, 0) = x$ for $x \in K$. The point x_* is the *base point* of the variation $\mathcal{V}$. If in addition there exists a linear map $L : X \times \mathbb{R}^k$ such that

$$\lim_{x \to x_*, \alpha \to 0, \alpha \in \mathbb{R}^k_+} \frac{\sup\{||y - x_* - L(x - x_*, \alpha)|| : y \in V(x, \alpha)\}}{||x - x_*|| + ||\alpha||} = 0\,, \tag{3.2}$$

then we call $\mathcal{V}$ *differentiable*, and the map L is the *differential* of $\mathcal{V}$. Clearly, L has to be of the form $L(z, \alpha) = z + \alpha_1 v_1 + \ldots + \alpha_k v_k$. We write $\mathcal{DV} \stackrel{\mathrm{def}}{=} (v_1, \ldots, v_k)$.

If $\Sigma = (\Omega, \mathcal{U}, f)$ is a control system on X, $\gamma_* = (\xi_*, \eta_*) \in \mathrm{CTraj}(\Sigma)$, $\mathrm{Dom}(\gamma_*) = [a, b]$, and $t_* \in [a, b]$, then a *Σ-admissible k-parameter variation of γ_* at time t_** is a k-parameter variation $\mathcal{V} = (x_*, V)$ in X with base point $x_* = \xi_*(t_*)$ such that

(ADM) for every $\varepsilon > 0$ there exist an $r(\varepsilon) > 0$, a $\tau_-(\varepsilon) \in [a, t_*]$, a $\tau_+(\varepsilon) \in [t_*, b]$, and a neighborhood $K(\varepsilon) \subseteq \Omega$ of $\xi_*(t_*)$ such that $K(\varepsilon) \times \sigma_k(r(\varepsilon)) \subseteq \mathrm{Dom}(V)$, for which the following holds:

(I) if $\xi_1 : [\tau_-(\varepsilon), t_*] \to \Omega$, $\xi_2 : [t_*, \tau_+(\varepsilon)] \to \Omega$ are trajectories of η_* such that $\xi_1(t_*) \in K(\varepsilon)$ and $\xi_2(t_*) \in V(\xi_1(t^*), \alpha)$ for some $\alpha \in \sigma_k(r(\varepsilon))$, if follows that $\xi_2(b) \in \mathcal{R}^{\Sigma}_{[\tau_-(\varepsilon), \tau_+(\varepsilon)]}(\xi_1(a))$.

Definition 3.3. A *variational vector* (VV) for a controlled trajectory γ_* of a control system Σ is a pair (v, t_*) such that there exists a Σ-admissible differentiable 1-parameter variation $\mathcal{V}$ of γ_* at time t_* such that $\mathcal{DV} = v$. A set $\mathbf{v}$ of VV's for γ_* is *compatible* if, whenever $\mathbf{w}$ is a finite subset of $\mathbf{v}$ all whose elements have the same variation time t_*, then there exist a $k \in \mathbb{N}$ and a Σ-admissible differentiable k-parameter variation $\mathcal{V}$ of γ_* at t_* such that, if $\mathcal{DV} = (v_1, \ldots, v_k)$, then $\{v : (v, t_*) \in \mathbf{w}\} \subseteq \{v_1, \ldots, v_k\}$. ∎

4 SKETCH OF THE PROOF: REDUCTION TO THE FINITE CASE

The simplest examples of VV's are the "needle variational vectors," defined as follows. Let η, η_* be open-loop controls such that f_η and f_{η_*} are LIB. Let $t_* \in \mathbb{R}$ be such that $[t_*, t_* + \varepsilon] \subseteq \mathrm{TDom}(f_\eta) \cap \mathrm{TDom}(f_{\eta_*})$ for some $\varepsilon > 0$. For $x \in \Omega$, $\alpha \geq 0$, define $V_{\eta, \eta_*, t_*}(x, \alpha)$ to be the set of all $y \in \Omega$ such that there exist $\delta \in \mathrm{Traj}(f_\eta)$, $\delta_* \in \mathrm{Traj}(f_{\eta_*})$ that satisfy (i) $\mathrm{Dom}(\delta) = \mathrm{Dom}(\delta_*) = [t_*, t_* + \alpha]$, (ii) $\delta(t_*) = x$, (iii) $\delta_*(t_*) = y$, and (iv) $\delta_*(t_* + \alpha) = \delta(t_* + \alpha)$. Then given any $x_* \in \Omega$ the pair $(x_*, V_{\eta, \eta_*, t_*})$ is a 1-parameter variation at x_*. (The compactness of the sets $V_{\eta, \eta_*, t_*}(x, \alpha)$ for (x, α) close to $(x_*, 0)$ and the upper semicontinuity of the SVM V_{η, η_*, t_*} near $(x_*, 0)$ follow from standard properties of solutions of ordinary differential equations. The existence of single-valued approximations follows by regularizing f_η and f_{η_*}.) Moreover, using standard techniques such as the Scorza-Dragoni theorem, one can prove that, if an interval $[a,b]$ is such that $[a,b] \subseteq \mathrm{TDom}(f_\eta) \cap \mathrm{TDom}(f_{\eta_*})$, then there exists a subset $E(\eta,\eta_*)$ of $[a,b]$ such that $[a,b] \backslash E(\eta,\eta_*)$ has measure zero, with the property that $(x_*, V_{\eta,\eta_*,t_*})$ is differentiable for every $t_* \in E(\eta,\eta_*)$ and every $x_* \in \Omega$, and the differential $\mathcal{D}(x_*, V_{\eta,\eta_*,t_*})$ is the vector $f_\eta(x_*,t_*) - f_{\eta_*}(x_*,t_*)$.

Now assume that $\gamma_* = (\xi_*,\eta_*)$ is a LIL-controlled trajectory of Σ with domain $[a,b]$, and η is such that $[a,b] \subseteq \mathrm{TDom}(\eta)$ and $f_\eta \in CVF_{LIB}(\Omega, I)$. If we pick $t_* \in E(\eta,\eta_*)$ and let $x_* = \xi_*(t_*)$, then it is easy to show that $(x_*, V_{\eta,\eta_*,t_*})$ is a Σ-admissible differentiable 1-parameter variation of γ_* at time t_*. (It suffices to pick $\tau_-(\varepsilon) = t_*$, $\tau_+(\varepsilon) = t_* + \varepsilon$.) Since $\mathcal{D}(x_*, V_{\eta,\eta_*,t_*}) = f_\eta(\xi_*(t_*),t_*) - f_{\eta_*}(\xi_*(t_*),t_*)$, we see that

$$\mathbf{v}_{\eta,\gamma_*} = \{(f_\eta(\xi_*(t_*),t_*) - f_{\eta_*}(\xi_*(t_*),t_*),t_*) : t_* \in E(\eta,\eta_*)\}$$

is a set of VV's for γ_*, which is obviously compatible since no two different members of $\mathbf{v}_{\eta,\gamma_*}$ can correspond to the same time t_*.

Using the above observations, it is easy to see that Theorem 2.2 will follow if we prove

Proposition 4.1. *Assume that (A1), (A2), (A3), (A4), (A5) hold. Let $\eta_* \in \mathcal{U}$ be such that $\gamma_* = (\xi_*,\eta_*)$ is a LIL-controlled trajectory of Σ. Let $\mathbf{v}$ be a finite compatible set of variational vectors for γ_*. Then, if $g(\mathcal{R}^{\Sigma}_{[a,b]}(S_0))$ is separated from S at $g(q_*)$, it follows that there exist: (I) a measurable function $A : [a,b] \to L(X,X)$ that satisfies (2.2), (II) a $B \in L(X,Y)$ such that $B \in \partial g(q_*)$, (III) a $w \in Y^* \backslash \{0\}$, and (IV) an absolutely continuous map $\zeta : [a,b] \to X^*$, such that (AE) holds, and in addition (i) $\zeta(b) = wB$, (ii) $wv \leq 0$ for all $v \in C$, (iii) $\zeta(a)v \geq 0$ for all $v \in C_0$, and (iv) $\zeta(t)v \geq 0$ for all $(v,t) \in \mathbf{v}$.*

The main difference between Proposition 4.1 and Theorem 2.2 is that in Proposition 4.1 the Hamiltonian minimization condition does not occur, and the set $\mathbf{v}$ of VV's is finite.

To prove that Proposition 4.1 implies Theorem 2.2, we consider first the case when all the f_η, $\eta \in \mathcal{U}$ are LIB. Fix a —not necessarily finite— compatible set $\mathbf{v}$ of VV's for γ_*. Let Q be the set of all triples (A,B,w) such that

(Q1) A is a measurable function from $[a,b]$ to $L(X,X)$ that satisfies (2.2),
(Q2) $B \in \partial g(q_*)$ (so in particular $B \in L(X,Y)$),
(Q3) $w \in Y^*$, $||w|| = 1$, and $wv \leq 0$ for all $v \in C$,
(Q4) if $\zeta_{A,B,w}$ is the unique absolutely continuous map $\zeta : [a,b] \to X^*$ such that (AE) holds and $\zeta(b) = wB$, then $\zeta_{A,B,w}(a)v \geq 0$ for all $v \in C_0$.

Then it is clear that Q is compact, if the set $\mathcal{A}$ of measurable selections A of $t \to \partial(f_{\eta_*})_t(\xi_*(t))$ is given the weak topology of $L^1([a,b], L(X,X))$. Moreover, $\zeta_{A,B,w}$ (regarded as a member of $L^\infty([a,b],X^*)$) depends continuously on (A,B,w). For each subset $\mathcal{S}$ of $\mathcal{U}$ and each subset $\mathbf{w}$ of $\mathbf{v}$, let $Q(\mathcal{S},\mathbf{w})$ be the set of those $(A,B,w) \in Q$ for which (i) the inequality

$$\zeta_{A,B,w}(t) f_{\eta_*}(\xi_*(t),t) \leq \zeta_{A,B,w}(t) f_\eta(\xi_*(t),t) \text{ for a.e. } t \in [a,b]\,, \tag{4.1}$$

holds for all $\eta \in \mathcal{S}$, and (ii) $\zeta_{A,B,w}(t)v \geq 0$ for all $(v,t) \in \mathbf{w}$. Our goal is to prove that $Q(\mathcal{U},\mathbf{v}) \neq \emptyset$. Suppose we show that $Q(\mathcal{S},\mathbf{w}) \neq \emptyset$ whenever $\mathcal{S}$, $\mathbf{w}$ are finite subsets of $\mathcal{U}$, $\mathbf{v}$. Then the conclusion that $Q(\mathcal{U},\mathbf{v}) \neq \emptyset$ follows by an elementary compactness argument.

Now let $\mathcal{S} \subseteq \mathcal{U}$, $\mathbf{w} \subseteq \mathbf{v}$, and assume that $\mathcal{S}$, $\mathbf{w}$ are finite. Let $\mathcal{S} = \{\eta_1, \ldots, \eta_m\}$. Set $\eta_0 = \eta_*$. Let $\mathbf{w} = \{(w_1,t_1), \ldots, (w_\ell,t_\ell)\}$. Let $k \in \mathbb{N}$. Using Lusin's theorem, find a compact subset F_k of $\cap_{i=1}^m E(\eta_i,\eta_*)$ such that $\text{meas}([a,b]\backslash F_k) < 2^{-k}$, with the property that the functions $t \to f_{\eta_i}(\xi_*(t),t)$ are continuous on F_k for $i = 0,1,\ldots,m$. If $N \in \mathbb{N}$, let $\mathcal{I}_N$ be the set of intervals $[a + \frac{j-1}{N}(b-a), a + \frac{j}{N}(b-a)]$, for $j = 1,\ldots,N$. Let $\mathcal{I}_N^+$ be the set of those $J \in \mathcal{I}_N$ such that $\text{meas}(F_k \cap J) > 0$. Let $F_{k,N} = \cup_{J\in\mathcal{I}_N^+}(F_k \cap J)$. Then $F_{k,N} \subseteq F_k$ and $\text{meas}(F_k\backslash F_{k,N}) = 0$. For fixed k and N, and $J \in \mathcal{I}_N^+$, find m different points $s_{k,N,J}^1, \ldots, s_{k,N,J}^m$ in $F_k \cap J$, making sure that none of the $s_{k,N,J}^i$ is an endpoint of J and that $s_{k,N,J}^i \neq t_j$ for all J, i, j. Let $\mathbf{z}(k,N)$ be the union of $\mathbf{w}$ and the set of all pairs $W_{k,N,J,i} = (f_{\eta_i}(\xi_*(s_{k,N,J}^i), s_{k,N,J}^i) - f_{\eta_*}(\xi_*(s_{k,N,J}^i), s_{k,N,J}^i), s_{k,N,J}^i)$, for all $J \in \mathcal{I}_N^+$ and all i. Then $\mathbf{z}(k,N)$ is a compatible set of VV's for γ_*. (The compatibility follows from the fact that the only way a set V of more that one VV in $\mathbf{z}(k,N)$ can correspond to the same time is if $V \subseteq \mathbf{w}$.) So Proposition 4.1 implies that there exists a triple $(A^{k,N}, B^{k,N}, w^{k,N}) \in Q(\emptyset, \mathbf{z}(k,N))$. Let $\tilde{F}_k = \cap_N F_{k,N}$, so $\text{meas}(F_k\backslash\tilde{F}_k) = 0$. Find a sequence $\{N(j)\}$ going to ∞ such that $\{(A^{k,N(j)}, B^{k,N(j)}, w^{k,N(j)})\}_{j=1}^\infty$ converges to a limit (A^k, B^k, w^k). Then $\zeta_{A^{k,N(j)},B^{k,N(j)},w^{k,N(j)}} \to \zeta_{A^k,B^k,w^k}$ uniformly. If $t \in \tilde{F}_k$, and $i \in \{1,\ldots,m\}$, then $t = \lim s_{k,N(j),J(j)}^i$ for some sequence $\{J(j)\}$. Since $W_{k,N(j),J(j),i} \in \mathbf{z}(k,N(j))$, we have

$$\zeta^{k,j}(s^{k,j,i})\Big(f_{\eta_i}(\xi_*(s^{k,j,i}), s^{k,j,i}) - f_{\eta_*}(\xi_*(s^{k,j,i}), s^{k,j,i})\Big) \geq 0\,, \tag{4.2}$$

where we have written $\zeta^{k,j} \stackrel{\text{def}}{=} \zeta_{A^{k,N(j)},B^{k,N(j)},w^{k,N(j)}}$, $s^{k,j,i} = s_{k,N(j),J(j)}^i$. Letting $j \to \infty$, and using the fact that the functions $t \to f_{\eta_i}(\xi_*(t),t)$ and $t \to f_{\eta_*}(\xi_*(t),t)$ are continuous on F_k, we get

$$\zeta_{A^k,B^k,w^k}(t)\Big(f_{\eta_i}(\xi_*(t),t) - f_{\eta_*}(\xi_*(t),t)\Big) \geq 0 \text{ for } t \in \tilde{F}_k\,. \tag{4.3}$$

Next let $\{k(j)\}$ be a sequence going to ∞ such that $\{(A^{k(j)}, B^{k(j)}, w^{k(j)})\}_{j=1}^\infty$ converges to a limit (A,B,w). Let $F = \cup_K \cap_{k\geq K} \tilde{F}_k$. Then $\text{meas}([a,b]\backslash\tilde{F}) = 0$. If $t \in \tilde{F}$, then $t \in \tilde{F}_{k(j)}$ for sufficiently large j, so $\zeta_{A^{k(j)},B^{k(j)},w^{k(j)}}(t)(f_{\eta_i}(\xi_*(t),t) - f_{\eta_*}(\xi_*(t),t)) \geq 0$ for large j, and then $\zeta_{A,B,w}(t)\Big(f_{\eta_i}(\xi_*(t),t) - f_{\eta_*}(\xi_*(t),t)\Big) \geq 0$. This shows that $(A,B,w) \in Q(\mathcal{S},\mathbf{w})$, so $Q(\mathcal{S},\mathbf{w}) \neq \emptyset$.

This completes the proof that Proposition 4.1 implies the conclusion of Theorem 2.2 for the case of a system Σ all whose CVF's are LIB. The other conclusion can then be proved by applying the compact exchange property, as we now show. Fix a relatively compact open subset $\hat{\Omega}$ of Ω such that ξ_* is entirely contained in $\hat{\Omega}$. Let $\hat{\Sigma} = (\hat{\Omega}, \mathcal{U}, \hat{f})$, where $\hat{f} = \{\hat{f}_\eta\}_{\eta\in\mathcal{U}}$ and $\hat{f}_\eta$ is the restriction of f_η to $\hat{\Omega} \times \text{TDom}(f_\eta)$. Then let $\tilde{\Sigma} = (\hat{\Omega}, \tilde{\mathcal{U}}, \tilde{f})$, where $\tilde{\mathcal{U}} = \{\eta \in \mathcal{U} : f_\eta \in CVF_{LIB}(\Omega, I)\}$ and $\tilde{f} = \{\hat{f}_\eta\}_{\eta\in\tilde{\mathcal{U}}}$. If we apply the result for LIB systems to the system $\tilde{\Sigma}$, we get the desired conclusion, except that

(WHM) is only known to hold for $\eta \in \tilde{\mathcal{U}}$. If $\eta \in \mathcal{U}$, let $\psi(t) = \sup\{||f_\eta(x,t)|| : x \in K\}$, where $K = \mathrm{Clos}(\hat{\Omega})$. Then ψ is measurable and finite. Let $L_N = \{t \in [a,b] : \psi(t) > N\}$. Fix $\varepsilon > 0$. Find N such that $\mathrm{meas}(L_N) < \frac{\varepsilon}{2}$. Find a compact subset F of $[a,b]\backslash L_N$ such that $\mathrm{meas}([a,b]\backslash F) < \varepsilon$. Using the compact exchange property, find a control $\eta_\#$ such that $f_{\eta_\#}$ agrees with f_η on $\Omega \times F$ and with f_{η_*} on $\Omega \times ([a,b]\backslash F)$. Then $\hat{f}_{\eta_\#} \in CVF_{LIB}(\hat{\Omega}, I)$, and therefore $\zeta(t)\Big(f_{\eta_\#}(\xi_*(t),t) - f_{\eta_*}(\xi_*(t),t)\Big) \geq 0$ for almost all $t \in [a,b]$. But then $\zeta(t)\Big(f_\eta(\xi_*(t),t) - f_{\eta_*}(\xi_*(t),t)\Big) \geq 0$ for a. e. $t \in F$, i.e. for all t in a set whose complement has measure $< \varepsilon$. Since ε is arbitrary, the conclusion follows.

5 SKETCH OF THE PROOF: SEMIDIFFERENTIABLE MAPS

To prove Proposition 4.1, it suffices to assume that there are regular maps $F_0 \in SVM(N_0, S_0)$, $F \in SVM(N,S)$, where N_0, N are relative neighborhoods of 0 in C_0, C, such that $F_0(0) = p_*$, $F(0) = g(q_*)$, and $F_0(x) = p_* + x + o(||x||)$, $F(y) = g(q_*) + y + o(||y||)$ as $x, y \to 0$ through values in N_0, N, respectively. Let $t_1 < t_2 < \ldots < t_m$ be the times of the VV's in $\mathbf{v}$. For each i, using the compatibility assumption, find a Σ-admissible differentiable variation $\mathcal{V}_i = (\xi_*(t_i), V_i)$ of γ_*, depending on k_i parameters, with the property that all the v's such that $(v,t_i) \in \mathbf{v}$ occur in $\mathcal{DV}_i$. Let Φ be the flow of the vector field f_{η_*}, so $t \to \Phi^{t,s}(x)$ is the maximal trajectory of f_{η_*} that goes through x at time s.

Let us define a set-valued map $M \in SVM(N_0 \times \mathbb{R}^{k_1} \times \ldots \times \mathbb{R}^{k_m}, Y)$ by letting $y \in M(z, \alpha^1, \ldots, \alpha^m)$ if there exist points $x_0, x_1, \hat{x}_1, x_2, \hat{x}_2, \ldots, x_m, \hat{x}_m, x_{m+1}$ such that $x_0 \in F_0(z)$, $x_1 = \Phi^{t_1,a}(x_0)$, $\hat{x}_1 \in V_1(x_1, \alpha^1)$, $x_2 = \Phi^{t_2,t_1}(\hat{x}^1)$, $\hat{x}_2 \in V_2(x_2, \alpha^2)$, $\ldots$, $x_{m+1} = \Phi^{b,t_m}(\hat{x}_m)$, $y = g(x_{m+1})$.

Then the hypothesis that $g(\mathcal{R}^{\Sigma}_{[a,b]}(S_0)) \cap S = \{g(q_*)\}$ implies that the images of the maps M and F are separated at $g(q_*)$. We want to show that this fact implies our conclusion. This requires using a "separation theorem," i.e. a necessary condition for two set-valued maps $\mu_i \in SVM(Q_i, Y)$ that map 0 to a point q to have the property that $\mathrm{Im}(\mu_1)$ and $\mathrm{Im}(\mu_2)$ are separated at q.

Notice that, if the μ_i are linear, the sets Q_i are convex and compact, and $q = 0$, then a necessary condition for $\mathrm{Im}(\mu_1) \cap \mathrm{Im}(\mu_2) = \{q\}$ is the existence of a nontrivial $w \in Y^*$ such that $w\mu_1 x \geq 0$ for all $x \in Q_1$ and $w\mu_2 x \leq 0$ for all $x \in Q_2$, *provided that at least one of the sets $\mu_i Q_i$ does not contain* 0 *in its interior relative to its linear span.* What we need is a generalization of this fact to the case when the μ_i are nonlinear and multivalued. Naturally, such a generalization should involve the "differentials" of the μ_i, so our first task is to single out a class of maps that can be differentiated in a reasonable sense and give rise to a good separation theorem. Moreover, since our map M is constructed by composing maps of two kinds —variations and Lipschitz flow maps— we should make sure that (I) our class contains all Lipschitz maps as well as our variations and (II) the "differential" of M can be computed in terms of the differentials of the variations and those of the flow maps, by means of some version of the chain rule.

For Lipschitz maps, one possible theory of generalized differentials is given by the Clarke generalized Jacobian. This theory, however, is not adequate here, because the chain rule for generalized Jacobians is not good enough, since all that can be said about the generalized gradient of a composite map is that it is contained in the convex hull of the composite of the generalized gradients. For Lipschitz flows, which are essentially iterates of a large number of Lipschitz maps, this gives a generalized gradient which is too large to be useful.

The correct theory for Lipschitz maps turns out to be that of *derivate containers*, introduced by J. Warga (cf. [11], [12], [13], [14]).

Definition 5.1. Let X, Y be FDRLS's, let Ω be an open subset of X, and let $F : \Omega \to Y$ be a map. A *Warga derivate container* of F at a point $p \in \Omega$ is a compact subset $\mathbf{D}$ of $L(X,Y)$ with the property that, if U is any neighborhood of $\mathbf{D}$ in $L(X,Y)$, then there exist a neighborhood V of p in Ω and a sequence $\{F_j\}$ of maps of class C^1 from V to Y such that (i) $F_j \to F$ uniformly on V as $j \to \infty$, and (ii) for every j, and every $x \in V$, the differential $DF_j(x)$ belongs to U. ▮

A function that has a Warga derivate container at p in the sense of Definition 5.1 is automatically Lipschitz near p, since it is a uniform limit of a sequence of functions that are uniformly Lipschitz on a neighborhood V of p. So the theory of derivate containers does not yet meet all our requirements, since we have to be able to differentiate maps such as the V_i, that need not be Lipschitz or even single-valued.

To produce a theory with all the desirable features we have to go beyond Lipschitz maps, and define the class of *semidifferentiable set-valued maps*:

Definition 5.2. If X, Y are FDRLS's, a set-valued map $F \in SVM(X,Y)$ is *semidifferentiable* at a point $p \in X$ if (i) F is regular, (ii) Dom(F) is a compact convex subset of X with nonempty interior, (iii) $p \in \mathrm{Dom}(F)$, (iv) F is single-valued at p, and (v) there exists a Lipschitz map $\Psi : V \to Y$, defined on a neighborhood V of p, such that $F(x) = \Psi(x) + o(||x-p||)$ as $x \to p$ through values in $V \cap \mathrm{Dom}(F)$ (i.e. $\lim_{x\to p, x\in \mathrm{Dom}(F), x\neq p} \frac{\sup\{||y-\Psi(x)||: y\in F(x)||\}}{||x-p||} = 0$). In that case, if $\mathbf{D}$ is a derivate container of Ψ at p, then $\mathbf{D}$ is a *semidifferential* of F at p. ▮

The following two results are the chain rule and the separation theorem for semidifferentiable maps. (The chain rule is trivial, but we call it a theorem because of the crucial role it plays in our proof of the Maximum Principle. The separation theorem is proved in [9].)

Theorem 5.1. *Let X^1, X^2, X^3 be FDRLS's, and let K^1, K^2 be compact convex subsets of X^1, X^2 with nonempty interior. Let $F^1 \in SVM(X^1, X^2)$, $F^2 \in SVM(X^2, X^3)$ be SVM's with domains K^1, K^2. Assume that $\mathrm{Im}(F^1) \subseteq K^2$. Define the composite map $F = F^2 \circ F^1 \in SVM(X_1, X_3)$ by*

$$(F^2 \circ F^1)(x^1) \stackrel{\mathrm{def}}{=} \{x^3 \in X^3 : (\exists x^2 \in F^1(x^1))(x^3 \in F^2(x^2))\} . \tag{5.1}$$

Then: (i) if F^1 and F^2 are compactly-valued and upper semicontinuous, it follows that F is compactly-valued and upper semicontinuous as well, (ii) if F^1 and F^2 are regular, then

F is regular, (iii) if $p^1 \in K^1$, $p^2 \in K^2$, $F^1(p^1) = p^2$, and F^1, F^2 are semidifferentiable at p^1, p^2 with semidifferentials $\mathbf{D}^1$, $\mathbf{D}^2$, then (a) F is semidifferentiable at p^1 and (b) if

$$\mathbf{D} = \mathbf{D}^2 \circ \mathbf{D}^1 = \{D^2 \circ D^1 : D^1 \in \mathbf{D}^1\,,\, D^2 \in \mathbf{D}^2\}\,, \tag{5.2}$$

then $\mathbf{D}$ is a semidifferential of F at p^1. ∎

Theorem 5.2. *Let X, Y be FDRLS's, let K be a compact convex subset of X with nonempty interior, and let $F \in SVM(X,Y)$ be such that $\mathrm{Dom}(F) = K$. Let $p \in K$ be such that F is single-valued at p, and write $q = F(p)$. Let $K_0 = \{x - p : x \in K\}$. Assume that F is semidifferentiable at p with semidifferential $\mathbf{D}$. Let $S \subseteq Y$ be such that $q \in S$, and assume that S has a generalized approximating cone C at q, such that either (i) C is not a linear subspace of Y, or (ii) $C = \{0\}$ and $q \in \mathrm{Clos}(S - \{q\})$. Finally, assume that $\mathrm{Im}(F)$ is separated from S at q. Then there exist $D \in \mathbf{D}$ and $w \in Y^*\backslash\{0\}$ such that $wDx \geq 0$ for every $x \in K_0$ and $wv \leq 0$ for every $v \in C$.* ∎

Notice that, if $X, Y, K, F, p, q, K, K_0, \mathbf{D}$ are as in the hypotheses of Theorem 5.2, and $q \notin \mathrm{Int}(\mathrm{Im}(F))$, then we can satisfy the remaining hypotheses of the theorem by taking $C = \{0\}$ and letting S be a sequence $\{q_j\}$ such that $q_j \to q$ but $q_j \notin \mathrm{Im}(F)$. We can then conclude that there is a $w \in Y^*\backslash\{0\}$ and a $D \in \mathbf{D}$ such that $wy \geq 0$ for every $y \in DK_0$, so $0 \notin \mathrm{Int}(DK_0)$. So Theorem 5.2 implies the *open mapping theorem*: if $X, Y, K, F, p, q, K, K_0, \mathbf{D}$ are as in the hypotheses of Theorem 5.2, and $0 \in \mathrm{Int}(DK_0)$ for every $D \in \mathbf{D}$, then $q \in \mathrm{Int}(\mathrm{Im}(F))$.

To prove Theorem 2.2 we use Theorems 5.1 and 5.2. The maps F_0, V_i, F, g and the flow maps are all semidifferentiable at the appropriate points, so if one knows how to compute semidifferentials for each of them, then the chain rule tells us how to get a semidifferential for M. Semidifferentials of F_0, the V_i, F and g are easily evaluated. For the flow maps, we use the following fact:

Proposition 5.1. *Let $a \leq b$, let X be a FDRLS, and let $\Omega \subseteq X$ be open. Let $f \in CVF_{LIL}(\Omega, [a,b])$. Let $\xi : [a,b] \to \Omega$ be a trajectory of f, and let $\mathcal{A}$ be the set of all measurable selections of $t \to \partial f_t(\xi(t))$. For each $A \in \mathcal{A}$, let $D_A : [a,b] \to L(X,X)$ be the solution of $\dot{D}(t) = A(t)D(t)$ such that $D(a) = \mathrm{id}_X$. Let Φ be the flow map corresponding to f over $[a,b]$ (that is, $\Phi(x) = y$ iff $(\exists \xi \in \mathrm{Traj}(f))(\xi(a) = x \wedge \xi(b) = y)$). Let $\mathbf{D} = \{D_A(b) : A \in \mathcal{A}\}$. Then $\mathbf{D}$ is a derivate container of Φ at $\xi(a)$.* ∎

We can now complete the proof of Theorem 2.2. For simplicity, assume that the number m of variation times is just 1. Write $k = k_1$, $\alpha = \alpha^1$, $V = V_1$, $\mathcal{V} = \mathcal{V}_1$, $\tau = t_1$, so M is the set-valud map given by

$$M(z, \alpha) = g(\Phi^{b,\tau}(V(\Phi^{\tau,a}(F_0(z)), \alpha)))\,. \tag{5.3}$$

Let $\mathcal{DV} = (v_1, \ldots, v_k)$. Let $\mathbf{D_1}$, $\mathbf{D}_2$ be, respectively, derivate containers of $\Phi^{\tau,a}$ at p_* and of $\Phi^{b,\tau}$ at $\xi_*(\tau)$. Use the fact that $\partial g(q_*)$ is a derivate container of g at q_*. The chain rule implies that, if we define $\hat{\mathbf{D}}$ to be the set of all maps

$$\hat{\mathbf{D}}_{D_1,D_2,B} : (z, \alpha) \to B(D_2(D_1 z + \sum_{i=1}^{k} \alpha_i v_i))\,, \tag{5.4}$$

for $D_1 \in \mathbf{D}_1$, $D_2 \in \mathbf{D}_2$, $B \in \partial g(q_*)$, then $\hat{\mathbf{D}}$ is a derivate container of M at $(0,0)$. If we now choose $\mathbf{D}_1$ and $\mathbf{D}_2$ to be the derivate containers given by Proposition 5.3, and apply Theorem 5.2, the conclusion follows.

6 HIGH ORDER VARIATIONS AND NONLIPSCHITZ RIGHT-HAND SIDES

We now give three examples, to illustrate the power of Theorem 2.2. A systematic discussion of high-order conditions in relation to Theorem 2.2 will be included in [10].

The first example illustrates how the Lojasiewicz Maximum Principle can apply in situations where the classical and nonsmooth versions do not:

Example 6.1. Consider a control system in $\mathbb{R}^2$ of the form $\dot{x} = y + u\mu(x,y)$, $\dot{y} = u\nu(x,y)$, $|u| \leq 1$, where μ, ν are continuous functions such that $\mu(0,0) = 0$ and $\nu(0,0) \neq 0$. Is Σ locally controllable from $(0,0)$? A formal application of the Maximum Principle yields a positive answer, except for the fact that Σ does not satisfy the technical hypotheses of the classical and nonsmooth versions, because the right-hand side need not be locally Lipschitz in x and y. On the other hand, the Lojasiewicz Maximum Principle provides a rigorous justification of the formal argument, since the reference trajectory $t \to \xi(t) = (0,0)$ corresponds to the control $u(t) \equiv 0$, and the right-hand side of Σ is of class C^1 when $u = 0$. ∎

The next two examples are well known in the literature on high-order conditions, but perhaps less so in the nonsmooth analysis community. They illustrate how, in very simple situations, it can happen that high-order conditions settle controllability or optimality questions for which the Maximum Principle alone does not suffice, and how these conditions involve certain Lie brackets. We first describe in each case how the new Lie bracket information solves the problem, and then state two theorems showing that, in fact, these solutions are special cases of Theorem 2.2. The applicability of Theorem 2.2 to the second example —which is a time-optimal control problem rather than one of controllability— presupposes the optimal control version of Theorem 2.2 which, as explained in §1, is a simple corollary of Theorem 2.2. (As is well known, the conclusion in this case —for autonomous systems— is the existence of an adjoint variable with the usual properties plus the fact that the constant value of the Hamiltonian is ≤ 0.)

Example 6.2. Consider the control system $\dot{x}_1 = u_1$, $\dot{x}_2 = u_2$, $\dot{x}_3 = u_2 x_1$ in $\mathbb{R}^3$, with control constraints $|u_1| \leq 1$, $|u_2| \leq 1$. If $\xi_* : [0,1] \to \mathbb{R}^3$ is the constant trajectory given by $\xi_*(t) \equiv (0,0,0)$, then $\xi_*(1)$ is an interior point of the reachable set from $\xi_*(0)$ —i.e. our system is locally controllable from $(0,0,0)$— but this fact cannot be detected by applying the usual Maximum Principle, since ξ_* is a Pontryagin extremal. The controllability along ξ_* can, however, be easily detected by an elementary Lie bracket calculation. It suffices to observe that, if we write our system in vector form as $\dot{x} = u_1 f_1(x) + u_2 f_2(x)$, where the vector fields f_1, f_2 have components $(1,0,0)$ and $(0,1,x_1)$, respectively, then $[f_1, f_2]$ has components $(0,0,1)$, so $f_1(0,0,0)$, $f_2(0,0,0)$

and $[f_1, f_2](0,0,0)$ are linearly independent, and then Chow's Theorem implies local controllability from 0. ∎

Example 6.3. Consider the control system $\dot{x}_1 = 1 + x_2^2$, $\dot{x}_2 = u$ in $\mathbb{R}^2$, with control constraint $|u| \leq 1$. Suppose we want to go from $(0,0)$ to $(1,0)$ in minimum time. One possible candidate is the trajectory $\xi_* : [0,1] \to \mathbb{R}^2$ given by $\xi_*(t) = (t,0)$, corresponding to the "singular" (i.e. not bang-bang) control $u_*(t) \equiv 0$. This trajectory passes the test of the usual Maximum Principle, but it is easy to see by direct inspection that ξ_* is not optimal, since it is clear that motion from left to right is much faster if one is far away from the x_1 axis, so a "bang-bang" control of the form $u(t) = 1$ for $0 \leq t \leq \tau$, $u(t) = -1$ for $\tau < t \leq 2\tau$ ought to do better than u_*. (The actual value of τ turns out to be the solution of $6\tau + 2\tau^3 = 3$, which obviously satisfies $0 < \tau < \frac{1}{2}$, and the optimal time is $2\tau \sim 0.93$.) The nonoptimality of u_* can be detected by means of a high-order test: we write our system in the form $\dot{x} = f(x) + ug(x)$, where f, g have components $(1 + x_2^2, 0)$ and $(0,1)$, respectively, and verify that the bracket $[g,[f,g]]$ has components $(-2,0)$. The well known Legendre-Clebsch condition says that, for our trajectory to be optimal, there must exist an adjoint vector $t \to \zeta(t)$ that satisfies the usual conditions of the Maximum Principle, as well as the "high-order condition" $\zeta(t)[g,[f,g]](\xi_*(t)) \leq 0$. In our case, if we write $\zeta = (\zeta_1, \zeta_2)$, the Hamiltonian is $H = \zeta_1(1 + x_2^2) + \zeta_2 u$. The adjoint equation says that $\dot{\zeta}_1 = 0$, $\dot{\zeta}_2 = -2\zeta_1 x_2$, which reduces to $\dot{\zeta}_1 = \dot{\zeta}_2 = 0$ along our trajectory ξ_*. The value of the Hamiltonian along (ξ_*, u_*) is ζ_1. The fact that u_* must minimize the Hamiltonian says that ζ_1 must vanish. The Maximum Principle also requires that $H \leq 0$ along (ξ_*, u_*), and that ζ be $\neq 0$. All these conditions can be satisfied by taking $\zeta = (c, 0)$, where $c < 0$ is a constant, and these are the only choices of ζ that meet all the requirements. However, the extra condition that $\zeta[g,[f,g]] \leq 0$, implies $-2c \leq 0$, which is a contradiction. So ξ_* is not optimal. ∎

We now show how the application of Chow's Theorem in Example 6.2 and the use of the Legendre-Clebsh condition in Example 6.3 are particular cases of Theorem 2.2. This follows from the following two facts, proved in [10]:

Theorem 6.1. *Let Π be a convex polyhedron in $\mathbb{R}^m$, and let $\mathcal{F}(\Pi)$ be the set of open faces of Π. Let $g_1, \ldots, g_m$ be vector fields of class C^1 on an open subset Ω of $\mathbb{R}^n$, and write $ug = u_1 g_1 + \ldots + u_m g_m$ for $u = (u_1, \ldots, u_m) \in \mathbb{R}^m$. Let Σ be the control system $\dot{x} = ug(x)$, $u \in \Pi$, where the open-loop controls are all measurable Π-valued functions defined on compact intervals. Let $\gamma_* = (\xi_*, \eta_*)$ be a controlled trajectory of Σ, and let $[a,b] = \mathrm{Dom}(\gamma_*)$. For each $F \in \mathcal{F}(\Pi)$, let $E_F(\eta_*) = \{t \in [a,b] : \eta_*(t) \in F\}$. Then, if $F \in \mathcal{F}(\Pi)$ and $v, w \in \mathbb{R}^m$ are parallel to F, it follows that $([vg, wg](\xi_*(t)), t)$ is a variational vector for γ_* for almost every $t \in E_F(\eta_*)$.* ∎

(Recall that an open face of Π is, by definition, a maximal member —with respect to inclusion— of the set $\{F : (\exists S \in Aff(\mathbb{R}^m))(F = \mathrm{Int}_S(\Pi \cap S))\}$, where $Aff(\mathbb{R}^m)$ is the set of affine subspaces of $\mathbb{R}^m$.)

Theorem 6.2. *Let f, g be vector fields of class C^1 on an open subset Ω of $\mathbb{R}^n$, and let Σ be the control system $\dot{x} = f(x) + ug(x)$, $u \in [-1, 1]$, where the open-loop controls are all measurable $[-1, 1]$-valued functions defined on compact intervals. Let $\gamma_* = (\xi_*, \eta_*)$ be a controlled trajectory of Σ, and let $[a, b] = \mathrm{Dom}(\gamma_*)$. Let $E(\eta_*) = \{t \in [a, b] : -1 < \eta_*(t) < 1\}$. Then $(-[g, [f, g]](\xi_*(t)), t)$ is a variational vector for γ_* for almost every $t \in E(\eta_*)$.* ∎

REFERENCES

1 R. M. Bianchini and G. Stefani, Local controllability about a reference trajectory, in *Analysis and Optimization of Systems*, A. Bensoussan and J.L. Lions Eds., Lect. Notes on Control 83, Springer-Verlag, Berlin, 1986.

2 F. H. Clarke, *Optimization and Nonsmooth Analysis,* Wiley Interscience, New York, 1983.

3 B. Kaskosz and S. Lojasiewicz Jr., "A Maximum Principle for generalized control systems," *Nonlinear Anal. TMA*, **9** (1985), pp. 109-130.

4 M. Kawski, "High-order small-time local controllability," in *Nonlinear Controllability and Optimal Control,* H.J. Sussmann ed., M. Dekker Inc. (1990), pp. 431-467.

5 H. Knobloch, *High Order Necessary Conditions in Optimal Control Theory*, Springer Lect. Notes in Control and Information Sciences, No. 34, New York (1981).

6 S. Lojasiewicz Jr., Local controllability of parametrized differential equations, in preparation.

7 L.S. Pontryagin, V.G. Boltyanskii, R.V. Gamkrelidze and E.F. Mischenko, *The Mathematical Theory of Optimal Processes*, Wiley, New York, 1962.

8 H. J. Sussmann, A general theorem on local controllability, *S.I.A.M. J. Control and Optimization* 25, No. 1: 158-194 (1987).

9 H.J. Sussmann, Open mapping and separation theorems for semidifferentiable maps, in preparation.

10 H.J. Sussmann, A strong Maximum Principle under weak hypotheses, in preparation.

11 J. Warga, Fat homeomorphisms and unbounded derivate containers, *J. Math. Anal. Appl.* **81** (1981), 545-560.

12 J. Warga, Controllability, extremality and abnormality in nonsmooth optimal control, *J. Optim. Theory Applic.* **41** (1983), 239-260.

13 J. Warga, Optimization and controllability without differentiability assumptions, *SIAM J. Control and Optimization* **21** (1983), 837-855.

14 J. Warga, Homeomorphisms and local C^1 approximations, *Nonlinear Anal. TMA* **12** (1988) 593-597.

20 On the Relationship Between the Optimal Quadratic Cost Problems on an Infinite Horizon, and on a Finite Horizon with Final Time Penalization: The Abstract Hyperbolic Case

Roberto Triggiani University of Virginia, Charlottesville, Virginia

Abstract

Given an abstract hyperbolic equation, we consider two optimal control problems with quadratic cost: the first over an infinite horizon, the second over a finite horizon with suitable finite state penalization. We then provide a new proof of the interesting relationship between them. In the simpler case (say, distributed control), where both Algebraic Riccati Equation (for the first problem) and Differential Riccati Equation (for the second problem) are available, such relationship is an immediate consequence of comparing these two equations. In the general case of boundary control and non-smoothing finite state penalization, however, no Differential Riccati Equation is available for the second problem. Our proof is Riccati equation-independent. Instead, it relies on intrinsic optimization properties of the two problems. Said relationship, besides being of interest in itself, plays a critical role in the numerical analysis of algebraic Riccati equations in the abstract hyperbolic case [L.1], [L-T.3].

Research partially supported by the National Science Foundation under Grant NSF-DMS-9204338.

1 Introduction. Statement of main result

1.1 Introduction and background material

Dynamical model. In this paper we consider the abstract differential equation

$$\dot{y} = Ay + Bu \text{ on, say,} [\mathcal{D}(A^*)]', \quad y(0) = y_0 \in Y, \tag{1.1}$$

A^* being the Y-adjoint of A, Y a Hilbert space, subject to the following assumptions

(H.1): $A : Y \supset \mathcal{D}(A) \to Y$ is the generator of a s.c. semigroup e^{At} on Y, $t \geq 0$, with (without loss of generality) $A^{-1} \in \mathcal{L}(Y)$;

(H.2): B is a (linear) continuous operator $U \to [\mathcal{D}(A^*)]'$: equivalently,

$$A^{-1}B \in \mathcal{L}(U; Y), \tag{1.2}$$

so that $B^* \in \mathcal{L}(\mathcal{D}(A^*); U)$ is defined by $(B^*x, u)_U = (x, Bu)_Y$, $u \in U$, $x \in \mathcal{D}(A^*)$;

(H.3): the (closable) operator $B^*e^{A^*t}$ can be extended as a map

$$B^*e^{A^*t} : \text{ continuous } Y \to L_2(0, T; U), \text{i.e.,} \tag{1.3a}$$

$$\int_0^T \|B^*e^{A^*t}x\|_U^2 dt \leq c_T \|x\|_Y^2, \quad x \in Y \tag{1.3b}$$

for all $0 < T < \infty$. Thus, by duality on (1.3), we obtain that [FLT, Appendix A]

$$(L_{0,T}u)(t) \doteq \int_0^t e^{A(t-\tau)}Bu(\tau)d\tau \tag{1.4a}$$

$$: \text{continuous } L_2(0, T; U) \to C([0, T]; Y) \tag{1.4b}$$

The solution to problem (1.1) is then given by

$$y(t) = y(t; y_0) = e^{At}y_0 + (L_{0,T}u)(t) \tag{1.5a}$$

$$: \text{continuous } \{y_0, u\} \in Y \times L_2(0, T; U) \to C([0, T]; Y) \tag{1.5b}$$

Remark 1.1. Assumption (H.3) = (1.3) is an abstract trace theory property. Over the past ten years, this property has been proved to hold true for many classes of partial differential equations (PDEs) by purely PDE methods (energy methods either in differential or in pseudodifferential form), including: second-order hyperbolic equations; Euler-Bernoulli, Kirchhoff, and Schrödinger equations; first-order hyperbolic systems, etc., all in arbitrary space dimensions and on explicitly identified spaces, see, e.g., class (H.2) of [L-T.1].

Optimal control problem. Interval $[0, T]$. We associate with the dynamics (1.1), or (1.5), the quadratic cost functional

$$J_{0,T;G}(u, y) = \int_0^T [\|Ry(t)\|_Z^2 + \|u(t)\|_U^2]dt + \|Gy(T)\|_{Z_f}^2, \tag{1.6}$$

where Z, Z_f are two other Hilbert spaces, and

(H.4)

$$R \in \mathcal{L}(Y;Z); \quad G \in \mathcal{L}(Y;Z_f). \tag{1.7}$$

The corresponding optimal control problem $O.C.P._{0,T;G}$ is:

Minimize

$$\begin{aligned} &J_{0,T;G}(u,y) \text{ over all } u \in L_2(0,T;U) \\ &\text{where } y \text{ is the solution of (1.1) due to } u. \end{aligned} \tag{1.8}$$

Assumptions (H.1) through (H.4) are in force throughout this paper.

Interval $[s,T]$. If the initial time for the dynamics (1.1) is $t = s \geq 0$, with corresponding initial condition $y(s) = y_0$, the resulting solution is now denoted by $y(t) \equiv y(t;s;y_0)$, so that $y(t;0;y_0)$ may be simply denoted by $y(t;y_0)$. The corresponding optimal control problem $O.C.P._{s,T;G}$ over the interval $[s,T]$, $s \leq T < \infty$, is then:

Minimize over all $u \in L_2(s,T;U)$ the functional cost

$$J_{s,T;G}(u,y) = \int_s^T \left[\|Ry(t)\|_Z^2 + \|u(t)\|_U^2\right] dt + \|Gy(T)\|_{Z_f}^2 \tag{1.9}$$

Optimal control problem. Interval $[0,\infty]$. We next consider the optimal control problem over the infinite horizon O.C.P$_\infty$: minimize

$$J_\infty(u,y) \equiv \int_0^\infty [\|Ry(t)\|_Z^2 + \|u(t)\|_U^2]dt \tag{1.10}$$

over all $u \in L_2(0,\infty;U)$, where $y(t) = y(t;y_0)$ is the solution of Equation (1.1) under the following control-theoretic assumptions.

F.C.C. (Finite Cost Condition): For every y_0, there exists a $\bar{u} \in L_2(0,\infty;U)$, such that the corresponding solution $\bar{y}$ of (1.1) satisfies $R\bar{y} \in L_2(0,\infty;Y)$, so that the corresponding cost is finite: $J_\infty(\bar{u},\bar{y}) < \infty$. (1.11)

Detectability Condition. [F-L-T.1] There exists a linear, densely defined operator $K : Z \supset \mathcal{D}(K) \to Y$ satisfying the following two conditions

(i)

$$\|K^*\|_Z^2 \leq c[\|B^*x\|_U^2 + \|x\|_Y^2], \quad \forall\, x \in \mathcal{D}(B^*) \subset Y; \tag{1.12a}$$

(ii) the s.c. semigroup $e^{A_K t}$ on Y, generated by

$$A_K = A + KR \tag{1.12b}$$

(as guaranteed by [F-L-T.1]), is uniformly stable on Y: there exist constants $M_k \geq 1$, $k > 0$, such that

$$\|e^{A_K t}\|_{\mathcal{L}(Y)} \leq M_k e^{-kt}, \quad t \geq 0. \tag{1.12c}$$

Under present assumptions, the above optimal control problems are studied in [F-L-T.1] and [L-T.2], see also [L-T.1], [L-T.4]. For clarity of exposition, we summarize below the relevant results from these references to be invoked in the sequel.

Summary of results of the $O.C.P_{s,T;G}$ in (1.9)

(1) There exists a unique optimal pair $\{u^0_{T,G}(\,\cdot\,;s;y_0), y^0_{T,G}(\,\cdot\,;s;y_0)\}$ of the optimal control problem $O.C.P._{s,T;G}$ in (1.9):

$$u^0_{T,G}(\,\cdot\,;s;y_0) \in L_2(s,T;U); \quad y^0_{T,G}(\,\cdot\,;s;y_0) \in C([s,T];Y). \tag{1.13}$$

(2) There exists an operator $P_{T,G}(t) \in \mathcal{L}(Y)$, $0 \le t \le T$, given explicitly by

$$P_{T,G}(t)x = \int_t^T e^{A^*(\tau-t)} R^* R y^0_{T,G}(\tau;t;x)d\tau + e^{A^*(T-t)} G^* G y^0_{T,G}(T;t;x) \tag{1.14a}$$

$$: \text{continuous } Y \to C([0,T];Y), \tag{1.14b}$$

such that

(i)
$$u^0_{T,G}(t;s;y_0) = -B^* P_{T,G}(t) y^0_{T,G}(t,s;y_0) \in L_2(s,T;U); \tag{1.15}$$

(ii)
$$P_{T,G}(t) \equiv P^*_{T,G}(t) \ge 0, \quad 0 \le t \le T; \tag{1.16}$$

(iii)
$$\begin{aligned} (P_{T,G}(t)x,x)_Y &= \int_t^T \left[\|Ry^0_{T,G}(\tau;t;x)\|^2_Z + \|u^0_{T,G}(\tau;t;x)\|^2_U\right] d\tau \\ &\quad + \|Gy^0_{T,G}(T;t;s)\|^2_{Z_f} \qquad (1.17a) \\ &= J^0_{t,T;G}(x) = J_{t,T;G}(u^0_{T,G}(\,\cdot\,;t;x), y^0_{T,G}(\,\cdot\,;t;x)). \qquad (1.17b) \end{aligned}$$

(3) Setting (we drop G)

$$\Phi_T(t,s)x \equiv y^0_{T,G}(t;s;x) \in C([s,T];Y), \tag{1.18}$$

$\Phi_T(t,s) \in \mathcal{L}(Y)$ is an evolution operator satisfying

$$\Phi_T(t,s) = \Phi_T(t,\tau)\Phi_T(\tau,s), \quad s \le \tau \le t. \tag{1.19}$$

Moreover,

$$\Phi_T(t+\sigma,t) = \Phi_{T-t}(\sigma,0). \tag{1.20}$$

Remark 1.2. We explicitly point out that in the present generality, no claim is made that $P_{T,G}(t)$ satisfies a Differential Riccati Equation, see [L-T.1, p. 28]. This is the case, instead, if minimal assumptions of smoothness are imposed on R and G [D-L-T.1], [L-T.2]. The lack of a Differential Riccati equation in the present setting is a key point in the problem investigated in this paper, see Remark 1.4 below. □

Summary of results of the $O.C.P_\infty$ in (1.10).

(a) There exists a unique optimal pair $\{u^0_\infty(\,\cdot\,;0;y_0), y^0_\infty(\,\cdot\,;0;y_0)\}$ of the optimal control problem $O.C.P_\infty$;

(b) there exists a non-negative, self-adjoint operator $0 \le P_\infty = P^*_\infty \in \mathcal{L}(Y)$, defined by

$$\lim_{T\uparrow\infty} P_{T,0}(t)x = P_\infty x,\ t \text{ fixed and arbitrary } < T,\ x \in Y, \tag{1.21}$$

where $P_{T,0}(\,\cdot\,)$ is the operator defined by (1.14a) with $G = 0$), such that:

(i) the unique optimal pair $\{u^0_\infty(t;0;y_0), y^0_\infty(t;0;y_0)\}$ of problem (1.10) is expressed in feedback form by

$$u^0_\infty(t;0;y_0) = -B^*P_\infty y^0_\infty(t;0;y_0) \in L_2(0,\infty;U); \tag{1.22}$$

(ii) the operator, with maximal domain

$$A_{P_\infty} = A - BB^*P_\infty : Y \supset \mathcal{D}(A_{P_\infty}) \to Y \tag{1.23}$$

is the generator of a s.c. semigroup $e^{A_{P_\infty}t}$ on Y, $t \ge 0$, and in fact,

$$\Phi_\infty(t)x = y^0_\infty(t;0;x) = e^{A_{P_\infty}t}x \in C([0,\infty];Y) \cap L_2(0,\infty;Y),\ x \in Y, \tag{1.24}$$

which, moreover, is uniformly stable: there exist constants $M_P \ge 1$ and $\omega_P > 0$ such that

$$\|e^{A_{P_\infty}t}\|_{\mathcal{L}(Y)} \le M_P e^{-\omega_P t},\ t \ge 0; \tag{1.25}$$

(iii) P_∞ possesses the following regularity properties

$$B^*P_\infty \in \mathcal{L}(\mathcal{D}(A);U) \cap \mathcal{L}(\mathcal{D}(A_{P_\infty});U); \tag{1.26}$$

(iv) P_∞ is the unique solution within the class of non-negative, self-adjoint solutions possessing the regularity properties (1.26), of the Algebraic Riccati Equation,

$$\begin{aligned}(P_\infty x, Az)_Y \; &+ \; (P_\infty Ax, z)_Y + (Rx, Rz)_Z = (B^*P_\infty x, B^*P_\infty z)_U \\ &\forall\, x, z \in \mathcal{D}(A);\ \text{or else}\ \forall\, x, z \in \mathcal{D}(A_{P_\infty});\end{aligned} \tag{1.27}$$

(v) the optimal cost is

$$J_\infty(u^0_\infty(\,\cdot\,;0;y_0), y^0_\infty(\,\cdot\,;0;y_0)) = (P_\infty y_0, y_0)_Y, \; y_0 \in Y. \tag{1.28}$$

(vi) The operator P_∞ satisfies the identity

$$P_\infty x = \int_0^{t_0} e^{A^*\sigma} R^* R \Phi_\infty(\sigma) x + e^{A^* t_0} P_\infty \Phi_\infty(t_0) x, \tag{1.29}$$

where t_0 is an arbitrary point $0 < t_0 < \infty$.

1.2 Coincidence on $[0,T]$ of the $O.C.P._\infty$ and of the $O.C.P._{T;G}$ with $G^*G = P_\infty$.

In this paper we compare two optimal control problems for the dynamics (1.1) under the standing assumptions (H.1)-(H.4): the $O.C.P._\infty$ in (1.10) over a infinite time horizon, whose theory—summarized above under the hypotheses (F.C.C.) = (1.11) and (D.C.) = (1.12)—produces in particular a unique non-negative, self-adjoint operator P_∞; and the $O.C.P._{0,T;G}$ in (1.6)-(1.8) over the finite time horizon $[0,T]$, with the special choice, however, of the operator G of final state penalization as given by $G^*G = P_\infty$. The claim is that these two problems coincide on $[0,T]$.

Theorem 1.1. Assume (H.1) through (H.4), as well as (F.C.C.) = (1.11) and (D.C.) = (1.12). Let $\{u^0_\infty(\,\cdot\,,0;y_0), y^0_\infty(\,\cdot\,,0;y_0)\}$ be the unique solution of the $O.C.P._\infty$ in (1.10) for the dynamics (1.1). Let $\{u^0_{T,G}(\,\cdot\,,0;y_0), y^0_{T,G}(\,\cdot\,,0;y_0)\}$ be the unique solution of the $O.C.P._{0,T;G}$ in (1.6)-(1.8) over the finite time $[0,T]$, with the operator G in (1.6) satisfying $G^*G = P_\infty$, where P_∞ is the unique algebraic Riccati operator of the $O.C.P._\infty$. Finally, let $P_{T,G}(t)$, $0 \le t \le T$, be the corresponding differential Riccati operator defined by Eqn. (1.14). Then, on $[0,T]$, we have

(i)

$$y^0_\infty(t,0;y_0) \equiv y^0_{T,G}(t,0;y_0), \; 0 \le t \le T, \text{ in } C([0,T];Y); \tag{1.30}$$

(ii)

$$u^0_\infty(t,0;y_0) = u^0_{T,G}(t,0;y_0), \; 0 \le t \le T, \text{ in } L_2(0,T;U); \tag{1.31}$$

(iii)

$$P_{T,G}(t) = P_\infty \equiv G^*G, \; 0 \le t \le T. \tag{1.32}$$

Remark 1.3. Theorem 1.1, besides being of interest in itself, plays a critical, and as yet irreplacable role, in the numerical approximation theory of the Algebraic Riccati Equation (1.27), see [G.1] in the case of B bounded, and [L.1] in the abstract unbounded framework of the present paper, suitable for boundary control problems for P.D.E.s as in Remark 1.1. □

Remark 1.4. In the finite-dimensional situation, or more generally, when the operator B is in $\mathcal{L}(U;Y)$ as in [G.1], the validity of (1.32) follows at once from comparing the respective algebraic Riccati equation and differential Riccati equation, as P_∞ is the unique stationary, non-negative, self-adjoint solution of the differential Riccati equation satisfying the terminal condition (at $t=T$): $G^*G = P_\infty$. However, under the present general assumptions, the theory—summarized above—does not claim that the operator $P_{T,G}(t)$, $0 \le t \le T$, defined by (1.14), satisfies a differential Riccati equation, see Remark 1.2. Thus, in the absence of a differential Riccati equation for the $O.C.P._{0,T;G}$, different techniques of proof must be devised. One such proof is given in [L.1, Lemma 4.2, p. 332]: it is a technical proof which uses formulas [L-T.1]-[L-T.3], [F-L-T.1], giving the optimal quantities of the optimal control problems, as well as a fixed point argument on the formula

$$\begin{aligned} P_{T,G}(t) &= \int_t^T e^{\hat{A}^*(\tau-t)}[R^*R + 2\omega P_{T,G}(\tau)]\hat{\Phi}_T(\tau,t)d\tau \\ &\quad + e^{\hat{A}^*(T-t)}G^*G\hat{\Phi}_T(T,t), \end{aligned} \tag{1.33}$$

where

$$\hat{A} = A - \omega I, \ \omega = \text{ fixed } > \omega_0 = \lim_{t\to\infty} \frac{\ell n\|e^{At}\|_{\mathcal{L}(Y)}}{t}, \tag{1.34}$$

$$\hat{\Phi}(\tau,t)x = e^{-\omega(\tau-t)}\Phi_T(\tau,t)x = e^{-\omega(\tau-t)}y^0_{T,G}(\tau,t;x). \tag{1.35}$$

A simple proof, different from the one in [L.1, p. 330], of formula (1.33) is given in the Appendix.

The goal of this paper is to provide a radically different proof of Theorem 1.1, which is based on intrinsic properties of the two optimization problems and makes no use of Riccati equations.

2 Proof of Theorem 1.1

Proof of (i) and (ii). Step 1. We shall establish that for $x \in Y$:

$$(P_\infty x, x)_Y = \int_0^\infty [\|Ry^0_\infty(t,0;x)\|^2_Z + \|u^0_\infty(t,0;x)\|^2_U]dt \tag{2.1}$$

$$\begin{aligned} &= \int_0^T [\|Ry^0_\infty(t,0;x)\|^2_Z + \|u^0_\infty(t,0;x)\|^2_U]dt \\ &\quad + (P_\infty y^0_\infty(T,0;x), y^0_\infty(T,0;x))_Y. \end{aligned} \tag{2.2}$$

Indeed, in view of the value formula (1.10) for an optimal process starting at $t=T$, we see that (2.2) is equivalent to establishing that

$$\int_T^\infty [\|Ry^0_\infty(t,0;x)\|^2_Z + \|u^0_\infty(t;0;x)\|^2_U]dt = (P_\infty y^0_\infty(T,0;x), y^0_\infty(T;0;x))_Y. \tag{2.3}$$

But (2.3) follows at once from observing that by (1.24),

$$y_\infty^0(T+\tau,0;x) = e^{(A-BB^*P_\infty)(T+\tau)}x \equiv e^{(A-BB^*P_\infty)\tau}e^{(A-BB^*P_\infty)T}x \quad (2.4)$$
$$= y_\infty^0(\tau,T;y_\infty^0(T;0;x)), \quad \tau \geq 0, \quad (2.5)$$

and by (1.22)

$$\begin{aligned} u_\infty^0(T+\tau,0;x) &= -B^*P_\infty y_\infty^0(T+\tau;0,x) \\ &= -B^*P_\infty e^{(A-BB^*P_\infty)(T+\tau)}x \\ &= -B^*P_\infty e^{(A-BB^*P_\infty)\tau}e^{(A-BB^*P_\infty)T}x \\ &= u_\infty^0(\tau,T;y_\infty^0(T,0;x)) \in L_2(0,\infty;U), \end{aligned} \quad (2.6)$$

so that, using (2.5) and (2.6), we obtain

$$\begin{aligned} &\int_T^\infty [\{\|Ry_\infty^0(t,0;x)\|_Z^2 + \|u_\infty^0(t,0;x)\|_U^2]dt \\ &= \int_0^\infty [\|Ry_\infty^0(\tau,T;y_\infty^0(T,0;x))\|_Z^2 + \|u_\infty^0(\tau,T;y_\infty^0(T,0;x))\|_U^2]dt \\ &= (P_\infty y_\infty^0(T,0;x), y_\infty^0(T,0;x))_Y, \end{aligned} \quad (2.7)$$

recalling (1.28) in the last step. Thus (2.3) is proved. [We note that the identity

$$y(T+\tau,0;x;u) = y(\tau,T;y(T,0;x;u);u) \quad (2.8)$$

holds true for all solutions of (1.1) due to u, not only for the optimal solution y_∞^0, through direct calculations.]

Step 2. By (1.17), the optimal pair $\{u_{T,G}^0(\,\cdot\,,0;x), y_{T,G}^0(\,\cdot\,,0;x)\}$ of problem $O.P.C._{\cdot 0,T;G}$ satisfies

$$\begin{aligned} (P_{T,G}(0)x,x)_Y &= \int_0^T [\|Ry_{T,G}^0(t,0;x)\|_Z^2 + \|u_{T,G}^0(t,0;x)\|_U^2]dt \\ &\quad + (G^*Gy_{T,G}^0(T,0;x), y_{T,G}^0(T,0;x))_Y. \end{aligned} \quad (2.9)$$

Since $G^*G = P_\infty$ by assumption, comparing the contribution (2.2) due to $\{u_\infty^0, y_\infty^0\}$ with the contribution (2.9) of the optimal pair $\{u_{T,G}^0, y_{T,G}^0\}$ on $[0,T]$, we conclude that

$$P_{T,G}(0) \leq P_\infty. \quad (2.10)$$

Step 3. We now prove that

$$P_{T,G}(0) \geq P_\infty. \tag{2.11}$$

In fact, we consider the optimal control problem on the infinite horizon $[T, \infty]$ with cost

$$\int_T^\infty [\{\|Ry(t)\|_Z^2 + \|u(t)\|_U^2]dt \tag{2.12}$$

initiating at the point $y^0_{T,G}(T, 0; x)$ at the initial time $t = T$. Call

$$\{u^0_\infty(\,\cdot\,; T; y^0_{T,G}(T, 0; x)), y^0_\infty(\,\cdot\,, T; y^0_{T,G}(T, 0; x))\} \tag{2.13}$$

the unique optimal solution. Define the functions

$$\tilde{u}(t, 0; x) = \begin{cases} u^0_{T,G}(t, 0; x) & 0 \leq t \leq T; \quad (2.14a) \\ u^0_\infty(t, T; y^0_{T,G}(T, 0; x)), & T \leq t < \infty; \quad (2.14b) \end{cases}$$

$$\tilde{y}(t, 0; x) = \begin{cases} y^0_{T,G}(t, 0; x) & 0 \leq t < T; \quad (2.15a) \\ y^0_\infty(t, T; y^0_{T,G}(T, 0; x)), & T \leq t < \infty; \quad (2.15b) \end{cases}$$

so that

$$\tilde{u} \in L_2(0, \infty; U), \quad R\tilde{y} \in L_2(0, \infty; Z), \tag{2.16}$$

and $\{\tilde{u}(t, 0; x), \tilde{y}(t, 0; x)\}$ is an admissible pair with respect to the dynamics (1.1) or (1.5a), in the sense that

$$\tilde{y}(t, 0; x) = e^{At}x + \{L_{0,T}\tilde{u}(\,\cdot\,, 0; x)\}(t). \tag{2.17}$$

Then by the value formula (1.28) (for an optimal process starting at $t = T$), we compute

$$\begin{aligned} (P_\infty y^0_{T,G}(T, 0; x), y^0_{T,G}(T, 0; x)) &= \int_T^\infty [\|Ry^0_\infty(t, T; y^0_{T,G}(T, 0; x))\|_Z^2 \\ &\quad + \|u^0_\infty(t, T; y^0_{T,G}(T, 0; x))\|_U^2]dt \quad (2.18) \\ \text{(using (2.14b) and (2.15b))} \quad &= \int_T^\infty [\|R\tilde{y}(t, 0; x)\|^2 + \|\tilde{u}(t, 0; x)\|^2]dt. \quad (2.19) \end{aligned}$$

Since $G^*G = P_\infty$ under present assumptions, we see that (2.19) is the second term at the right-hand side of (2.9), while in the integrand of the first term in (2.9) we recall (2.14a) and (2.15a). We thus obtain that (2.9) can be rewritten as

$$(P_{T,G}(0)x, x)_Y = \int_0^\infty [\|R\tilde{y}(t, 0; x)\|_Z^2 + \|\tilde{u}(t, 0, x)\|_U^2]dt \tag{2.20}$$

for the admissible pair $\{\tilde{u}, \tilde{y}\}$ in (2.17). Since $\{u^0_\infty(\cdot, 0; x), y^0_\infty(\cdot, 0, x)\}$ is the optimal pair for $O.P.C._{\cdot\infty}$, a comparison between (2.2) and (2.20) yields (2.11), as desired.

Step 4. Eqn. (2.10) and (2.11) then yields

$$P_{T,G}(0) = P_\infty. \tag{2.21}$$

By (2.9) and (2.2) where $G^*G = P_\infty$, identity (2.21) allows us to conclude that identities (1.30) and (1.31) must hold true on $[0, T]$, since $\{u^0_{T,G}(\cdot, 0; x), y^0_{T,G}(\cdot, 0; x)\}$ is the unique optimal pair of $O.C.P._{\cdot 0,T;G}$.

(iii) So far we have obtained $P_{T,G}(0) = P_{T,G}(G) = P_\infty = G^*G$. To show (1.32) in full, we return to the defining formula (1.14) for $P_{T,G}(t)$, which we rewrite here after the change of variable $\tau - t = \sigma$, and after using $G^*G = P_\infty$ as

$$\begin{aligned} P_{T,G}(t)x &= \int_0^{T-t} e^{A^*\sigma} R^* R y^0_{T,G}(t+\sigma, t; x) d\sigma \\ &\quad + e^{A^*(T-t)} P_\infty y^0_{T,G}(T, t; x). \end{aligned} \tag{2.22}$$

But, recalling (1.20) via (1.18), we have

$$y^0_{T,G}(t+\sigma, t; x) = y^0_{T-t,G}(\sigma, 0; x), \quad 0 \le \sigma \le T - t, \tag{2.23}$$

so that, since $0 \le \sigma \le T - t$ and $G^*G = P_\infty$, (1.30) of part (i) gives also

$$y^0_{T-t,G}(\sigma, 0; x) = y^0_\infty(\sigma, 0; x), \quad 0 \le \sigma \le T - t. \tag{2.24}$$

Using (2.23), (2.24) into (2.22) yields for $0 \le t \le T$:

$$P_{T,G}(t)x = \int_0^{T-t} e^{A^*\sigma} R^* R y^0_\infty(\sigma, 0; x) d\sigma + e^{A^*(T-t)} P_\infty y^0_\infty(T-t, 0; x) \tag{2.25}$$

$$= P_\infty, \tag{2.26}$$

where in going from (2.25) to (2.26) we have recalled (1.29), with (arbitrary) $t_0 = T - t$. □

Appendix: Proof of (1.33)

Proposition A. Assume the standing hypotheses (H.1), (H.2), (H.3) on the dynamics, as well as (H.4) on R, G. Then, with reference to (1.34),(1.35), the operator $P_{T,G}(t)$ defined by (1.14) satisfies the identity:

$$\begin{aligned} P_{T,G}(t)x &= \int_t^T e^{\hat{A}^*(\tau-t)} [R^*R + 2\omega P_{T,G}(\tau)] \hat{\Phi}_T(\tau, t)x \, d\tau \\ &\quad + e^{\hat{A}^*(T-t)} G^*G \hat{\Phi}_T(T, t)x, \quad x \in Y. \end{aligned} \tag{A.1}$$

Remark A.1. Identity (A.1) coincides with the identity of [L-T.5, Eqn. (2.13), p. 155], [L-T.3, Chapter 3, Eqn. (3.1.9)] (with $G = 0$ and $-\hat{A}$ there replaced by $\hat{A}$ here). In these references, proof of this identity was based on the property that, in that context, where the semigroup e^{At} is, moreover, analytic, $P_{T,G}(t)$ satisfies a Differential Riccati Equation. In the generality of the present setting (dealing with hyperbolic problems), $P_{T,G}(t)$ is simply defined by (1.14) and no claim is now made that such $P_{T,G}(t)$ is a solution of a Differential Riccati Equation (unless additional regularity is imposed on R and G, see Remarks 1.2 and 1.4). Thus, a different proof, independent of the D.R.E., will be given to establish (A.1).

One such proof is given in [L.1, p. 330]. We provide here a simple different proof.

Proof of Proposition A.1. We shall drop the subscript G in this proof. We shall essentially verify (A.1) by using (1.14). Recalling (1.35) and the evolution properties $\Phi_T(\sigma,\tau)\Phi_T(\tau,t) = \Phi_T(\sigma,t)$ from (1.19) we obtain from (1.14):

$$\begin{aligned} P_T(\tau)\hat{\Phi}(\tau,t)x &= P_T(\tau)e^{-\omega(\tau-t)}\Phi_T(\tau,t)x \\ &= \int_\tau^T e^{-\omega(\tau-t)}e^{A^*(\sigma-\tau)}R^*R\Phi_T(\sigma,\tau)\Phi_T(\tau,t)x\,d\sigma \\ &\quad + e^{-\omega(\tau-t)}e^{A^*(T-t)}G^*G\Phi_T(T,\tau)\Phi_T(\tau,t)x \\ &= \int_\tau^T e^{-2\omega(\tau-\sigma)}e^{-\omega(\sigma-\tau)}e^{A^*(\sigma-\tau)}R^*Re^{-\omega(\sigma-t)}\Phi_T(\sigma,t)x\,d\sigma \\ &\quad + e^{-2\omega(\tau-T)}e^{-\omega(T-\tau)}e^{A^*(T-\tau)}G^*Ge^{-\omega(T-t)}\Phi_T(T,t)x. \end{aligned} \tag{A.2}$$

Recalling (1.34), (1.35), we thus obtain from (A.2):

$$\begin{aligned} P_T(\tau)\hat{\Phi}(\tau,t)x &= \int_\tau^T e^{2\omega(\sigma-\tau)}e^{\hat{A}^*(\sigma-\tau)}R^*R\hat{\Phi}_T(\sigma,t)x\,d\sigma \\ &\quad + e^{2\omega(T-\tau)}e^{\hat{A}^*(T-\tau)}G^*G\hat{\Phi}_T(T,t)x. \end{aligned} \tag{A.3}$$

From (A.3), using the semigroup property of $e^{\hat{A}^*t}$, we write

$$\begin{aligned} e^{\hat{A}^*(\tau-t)}P_T(\tau)\hat{\Phi}_T(\tau,t)x &= \int_\tau^T e^{2\omega(\sigma-\tau)}e^{\hat{A}^*(\sigma-t)}R^*R\hat{\Phi}_T(\sigma,t)x\,d\sigma \\ &\quad + e^{2\omega(T-\tau)}e^{\hat{A}^*(T-t)}G^*G\hat{\Phi}_T(T,t)x. \end{aligned} \tag{A.4}$$

Next, from (A.4), changing the order of integration we obtain

$$\int_t^T e^{\hat{A}^*(\tau-t)} P_T(\tau)\hat{\Phi}_T(\tau,t)x\,d\tau$$

$$= \int_t^T \int_\tau^T e^{2\omega(\sigma-\tau)} e^{\hat{A}^*(\sigma-t)} R^*R\hat{\Phi}_T(\sigma,t)x\,d\sigma\,d\tau$$

$$+ \left(\int_t^T e^{2\omega(T-\tau)}d\tau\right) e^{\hat{A}^*(T-t)} G^*G\hat{\Phi}_T(T,t)x$$

$$= \int_t^T \left(\int_t^\sigma e^{2\omega(\sigma-\tau)}d\tau\right) e^{\hat{A}^*(\sigma-t)} R^*R\hat{\Phi}_T(\sigma,t)x\,d\sigma$$

$$+ \left(\int_t^T e^{2\omega(T-\tau)}d\tau\right) e^{\hat{A}^*(T-t)} G^*G\hat{\Phi}_T(T,t)x. \tag{A.5}$$

Performing the integrations in τ in (A.5), we obtain

$$2\omega \int_t^T e^{\hat{A}^*(\tau-t)} P_T(\tau)\hat{\Phi}_T(\tau,t)x\,d\tau$$

$$= \int_t^T [e^{2\omega(\sigma-t)} - 1]e^{\hat{A}^*(\sigma-t)} R^*R\hat{\Phi}_T(\sigma,t)x\,d\sigma$$

$$[e^{2\omega(T-t)} - 1]e^{\hat{A}^*(T-t)} G^*G\hat{\Phi}_T(T,t)x, \tag{A.6}$$

or moving the terms with the minus sign to the left and noticing that by (1.34), (1.35),

$$e^{2\omega(\sigma-t)} e^{\hat{A}^*(\sigma-t)} R^*R\hat{\Phi}_T(\sigma,t)x = e^{A^*(\sigma-t)} R^*R\Phi_T(\sigma,t)x, \tag{A.7}$$

we arrive at

$$\int_t^T e^{\hat{A}^*(\tau-t)}[R^*R + 2\omega P_T(\tau)]\hat{\Phi}_T(\tau,t)x\,d\tau + e^{\hat{A}^*(T-t)} G^*G\hat{\Phi}_T(T,t)x$$

$$= \int_t^T e^{A^*(\sigma-t)} R^*R\Phi_T(\sigma,t)x\,d\sigma$$

$$= P_T(t)x, \tag{A.8}$$

where in the last step we have recalled (1.14). Thus, (A.8) proves (A.1). □

References

[D-L-T.1] G. DaPrato, I. Lasiecka, and R. Triggiani, A direct study of Riccati equations arising in boundary control problems for hyperbolic equations, *J. Diff. Eqns.* 64 (1986), 26-47.

[F-L-T.1] F. Flandoli, I. Lasiecka, and R. Triggiani, Algebraic Riccati equations with non-smoothing observation arising in hyperbolic and Euler-Bernoulli equations, *Ann. di Matem. Pura et Applic.* IV Vol. CLIII (1988), 307-382.

[G.1] J. S. Gibson, The Riccati integral equations for optimal control problems in Hilbert spaces, *SIAM J. on Control and Optimiz.* 17 (1979),537-565.

[L.1] I. Lasiecka, Approximations of solutions to infinite-dimensional algebraic Riccati equations with unbounded input operators, *Num. Funct. Anal. and Optimiz.* 11 (1990), 303-378.

[L-T.1] I. Lasiecka and R. Triggiani, *Differential and Algebraic Riccati Equations with Application to Boundary/Point Control Problems: Continuous Theory and Approximation Theory*, Volume 164 in the Springer-Verlag Lecture Notes LNCIS Series (1991), p. 160.

[L-T.2] I. Lasiecka and R. Triggiani, Riccati equations for hyperbolic partial differential equations with $L_2(0,T;L_2(\Gamma))$-Dirichlet boundary terms, *SIAM J. Control and Optimiz.* 24 (1986), 884-925.

[L-T.3] I. Lasiecka and R. Triggiani, Volume for Encyclopedia of Mathematics and its Applications, Cambridge University Press, in preparation.

[L-T.4] I. Lasiecka and R. Triggiani, Differential Riccati equations with unbounded coefficients: Applications to boundary control/boundary observation hyperbolic problem, *J. of Nonlinear Analysis* 17, no. 7 (1991), 655-682.

[L-T.5] I. Lasiecka and R. Triggiani, The regulator problem for parabolic equations with Dirichlet boundary control. Part I: Riccati's feedback synthesis and regularity of optimal solutions, *Appl. Math. and Optimiz.* 16 (1987), 147-168.

21 A Sharp Result on the Exponential Operator-Norm Decay of a Family $T_h(t)$ of Strongly Continuous Semigroups Uniformly in h

Roberto Triggiani University of Virginia, Charlottesville, Virginia

Abstract

Given a family $T_h(t)$ of s.c. semigroups on a Banach space X, we give a sharp result on the exponential decay in $\mathcal{L}(X)$ of the family, which is uniform in the parameter h. Sharpness is shown by means of counterexamples. This result generalizes a well-known and very useful criterion of stability of a single s.c. semi-group. The obtained generalization plays a critical role in a variety of applications, from approximation theory to regularization theory.

1. Introduction. Statement of main results

We first recall a well-known stability criterion for a single strongly continuous (s.c.) semigroup, and next we present its (sharp) generalization to a family of s.c. semigroups, depending on a parameter. Sharpness is shown by means of counterexamples.

Case of a single semigroup T(t). Let X be a Banach space and let $T(t) \in \mathcal{L}(X)$ be a s.c. semigroup of bounded operators on X, $t \geq 0$. The following result is well known.

Theorem A. T(t) is (exponentially) stable in $\mathcal{L}(X)$:

$$\|T(t)\|_{\mathcal{L}(X)} \leq M e^{-\mu t}, \quad t \geq 0 \text{ for some } M \geq 1, \mu > 0 \tag{1.1}$$

if (and only if) for some p, $1 \leq p < \infty$

$$\int_0^\infty \| T(t)x \|_X^p \, dt < \infty \text{ for every } x \in X \; \square \tag{1.2}$$

The "if part" of this result was first established for p = 2 and for X a Hilbert space in [D.1], and was later generalized to the above statement in [P.1] (see also [P.2, Thm 4.1, p. 116]), with a different proof. Since then, this result has seen (in the case p = 2) a remarkable range of applications, from optimal control theory with quadratic cost functional, to uniform stabilization of conservative hyperbolic and plate-like (Petrovski type) partial differential equations. Indeed, further generalizations of Theorem A are possible: one may e.g. relax the p-th power of the norm in X in condition (1.2) to other functions (work of J. Zabczyk (1974), S. Rolewicz (1986), W. Littman (1989)) but for further details we refer to [B-D-D-M.1 p. 21], since these extensions are beyond the main scope of the present note.

Case of a family $T_h(t)$ of semigroups. The aim of this paper is to give a sharp generalization of Theorem A to the case of a family $T_h(t)$ of s.c. semigroups on X, depending on the parameter h. This extension plays an equally important role in approximation theory e.g. of optimal control problems and related (operator) algebraic Riccati equations, where T(t) is the original s.c. semigroups of the "continuous" problem, and $T_h(t)$ is the family of approximating s.c. semigroups depending on the parameter of discretization h [L-T.1-2], [La.1]. Other areas of critical application of our generalization include: adaptive control [L-T.3]; parabolic regularization of wave equations, where now h is the parameter of regularization [B.1], etc. Thus, let $h \downarrow 0$, $0<h \leq h_0 < \infty$ be a parameter and let $T_h(t)$, $t \geq 0$, be a family of s.c. semigroups on X. The generalization of Theorem A is the following criterion

Theorem 1. Assume that

(i) there exist constants $M \geq 1$ and $\omega > 0$, independent of h, such that

$$\| T_h(t) \|_{\mathcal{L}(X)} \leq M e^{\omega t}, \quad t \geq 0, \text{ uniformly in h} \tag{1.3}$$

(ii) there exists a constant $c > 0$, independent of h, such that for some p, $1 \leq p < \infty$

$$\int_0^\infty \| T_h(t)x \|_X^p \, dt \leq c \| x \|_X^p \quad \forall x \in X, \text{ uniformly in h} \tag{1.4}$$

Then, there exist constants $K \geq 1$ and $\alpha > 0$, independent of h, such that

$$\| T_h(t) \|_{\mathcal{L}(X)} \leq K e^{-\alpha t}, \quad t \geq 0, \ \alpha > 0 \text{ uniformly in h}. \ \square \tag{1.5}$$

To be sure, the statement of this Theorem has already appeared in the literature, [L-T.1, p. 202], where for p = 2 played a key role in establishing the decay in $\mathcal{L}(X)$ of various approximating semigroups, uniformly in h, which arise in the numerical analysis of the (operator) algebraic Riccati equation, associated with optimal control problems. However, because of space limitations, [L-T.1] did not provide a proof. Rather, [L-T.1, p. 202, bottom] only noted that a proof of Theorem 1 can be given by proceeding as in the proof of Theorem A, as given in [P.2] in the case of a single s.c. semigroup. The authors of [L-T.1] have since used the statement of Theorem 1 in related work on the numerical analysis of abstract parabolic problems [L-T.2, p. 22 in Supplement (proof of Thm. 4.6)] as well as abstract hyperbolic problems [La.1, p. 340]; in all cases, however, with no specific proof, and always making reference to the proof of [P.2] for a single s.c. semigroup for justification. Recently, various circles, with motivation coming from areas other than the numerical analysis of Riccati equations and unaware of the validity of Theorem 1, have wondered about the possibility of giving an extension of Theorem A for a family of s.c. semigroups depending on a parameter. The first guess is to attempt to prove the desired uniform decay (1.5) by using only assumption (1.4), which is—via the closed graph theorem—the perfect counterpart of assumption (1.2) in the single semigroup case. We shall see in section 3, that this guess is erroneous. In effect, even upon stumbling on the statement of Theorem 1 as in [L-T.1-2], [La.1], the question of its proof still remains unsettled, as reference to the proof of a single semigroup as given in [P.2] provides only a model, which however needs some appropriate modifications before succeeding in the case of a family of semigroups. It was, therefore, suggested to write up a note, to put the issue to rest. Accordingly, the aim of this note is twofold:

(i) to give (in section 2) a proof of Theorem 1, by providing the appropriate modifications which are required over the proof of [P.2] for a single semigroup as in Theorem A;

(ii) to show (in section 3) that Theorem 1 is sharp; i.e. that assumption (1.3) cannot be dispensed with, as it is not implied by assumption (1.4). (This then shows that the first guess, referred to above, in extending Theorem A is indeed erroneous.)

2. Proof of Theorem 1.

The main point of the proof, which requires modifications over the proof in [P.2] for a single s.c. semigroup, is singled out in the following

Step 1. Lemma 2.1. Under the assumptions of Theorem 1 we have: there exists a constant $k > 0$, independent of h, such that

$$\|T_h(t)\|_{\mathcal{L}(X)} \leq k \quad \forall\, t \geq 0, \text{ uniformly in h} \tag{2.1}$$

Proof. It suffices to show that: for each $x \in X$, there exists a constant C_x (independent of h), such that

$$\|T_h(t)x\|_X \leq C_x \quad \forall\, t \geq 0, \ \forall x \in X, \ \forall h \tag{2.2}$$

for then (2.2) implies (2.1) by the Principle of Uniform Boundedness. We then prove (2.2). Suppose (2.2) is false. Then, there exist: an $x \in X$, a sequence $h_j \downarrow 0$ and a sequence $t_j \uparrow +\infty$, as $j \to +\infty$, such that

$$\delta_j \equiv \| T_{h_j}(t_j)\, x \|_X \to +\infty \quad \text{as } j \to +\infty \tag{2.3}$$

Without loss of generality we may take $t_{j+1} - t_j > \frac{1}{\omega}$, where ω is given by (1.3). Define

$$\Delta_j = [t_j - \frac{1}{\omega},\ t_j] \tag{2.4}$$

as in [P.2], so that the intervals Δ_j do not overlap. Let now $t \in \Delta_j$, then

$$\delta_j = \| T_{h_j}(t_j)\, x \|_X = \| T_{h_j}(t_j - t)\, T_{h_j}(t)\, x \|_X \le \| T_{h_j}(t_j - t) \|_{\mathcal{L}(X)} \| T_{h_j}(t)\, x \|_X \tag{2.5}$$

But, by (1.3), the choice of t and the size of Δ_j, we obtain

$$\| T_{h_j}(t_j - t) \|_{\mathcal{L}(X)} \le M\, e^{\omega (t_j - t)} \le M\, e^{\omega \frac{1}{\omega}} = M\, e \tag{2.6}$$

which inserted into (2.5) yields

$$\| T_{h_j}(t)\, x \|_X \ge \frac{\delta_j}{Me}, \quad t \in \Delta_j\,. \tag{2.7}$$

Hence, using (2.7) and the size $1/\omega$ of Δ_j from (2.4), we obtain from (2.3)

$$\sup_h \int_0^\infty \| T_h(t)\, x \|_X^p\, dt \ge \int_{\Delta_j} \| T_{h_j}(t)\, x \|_X^p\, dt \ge \left[\frac{\delta_j}{Me} \right]^p \frac{1}{\omega} \to +\infty \quad \text{as } j \to \infty \tag{2.8}$$

and (2.8) contradicts assumption (1.4). Thus, (2.2) holds true □

Step 2. Once Lemma 2.1 is proved, we may finish off the proof of Theorem 1, by paralleling [P.2]. For each h fixed and each $x \in X$ fixed, we have

$$T_h(t)\, x \to 0 \quad \text{as } t \to +\infty \tag{2.9}$$

a result established in [P.2], since h is fixed. Next we define

$$t_{x,h}(\rho) = \max\{t: \|T_h(s)\,x\|_X \geq \rho\,\|x\|_X, \text{ for } 0 \leq s \leq t\} \tag{2.10}$$

By strong continuity

$$\|T_h(t_{x,h}(\rho))\,x\| = \rho\,\|x\| \tag{2.11}$$

Moreover, by (2.9), $t_{x,h}(\rho)$ is finite and positive for every h and x. By (2.10)

$$t_{x,h}(\rho)\,\rho^p\,\|x\|_X^p \leq \int_0^{t_{x,h}(\rho)} \|T_h(t)\,x\|_X^p\,dt \leq \int_0^{\infty} \|T_h(t)\,x\|_X^p\,dt \leq c\,\|x\|_X^p \tag{2.12}$$

recalling assumption (1.4) in the last step of (2.12), whereby

$$t_{x,h}(\rho) \leq \frac{c}{\rho^p} = \text{independent of } h \equiv t_0 \tag{2.13}$$

For $t > t_0$, in the norm of X

$$\|T_h(t)\,x\| \leq \|T_h(t - t_{x,h}(\rho))\|\,\|T_h(t_{x,h}(\rho))\,x\| \leq k\,\rho\,\|x\|, \quad t > t_0, \text{ uniformly in } h \tag{2.14}$$

where in the last step of (2.14), we have recalled (2.1) of Lemma 2.1, as well as (2.11). Given c from (1.4), we next select $\rho > 0$ so that via (2.14)

$$\|T_h(t)x\| \leq \beta\,\|x\|, \; t > t_0, \text{ uniformly in } h, \; \beta = k\rho < 1. \tag{2.15}$$

Starting from (2.15), the remaining of the proof to obtain the desired conclusion (1.5), with

$$\mu = \frac{-1}{t_1} \ln \beta > 0, \; t_1 > t_0; \; K = \frac{c}{\beta} \tag{2.16}$$

is now standard, [P.2].

Remark 2.1. G. Da Prato [D.2] has observed an alternative route to Step 2, once Lemma 2.1 is established. It is based on the proof of Theorem A for a single s.c. semigroup, as given in [B-D-D-M.1, p. 23]. One writes, in the X-norm:

$$t\|T_h(t)\,x\|^p = \int_0^t \|T_h(t)\,x\|^p\,dr \le \int_0^t \|T_h(r)\|^p\,\|T_h(t-r)\,x\|^p\,dr \tag{2.17}$$

$$\text{(by (2.1),(1.4))} \quad \le k^p \int_0^t \|T_h(t-r)\,x\|^p\,dr \le k^p \int_0^\infty \|T_h(\sigma)\,x\|^p\,d\sigma \le k^p\,c\,\|x\|^p \tag{2.18}$$

where in going from (2.17) to (2.18) we have invoked (2.1) of Lemma 2.1, while in the last step we have recalled assumption (1.4). The rest of the proof proceeds as in [B-D-D-M.1] □

3. Assumption (1.4) does not imply (1.3): counterexamples

The goal of this section is to provide (classes of) counterexamples, which show that assumption (1.4) does not imply (1.3). Thus, the extension of Theorem A to a family of s.c. semigroups, based only on assumption (1.4), is false.

(i) Let $X = L_p(0,1)$, $1 \le p < \infty$. Let $a_h \ge 1$ be a function of the parameter h such that

$$a_h \uparrow +\infty \text{ as } h \downarrow 0 \quad (\text{e.g. } a_h = \frac{1}{h}) \tag{3.1}$$

and define g_h by

$$(a_h)^{g_h} = \ln a_h; \quad \text{or } 1 \ge g_h = \frac{\ln \ln a_h}{\ln a_h} \downarrow 0 \quad \text{as } h \downarrow 0 \tag{3.2}$$

(ii) We now define a family $T_h(t)$ of s.c. semigroups on X by setting for $f \in X$: (Fig. 1)

$$\{T_h(t)\,f\}(x) = \begin{cases} (a_h)^{t/p}\, f\left(x + \dfrac{t}{g_h}\right) & 0 \le x \le 1 - \dfrac{t}{g_h} \quad (3.3a) \\[2ex] 0 & 1 - \dfrac{t}{g_h} < x \le 1 \quad (3.3b) \end{cases}$$

(We omit verification of the semigroup property and of the strong continuity at the origin.) Thus, the action of T_h on $f \in X$ consists of two main features: translation of f to the left as t increases with speed $1/g_h$, while modulating its amplitude by $(a_h)^{t/p}$.

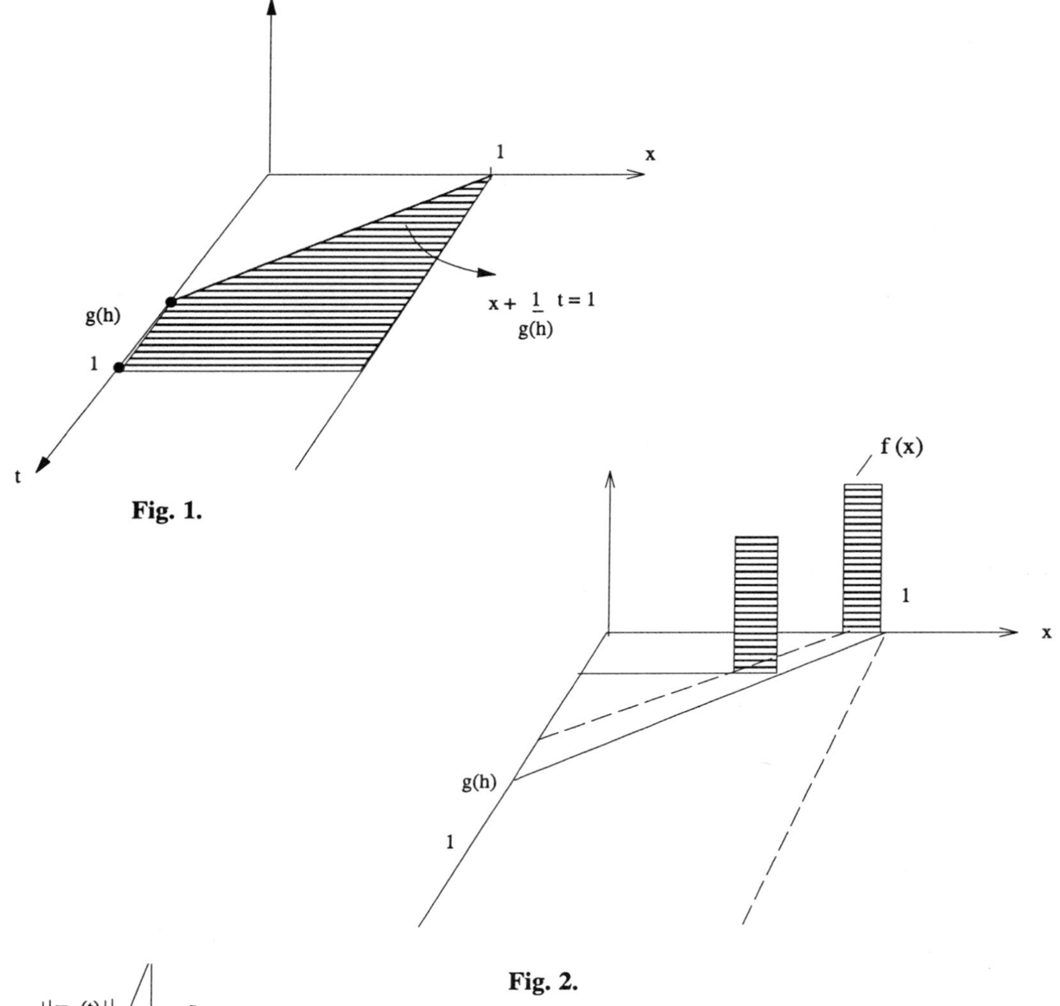

Fig. 1.

Fig. 2.

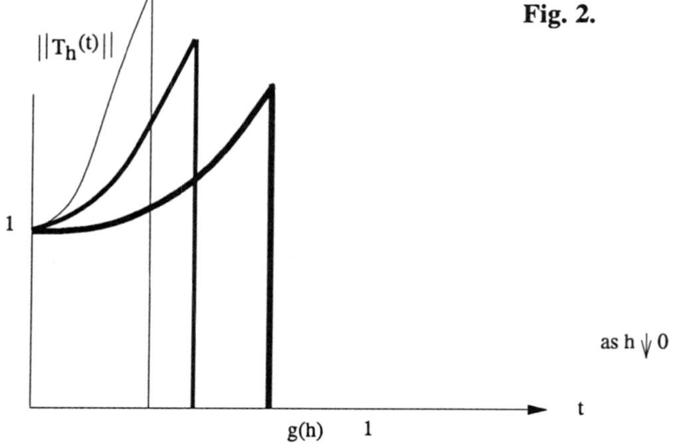

Fig. 3

(iii) We note that the s.c. semigroups $T_h(t)$ are all nilpotent for $t \geq g_h$ with threshold of nilpotency $g_h \downarrow 0$ as $h \downarrow 0$ from (3.2). Moreover, the uniform norm is given by

$$\|T_h(t)\|_{\mathcal{L}(X)} = \begin{cases} (a_h)^{t/p} & 0 \leq t \leq g_h \qquad (3.4a) \\ 0 & g_h < t \qquad (3.4b) \end{cases}$$

as can be readily seen by taking $L_p(0,1)$—functions f of norm 1, identically 1 near x = 1, on progressively smaller intervals with end-point x = 1, and zero elsewhere (Fig. 2).

(iv) We now show that: the peak (maximum) of $\|T_h(t)\|$, which occurs at $t = g_h \downarrow 0$ by (3.4), is monotone increasing to $+\infty$ as a function of $h \downarrow 0$, so that—as a consequence—the uniform bound

$$\|T_h(t)\|_{\mathcal{L}(X)} \leq C\, e^{at} \quad t \geq 0 \text{ for some } C \geq 1 \text{ and } a > 0 \tag{3.5}$$

is violated. In fact, the maximum of $\|T_h(t)\|$ at $t = g_h$ by (3.4) equals

$$F(h) \equiv \|T_h(g_h)\|_{\mathcal{L}(X)} = (a_h)^{g_h/p} \tag{3.6}$$

Thus, recalling (3.2), we see that (3.6) implies

$$\ln F(h) = \frac{g_h}{p} \ln a_h = \frac{1}{p} \ln \ln a_h \uparrow +\infty \quad \text{as } h \downarrow 0 \tag{3.7}$$

so that, as desired

$$\lim_{h \downarrow 0} F(h) = \|T_h(g_h)\|_{\mathcal{L}(X)} = +\infty \tag{3.8}$$

and then condition (3.5) is impossible. Thus, as $h \downarrow 0$, the family $T_h(t)$ has the maximum of its uniform norm which occurs at $t = g_h \downarrow 0$ and which explodes to $+\infty$, (Fig. 3).

(v) Finally, we now prove the following uniform bound

$$\int_0^\infty \| T_h(t)\, f \|_X^p \, dt \le \| f \|_X^p , \quad \forall\, f \in X , \text{ uniformly in } h \tag{3.9}$$

In fact, returning to (3.4), we compute for $f \in X$, by (3.3):

$$\int_0^\infty \| T_h(t) f \|_X^p \, dt = \int_0^{g_h} \| T_h(t)\, f \|_X^p \, dt = \int_0^{g_h} \left\{ \int_0^{1 - t/g_h} (a_h)^t \, | f(x + \frac{t}{g_h}) |^p \, dx \right\} dt \tag{3.10}$$

Thus, setting $\sigma = x + t/g_h$ in (3.10)

$$\int_0^\infty \| T_h(t)\, f \|_X^p \, dt = \int_0^{g_h} (a_h)^t \int_{t/g_h}^{1} | f(\sigma) |^p \, d\sigma \, dt \le \left(\int_0^{g_h} (a_h)^t \, dt \right) \int_0^1 | f(\sigma) |^p \, d\sigma$$

$$= \left[\frac{(a_h)^t}{\ln a_h} \right]_{t=0}^{t = g_h} \| f \|_X^p \le \frac{(a_h)^{g_h}}{\ln a_h} \| f \|_X^p \le \| f \|_X^p \tag{3.11}$$

after recalling (3.2) in the last step in (3.11). Then (3.11) proves (3.9).

In conclusion: the family of s.c. semigroups $T_h(t)$ in (3.3) satisfies condition (1.4) with $c = 1$ (see (3.9)), but does not satisfy condition (1.3) (see point (iv), violation of (3.5)).

References

[B-D-D-M.1] A. Bensoussan, G. Da Prato, M.Delfour, S. Mitter, *Representation and Control of Infinite Dimensional Systems,* Vol. I Birkhausr 1992.

[B.1] F. Bucci, Singular perturbation for controlled wave equations, *J. Math. Signals & Control,* to appear. Also, Ph.D. dissertation, Scuola Normale Superiore, Pisa, Italy.

[D.1] R. Datko, Extending a theorem of Lyapunov to Hilbert spaces, *J. Math. Anal. & Appl.,* 32 (1970), 610-616.

[D.2] G. Da Prato, private communication, July 1993.

[La. 1] I. Lasiecka, Approximations of solutions to infinite-dimensional algebraic Riccati equations with unbounded input-operators *Numer. Funct. Anal. & Optimiz.,* 11 (1990) 303-378.

[L-T.1] I. Lasiecka and R. Triggiani, The regulator problem for parabolic equations with

Dirichlet boundary control, Part II, *Appl. Math. & Optim.,* 16 (1987), 187-216.

[L-T.2] I. Lasiecka and R. Triggiani, Numerical approximations of algebraic Riccati equations for abstract systems modelled by analytic semigroups and applications *Mathem. of Computation,* Vol. 57 (1991), 639-662, and Supplement p. 513-537.

[L-T.3] I. Lasiecka and R. Triggiani, Uniform convergence to the solutions of Riccati equations arising in boundary/point control problems, LNCIS 184, *Proceedings of Stochastic Theory and Adaptive Control,* held at University of Kansas, September 26-28, 1991, Editors: T.E. Duncan and B. Pasik-Duncan.

[P.1] A. Pazy, On the applicability of Lyapumov's theorem in Hilbert space, *SIAM J. Math. Anal.,* 3 (1972), 291-294.

[P.2] A. Pazy, *Semigroups of Linear Operators and Applications to Partial Differential Equations,* Springer-Verlag 1983.